The Fragile Earth

The Fragile Earth

Writing from *The New Yorker* on Climate Change

Edited by David Remnick and Henry Finder

HARPER LARGE PRINT

An Imprint of HarperCollins*Publishers*

HarperCollins books may be purchased for educational, business, or sales promotional use. For information, please e-mail the Special Markets Department at SPsales@harpercollins.com.

FIRST HARPER LARGE PRINT EDITION

ISBN: 978-0-06-302921-7

Library of Congress Cataloging-in-Publication Data is available upon request.

20 21 22 23 24 LSC 10 9 8 7 6 5 4 3 2 1

All pieces in this collection were originally published in *The New Yorker.* The publication date is given at the beginning of each piece.

Contents

Part II—Hell and High Water: *Where We Are*

Part III— Changing the Weather: *What We Can Do Now*

Foreword

David Remnick

In the nineteen-eighties, a writer named Bill McKibben was regarded around *The New Yorker* office as something of a prodigy. The pieces in the Talk of the Town section in those days were unsigned, but everyone at 25 West Forty-third Street knew that McKibben was often writing half of them, sometimes more. Still in his twenties, he went wherever William Shawn, the editor of the magazine, sent him—to trade shows, ballgames, political rallies, the piers—and, in no time at all, he returned to the office and bashed out something charming or funny or sharp. Some writers in that era did not know what to make of him. McKibben had the laconic bearing of an Episcopal novitiate but worked with the metabolism of a hummingbird.

And, as addicted as he had become to turning out his metropolitan dispatches on a snug weekly deadline, he was also eager to write something longer, more deeply reasoned, more far-flung. He had a particular passion for environmental matters and was a devoted reader of Henry David Thoreau and John Burroughs, Wendell Berry and Rachel Carson. Perhaps he'd do something in that mode. After giving the matter considerable thought, he approached Shawn with a proposal.

"Can I write about my apartment?" he asked.

What McKibben had in mind was rather high-concept. He lived in a nothing-special apartment at Bleecker Street and Broadway. What would he learn, he wondered, if he followed every pipe and wire and chute that connected his apartment to the greater world to see where it all led? Where exactly did his water and electricity come from? Where did all his coffee grounds and apple cores go?

With a notebook and a credit card, McKibben flew off to Brazil to see where Con Edison was getting its oil. He visited the La Grande hydroelectric dams in sub-arctic Quebec; water flowed from there into the James Bay, into the Hudson Bay, and, eventually, into Bill McKibben's faucet. He saw uranium being extracted at the Hack Canyon mine, in Arizona—and eventually brought to the Indian Point nuclear power plant, the better to provide electricity for the lights and air-conditioners at Bleecker and Broadway. McKibben's essay on the exploitation of nature and its resources, titled "Apartment," ran in the March 17, 1986, issue of the magazine.

An important outcome of that piece, for its author, was that it suggested further work. "'Apartment' had the effect of reminding me, or maybe teaching me for the first time, that the world was physical," McKibben tells me. "Somehow, I became attuned to the idea that the world and human arrangements were more vulnerable than I had ever thought. I could see what these lifelines looked like. And, at that moment, I also started reading the emerging literature of climate science."

One voice in particular stood out. In June, 1988, James Hansen, a scientist best known in his early career for the study of Venus and its atmospheric conditions, came before a Senate committee and testified that the Earth was now warmer than it had ever been in recorded history, and that the chief cause was our heedless consumption of fossil fuels. This warming, he said, was intensifying all the time and would soon lead to rising sea levels and extreme weather conditions—powerful hurricanes, ruinous droughts and fires and floods. It would unavoidably threaten natural systems and the social order.

McKibben was paying attention. The next year, he published a long, scrupulously reported, yet meditative piece of writing on climate change and what it could mean for the planet. It was called "Reflections: The End of Nature," and it was the first truly extensive exploration of climate change in the nonscientific press. At the time, some people viewed the piece as exceedingly pious and comically apocalyptic. Lewis Lapham, the editor of *Harper's Magazine*, wrote: "'The End of Nature' plays to the superstitions of the environmental left. The author, a young man of sensibility named Bill McKibben, strives for sanctimonious effect that is earnest, doom-ridden, precious, and tear-stained."

Still, the charges levelled against McKibben—of self-righteousness, fear-mongering, exaggeration—did not yet come from more organized and ominous forces. The oil-and-gas industry and right-wing, anti-environmentalist groups would eventually mobilize in an effort to discredit the work of writers like McKibben. Young men with video cameras would try to capture him in acts of hypocrisy, like getting into his car. He would be on the receiving end of death threats. But all of that came later.

When "The End of Nature" first appeared, it felt like speculative literature. The more alarming signs of climate change—melting ice sheets, species disappearing by the thousands—were not yet a matter of constant coverage in the press. Which makes it all the more remarkable that McKibben managed to bring both the science and the politics to a general audience. No corner of the earth, he made plain, could now be considered wild; everything, every ecosystem, was affected by human civilization. He studied the ramifications of a global crisis precipitated by the "awesome power of Man, who has overpowered in a century the processes that have been slowly evolving and changing of their own accord since the earth was born." McKibben's classic essay was, in effect, an echo of James Hansen's lonely testimony, the beginning of the literature on climate change.

The New Yorker has a long tradition of publishing pieces about the natural world and its vulnerability, among them Rachel Carson's "The Sea" (1951) and "Silent Spring" (1962), John McPhee's "Encounters with the Archdruid" (1971), and Jonathan Schell's "The Fate of the Earth" (1982). Carson's work had a distinct political impact, changing the national mindset about the dangers of pesticides. McPhee's work explored the clashing values of those who sought to protect the environment and those who sought to exploit it. Schell dramatized, at the height of the Cold War, the costs to the earth of nuclear conflict. What this anthology aims to represent is the magazine's efforts, over the past three decades, to examine what is now referred to as the climate emergency—its explanation and origins; the conditions it has already created; the likely consequences if we go on as we have; and (with some measure of hope) the possibilities of mitigation or adaptation.

In the worst sense, we have come a long way since "The End of Nature." If McKibben presaged a speculative dystopia, we now live in its opening chapters. In the intervening years, the world's governments have done precious little of the work and the coördination needed to ease the overheating of the atmosphere. The oil companies have conspired to buy off opposition through vast lobbying efforts in Washington and around the world. Awareness of climate change has certainly intensified, but, so far, with little to show by way of reform or results.

Hansen's early alarms have largely proved accurate or have been grimly overtaken. In September, 2019, the Intergovernmental Panel on Climate Change, a U.N. commission led by a hundred scientists from around the world, issued a report about a crisis that was already upon us—record heat waves, catastrophic fires, devastated coral reefs, monster storms, disappearing glaciers and thinning ice sheets, rising sea levels—with far worse to come if CO_2-emission levels were not radically reduced. Floods that were, in the past, freakish, once-in-a-century events will, by 2050, be an ordinary occurrence. Sea levels could rise more than three feet by the end of the century. Coastal cities all over the world will be endangered. Los Angeles, to take but one example, will face not only constant fire seasons but economic devastation along the shore; Marina del Rey and Venice Beach could be swamped. Other jeopardized cities, according to the U.N. report, include Bangkok, Barcelona, Honolulu, Jakarta, Lima, Manila, Miami, San Diego, San Juan, and Sydney. The state of the world's oceans is perilous, the report said: the water is growing warmer and more acidic, a deadly combination for marine ecosystems. More than five hundred million people already live in areas that are turning into deserts. Certain cities and regions in the Middle East, Africa, and South Asia will become so hot that they will be uninhabitable.

The social and economic concerns are vast. The report noted that, in a world in which more than ten percent of the population is undernourished, climate change could lead to food shortages, mass migrations, and political upheaval. As Jonathan Blitzer has reported in *The New Yorker*, climate change in Central America, which has devastated farming, is a leading cause of migration north to Mexico and the United States. The implications of all these current and impending conditions beggar the imagination. Profound changes in political and human behavior—in the way we eat, the way dispose of our trash, the way we heat and cool our homes, the way we travel—are necessary.

Citizen activists around the world, many of them very young, have lost patience with the indifference of their elders and the floundering efforts of the political élites. Time, they know, is slipping away. "If you choose to fail us," Greta Thunberg told a gathering of world leaders at the U.N., "we will never forgive you."

When I was a young reporter at the Washington *Post*, I worked under Leonard Downie, a gifted editor who was a consistent champion of investigative reporting. He also had a peculiar fondness for weather stories. Rainstorms, blizzards, heat waves—he was enthralled by them all. On the eve of a storm, he liked nothing more than to deploy a huge cast of reporters and editors to cover it. Then he would give the weather immense front-page play. He was often mocked for this, but, in fact, he understood something significant. The weather is what we have in common; it affects everyone's life in the most immediate ways.

Weather stories are now of a radically different order and frequency. We already live in a landscape of increasingly common hurricanes, tornadoes, droughts, and fires. For most of us, the penny has dropped: the adverse weather we are facing is no longer just a matter of nature acting on its own. Our behavior—our heedlessness, our inability to alter our patterns of consumption and greed—is the dominant factor. Unless this planet's human inhabitants change our ways, and swiftly, weather conditions will only worsen. On this score, climate experts are happy to show us their finely calibrated models. Yet how can their findings be made vivid and comprehensible to a broader audience?

In the twenty-first century, *The New Yorker*'s leading voice on the environment has been Elizabeth Kolbert. Bill McKibben continues to write for this magazine and for other publications; he also publishes books and teaches. Much of his time, however, has been taken up with organizing and activism. As the leader of 350.org, he has led hundreds of climate-related demonstrations around the world. But the beat he established at *The New Yorker* could not remain empty. Kolbert came to the magazine from the New York *Times*, in 1999, with the intention of covering

City Hall. When she had had her fill of Rudy Giuliani and Michael Bloomberg, though, she turned to climate change; there could be nothing more important for her to pursue. She announced her ambitions, in 2005, with a three-part series called "The Climate of Man." Since then, she has extensively travelled the world—visiting Arctic glaciers, tropical rain forests, United Nations offices—and charted the forms and forces of its depredation. Her prose is razor-sharp, unforgiving, unillusioned. She aims not to soothe but to state things plainly, unmistakably, with the trenchancy of precision. I remember coming across this paragraph in her 2009 article "The Sixth Extinction" and feeling as if I'd had the wind knocked out of me:

> Once a mass extinction occurs, it takes millions of years for life to recover, and when it does it generally has a new cast of characters; following the end-Cretaceous event, mammals rose up (or crept out) to replace the departed dinosaurs. In this way, mass extinctions, though missing from the original theory of evolution, have played a determining role in evolution's course; as Richard Leakey has put it, such events "restructure the biosphere" and so "create the pattern of life." It is now generally agreed among biologists that another mass extinction is under way.

> Though it's difficult to put a precise figure on the losses, it is estimated that, if current trends continue, by the end of this century as many as half of earth's species will be gone.

In this anthology, you will find different approaches to climate change, across a range of geographies. To tell the story of the crisis—its past, present, and future—the book will take you from Greenland to the Great Plains, into sepulchral laboratories and emerald rain forests. It will tell you how human beings created this epochal condition, even as early warnings were sounded; it will assess the consequences that climate change has already wrought, describe what the very near future portends; and explore what can be done to either forestall or try to cope with a looming cataclysm. You'll find science writers, foreign correspondents, essayists, and more engaged in the critical task of trying to think through our predicament. Collectively, their work will underscore what many people have, at long last, come to recognize: that climate change isn't an "issue" to be considered among a list of others. Rather, it concerns the very preconditions for all species to go on living on this planet. "The Fragile Earth" seeks to illuminate the emergency from a multitude of perspectives. We hope that it contributes to a shared sense of urgency—and to a shared spirit of change.

PART I

A Crack in the Ice

How We Got Here

Reflections: The End of Nature

Bill McKibben

September 11, 1989

Nature, we believe, takes forever. It moves with infinite slowness through the many periods of its history, whose names we can dimly recall from high-school biology—the Cambrian, the Devonian, the Triassic, the Cretaceous, the Pleistocene. At least since Darwin, nature writers have taken pains to stress the incomprehensible length of this path. "So slowly, oh, so slowly, have the great changes been brought about," John Burroughs wrote in 1912. "The Orientals try to get a hint of eternity by saying that when the Himalayas have been ground to powder by allowing a gauze veil to float against them once in a thousand years, eternity will only have just begun.

Our mountains have been pulverized by a process almost as slow." We have been told that man's tenure is as a minute to the earth's day, but it is that vast day that has lodged in our minds. The age of the trilobites began six hundred million years ago. The dinosaurs lived for a hundred and fifty million years. Since even a million years is utterly unfathomable, the message is: Nothing happens quickly. Change takes unimaginable—"Geologic"—time.

This idea about time is essentially misleading, for the world as we know it, the world with human beings formed into some sort of civilization, is of quite comprehensible duration. People began to collect in a rudimentary society in the north of Mesopotamia some twelve thousand years ago. Using twenty-five years as a generation, that is four hundred and eighty generations ago. Sitting here at my desk, I can think back five generations—I have photographs of four. That is, I can think back one-ninety-sixth of the way to the start of civilization. A skilled genealogist could easily get me one fiftieth of the distance back. And I can conceive of how most of those forebears lived. From the work of archeologists and from accounts like those in the Bible I have some sense of daily life at least as far back as the time of the Pharaohs, which is almost half the way. Three hundred and twenty generations ago, Jericho was a walled city of three thousand souls. Three hundred and twenty is a large number, but not in the way that six hundred million is a large number, not inscrutably large. And within those twelve thousand years of civilization time is not uniform. The world as we *really* know it dates back to the Renaissance. The world as we really know it dates back to the Industrial Revolution. The world as we feel comfortable in it dates back to perhaps 1945.

In other words, our sense of an unlimited future, which is drawn from that apparently bottomless well of the past, is a delusion. True, evolution, grinding on ever so slowly, has taken billions of years to create us from slime, but that does not mean that time always moves so ponderously. Over a lifetime or a decade or a year, big and impersonal and dramatic changes can take place. We have accepted the idea that continents can drift in the course of aeons, or that continents can die in a nuclear second. But normal time seems to us immune from such huge changes. It isn't, though. In the last three decades, for example, the amount of carbon dioxide in the atmosphere has increased more than ten per cent, from about three hundred and fifteen parts per million to about three hundred and fifty parts per million. In the last decade, an immense "hole" in the ozone layer has opened up above the South Pole each fall, and, according to the Worldwatch Institute, the percentage of West German forests damaged by acid rain has risen from less than ten per cent to more than fifty per cent. Last year, for perhaps the first time since that starved Pilgrim winter at Plymouth, America consumed more grain than it grew. Burroughs again: "One summer day, while I was walking along the country road on the farm where I was born, a section of the stone wall opposite me, and not more than three or four yards distant, suddenly fell down. Amid the general stillness and immobility about me, the effect was quite startling. . . . It was the sudden summing-up of half a century or more of atomic changes in the material of the wall. A grain or two of sand yielded to the pressure of long years, and gravity did the rest."

In much the same comforting way that we think of time as imponderably long, we consider the earth to be inconceivably large. Although with the advent of space flight it became fashionable to picture the planet as a small orb of life and light in a dark, cold void, that image never really took hold. To any one of us, the earth is enormous, "infinite to our senses." Or, at least, it is if we think about it in the usual horizontal dimensions. There is a huge distance between my house, in the Adirondack Mountains, and Manhattan—it's a five-hour drive through one state in one country of one continent. But from my house to Allen Hill, near town, is a trip of five and a half miles. By bicycle it takes about twenty minutes, by car seven or eight. I've walked it in an hour and a half. If you turned that trip on its end, the twenty-minute pedal past Bateman's sandpit and the graveyard and the waterfall would take me to the height of Mt. Everest—almost precisely to the point where the air is too thin to breathe without artificial assistance. Into that tight space, and the layer of ozone above it, are crammed all that is life and all that maintains life.

This, I realize, is a far from novel observation. I repeat it only to make the case I made with regard to time. The world is not as large as we intuitively believe—space can be as short as time. For instance, the average American car driven the average American distance—ten thousand miles—in an average American year releases its own weight in carbon into the atmosphere. Imagine every car on a busy freeway pumping a ton of carbon into the atmosphere, and the sky seems less infinitely blue.

Along with our optimistic perceptions of time and space, other, relatively minor misunderstandings distort our sense of the world. Consider the American failure to convert to the metric system. Like all schoolchildren of my vintage, I spent many days listening to teachers explain litres and metres and hectares and all the other logical units of measurement, and then promptly forgot about it. All of us did, except the scientists, who always use such units. As a result, if I read that there will be a rise of 0.8 degrees Celsius in the temperature between now and the year 2000, it sounds less ominous than a rise of a degree and a half Fahrenheit. Similarly, a ninety-centimetre rise in sea level sounds less ominous than a one-yard rise—and neither of them sounds all that ominous until one stops to think that over a beach with a normal slope such a rise would bring the ocean ninety metres (that's two hundred and ninety-five feet) above its current tideline. In somewhat the same way, the logarithmic scale we use to determine the acidity or alkalinity of our soils and our waters—pH—distorts reality for anyone who doesn't use it on a daily basis. Normal rainwater has a pH of 5.6. But the acidified rain that falls on Buck Hill, behind my house, has a pH of 4.6 to 4.2, which is from ten to fourteen times as acid as normal.

Of all such quirks, though, probably the most significant is an accident of the calendar: we live too close to the year 2000. Forever we have read about the year 2000. It has become a symbol of the bright and distant future, when we will ride in air cars and talk on video phones. The year 2010 still sounds far off, almost unreachably far off, as if it were on the other side of a great body of water. But 2010 is as close as 1970—as close as the breakup of the Beatles—and the turn of the century is no farther in front of us than Ronald Reagan's election to the Presidency is behind. We live in the shadow of a number, and that makes it hard to see the future.

Our comforting sense, then, of the permanence of our natural world—our confidence that it will change gradually and imperceptibly, if at all—is the result of a subtly warped perspective. Changes in our world which can affect us can happen in our lifetime—not just changes like wars but bigger and more sweeping events. Without recognizing it, we have already stepped over the threshold of such a change. I believe that we are at the end of nature.

By this I do not mean the end of the world. The rain will still fall, and the sun will still shine. When I say "nature," I mean a certain set of human ideas about the world and our place in it. But the death of these ideas begins with concrete changes in the reality around us, changes that scientists can measure. More and more frequently, these changes will clash with our perceptions, until our sense of nature as eternal and separate is finally washed away and we see all too clearly what we have done.

Svante Arrhenius took his doctorate at the University of Uppsala in 1884. His thesis earned him the lowest possible grade short of outright failure. Nineteen years later, the same thesis, which was on the conductivity of solutions, earned him a Nobel Prize. He later explained the initial poor reception: "I came to my professor, Cleve, whom I admired very much, and I said, 'I have a new theory of electrical conductivity as a cause of chemical reactions.' He said, 'This is very interesting,' and then he said, 'Goodbye.' He explained to me later that he knew very well that there are so many different theories formed, and that they are almost all certain to be wrong, for after a short time they disappeared; and therefore, by using the statistical manner of forming his ideas, he concluded that my theory also would not exist long."

Arrhenius's understanding of electrolytic conduction was not his only shrug-provoking new idea. As he surveyed the first few decades of the Industrial Revolution, he realized that man was burning coal at an unprecedented rate—"evaporating our coal mines into the air." Scientists already knew that carbon dioxide, a by-product of fossil-fuel combustion, trapped solar infrared radiation that would otherwise have been reflected back to space. The French polymath Jean-

Baptiste Joseph Fourier had speculated about the effect nearly a century before, and had even used the hothouse metaphor. But it was Arrhenius, employing measurements of infrared radiation from the full moon, who did the first calculations of the possible effects of man's stepped-up production of carbon dioxide. The average global temperature, he concluded, would rise as much as nine degrees Fahrenheit if the amount of carbon dioxide in the air doubled from its pre-industrial level; that is, heat waves in mid-American latitudes would run as high as a hundred and thirty degrees, the seas would rise several metres, crops would wither in the fields.

This idea floated in obscurity for a very long time. Now and then, a scientist took it up—the British physicist G. S. Callendar speculated in the nineteen-thirties that rising carbon-dioxide levels could account for the warming of North America and northern Europe which meteorologists had begun to observe in the eighteen-eighties. But that warming seemed to be replaced by a decline, beginning in the nineteen-forties; in any case, we were too busy creating better living through petroleum to be bothered with such long-term speculation. And the few scientists who did consider the matter concluded that the oceans, which hold much more carbon dioxide than the atmosphere, would soak up any excess that man churned out—that the oceans were an infinite sink down which to pour the problem.

Then, in 1957, two scientists at the Scripps Institution of Oceanography, in California, Roger Revelle and Hans Suess, published a paper in the journal *Tellus* on this question of the oceans. What they found may turn out to be the single most important limit in an age of limits. They found that the conventional wisdom was wrong: the upper layer of the oceans, where the air and sea meet and transact their business, would absorb less than half of the excess carbon dioxide produced by man. "A rather small change in the amount of free carbon dioxide dissolved in seawater corresponds to a relatively large change in the pressure of carbon dioxide at which the oceans and atmosphere are at equilibrium," they wrote. That is to say, most of the carbon dioxide being pumped into the air by millions of smokestacks, furnaces, and car exhausts would stay in the air, where, presumably, it would gradually warm the planet. "Human beings are now carrying out a large-scale geophysical experiment of a kind that could not have happened in the past nor be repeated in the future," they concluded, adding, with the morbid dispassion of true scientists, that this experiment, "if adequately documented, may yield a far-reaching insight into the processes of weather and climate." While there are other parts to this story—the depletion of the ozone, acid rain, genetic engineering—the story of the end of nature centers on this greenhouse experiment, with what will happen to the weather.

When we drill into an oil field, we tap into a vast reservoir of organic matter—the fossilized remains of aquatic algae. We unbury it. When we burn oil—or coal, or methane (natural gas)—we release its carbon into the atmosphere in the form of carbon dioxide. This is not pollution in the conventional sense. Carbon monoxide is pollution—an unnecessary by-product; a clean-burning engine releases less of it. But when it comes to carbon dioxide a clean-burning engine is no better than the motor in a Model T. It will emit about five and a half pounds of carbon in the form of carbon dioxide for every gallon of gasoline it consumes. In the course of about a hundred years, our various engines and industries have released a very large portion of the carbon buried over the last five hundred million years. It is as if someone had scrimped and saved his entire life and then spent everything on one fantastic week's debauch. In this, if in nothing else, wrote the great biologist A. J. Lotka, "the present is an eminently atypical epoch." We are living on our capital, as we began to realize during the oil crises of the nineteen-seventies. But it is more than waste, more than a binge. We are spending that capital in such a way as to alter the atmosphere.

There has always been, at least since the start of life, a certain amount of carbon dioxide in the atmosphere, and it has always trapped a certain amount of the sun's radiation to warm the earth. If there were no atmospheric carbon dioxide, our world might resemble Mars: it would probably be so cold as to be lifeless. A little greenhouse effect is a good thing—life thrives in its warmth. The question is: How much? On Venus, the atmosphere is ninety-seven per cent carbon dioxide. As a result, it traps infrared radiation a hundred times as efficiently as the earth's atmosphere, and keeps the planet a toasty seven hundred degrees warmer than the earth. The earth's atmosphere is mostly nitrogen and oxygen; it is only about .035 per cent carbon dioxide, which is hardly more than a trace. The worries about the greenhouse effect are worries about raising that figure to .055 or .06 per cent, which is not very much. But enough, it turns out, to make everything different.

In 1957, when Revelle and Suess wrote their paper, no one even knew for certain whether carbon dioxide was increasing. The Scripps Institution hired a young researcher, Charles Keeling, and he set up monitoring stations at the South Pole and on the side of Mauna Loa, in Hawaii, eleven thousand feet above the Pacific. His data soon confirmed their hypothesis: more and more carbon dioxide was entering the atmosphere. When the first readings were taken, in 1958, the atmosphere at Mauna Loa contained about three hundred and fifteen parts per million of carbon dioxide. Subsequent readings showed that each year the amount increased, and at a steadily growing rate. Initially, the annual increase was about seven-tenths of a part per million; in recent years, the rate has doubled, to one and a half parts per million. Admittedly, one and a half parts per million sounds absurdly small. But scientists, by drilling holes in glaciers and testing the air trapped in ancient ice, have calculated that the carbon-dioxide level in the atmosphere prior to the Industrial Revolution was about two hundred and eighty parts per million, and that this was as high a level as had been recorded in the past hundred and forty thousand years. At a rate of one and a half parts per million per year, the pre-Industrial Revolution concentration of carbon dioxide would double in the next hundred and forty years. Since, as we have seen, carbon dioxide at a very low level largely determines the climate, carbon dioxide at double that very low level, small as it is in absolute terms, could have an enormous effect.

And the annual increase seems nearly certain to go higher. The essential facts are demographic and economic, not chemical. The world's population has more than tripled in this century, and is expected to double, and perhaps triple again, before reaching a plateau in the next century. Moreover, the tripled population has not contented itself with using only three times the resources. In the last hundred years, industrial production has grown fiftyfold. Four-fifths of that growth has come since 1950, almost all of it based on fossil fuels. In the next half century, a United Nations commission predicts, the planet's thirteen-trillion-dollar economy will grow five to ten times larger.

These facts are almost as stubborn as the chemistry of infrared absorption. They mean that the world will use more energy—two to three per cent more a year, by most estimates. And the largest increases may come in the use of coal—which is bad news, since coal spews more carbon dioxide into the atmosphere than any other fuel. China, which has the world's largest hardcoal reserves and recently passed the Soviet Union as the world's largest coal producer, has plans to almost double coal consumption by the year 2000. A model devised by the World Resources Institute predicts that if energy use and other contributions to carbon-dioxide levels continue to grow very quickly, the amount of atmospheric carbon dioxide will have doubled from its pre-Industrial Revolution level by about 2040; if they grow somewhat more slowly, as most estimates have it, the amount will double by about 2070. And, unfortunately, the solutions are neither obvious nor easy. Installing some kind of scrubber on a power-plant smokestack to get rid of the carbon dioxide might seem an obvious fix, except that a system that removed ninety per cent of the carbon dioxide would reduce the effective capacity of the plant by eighty per cent. One often heard suggestion is to use more nuclear power. But, because so much of our energy is consumed by automobiles and the like, even if we mustered the political will and the economic resources to quickly replace each of our non-nuclear electric plants with nuclear ones our carbon-dioxide output would fall by only about thirty per cent. The same argument would apply, at least initially, to fusion or any other clean method of producing electricity.

Burning fossil fuels is not the only method human beings have devised to increase the level of atmospheric carbon dioxide. Burning down a forest also sends clouds of carbon dioxide into the air. Trees and shrubby forests still cover forty per cent of the land on earth, but the forests have shrunk by about a fifth since pre-agricultural times, and the shrinkage is accelerating. In the Brazilian state of Pará, for instance, nearly seventy thousand square miles were deforested between 1975 and 1986; in the hundred years preceding that decade, settlers had cleared about seven thousand square miles. The Brazilian government has tried to slow the burning, but it employs fewer than nine hundred forest wardens in an area larger than Europe.

This is not news; it is well known that the rain forests are disappearing, and are taking with them a majority of the world's plant and animal species. But forget for a moment that we are losing a unique resource, a cradle of life, irreplaceable grandeur, and so forth. The dense, layered rain forest contains from three to five times as much carbon per acre as an open, dry forest—an acre of Brazil in flames equals between three and five acres of Yellowstone. Deforestation currently adds about a billion tons of carbon to the atmosphere annually, which is twenty per cent or more of the amount produced by the burning of fossil fuels. And that acre of rain forest, which has poor soil and can support crops for only a few years, soon turns to desert or to pastureland. And where there's pasture there are cows. Cows support in their stomachs huge numbers of anaerobic bacteria, which break down the cellulose that cows chew. That is why cows, unlike people, can eat grass. The bugs that digest the cellulose excrete methane, the same natural gas we use as fuel. And unburned methane, like carbon dioxide, traps infrared radiation and warms the earth. In fact, methane is twenty times as efficient as carbon dioxide at warming the planet, so even though it makes up less than two parts per million of the atmosphere it can have a significant effect. Though it may come from seemingly "natural" sources—the methanogenic bacteria—the present huge numbers of these bacteria are man's doing. Mankind owns well over a billion head of cattle, not to mention a large number of camels, horses, pigs, sheep, and goats; together, they belch about seventy-three million metric tons of methane into the air each year—a four-hundred-and-thirty-five-per-cent increase in the last century.

We have raised the number of termites, too. Like cows, termites harbor methanogenic bacteria, which is why they can digest wood. We tend to think of termites as house-wreckers, but in most of the world they are house-builders, erecting elaborate, rock-hard mounds twenty or thirty feet high. If a bulldozer razes a mound, worker termites can rebuild it in hours. Like most animals, they seem limited only by the supply of food. When we clear a rain forest, all of a sudden there is dead wood everywhere—food galore. As deforestation has proceeded, termite numbers have boomed; Patrick Zimmerman, of the National Center for Atmospheric

Research, in Boulder, Colorado, estimates that there is more than half a ton of termites for every man, woman, and child on earth. Termites excrete phenomenal amounts of methane: a single mound may give off five litres a minute.

Researchers differ on the importance of termites as a methane source, but they agree about rice paddies. The oxygenless mud of marsh bottoms has always sheltered the methane-producing bacteria. (Methane is sometimes known as swamp gas.) But rice paddies may be even more efficient; the rice plants themselves act a little like straws, venting as much as a hundred and fifteen million tons of methane annually. And rice paddies must increase in number and size every year, to feed the world's growing population. Then, there are landfills. Twenty per cent of a typical landfill is putrescible: it rots, creating carbon dioxide and methane. At the main New York City landfill, on Staten Island, the methane is pumped from under the trash straight to the stoves of thousands of homes, but at most landfills it just seeps out.

What's more, some scientists have begun to think that these sources by themselves may not account for all the methane. For one thing, an enormous amount of methane is locked up as hydrates in the tundra and in the mud of the continental shelves. These are, in essence, methane ices; the ocean muds alone may hold ten trillion tons of methane. If the greenhouse effect warms the oceans, if it begins to thaw the permafrost, then those ices could start to melt. Some estimates of the potential methane release from the ocean muds run as high as six hundred million tons a year—an amount that would more than double the present atmospheric concentration. This would be a nasty example of a feedback loop: warm the atmosphere and release methane; release methane and warm the atmosphere; and so on.

When all the sources of methane are combined, we have done an even more dramatic job of increasing methane than of increasing carbon dioxide. Samples of ice from Antarctic glaciers show that the concentration of methane in the atmosphere has fluctuated between 0.3 and 0.7 parts per million for the last hundred and sixty thousand years, reaching its highest levels during the earth's warmest periods. In 1987, methane composed 1.7 parts per million of the atmosphere; that is, there is now two and a half times as much methane in the atmosphere as there was at any time since the onset of the ice age preceding the most recent one. The level is now increasing at a rate of one per cent a year.

Man is also pumping smaller quantities of other greenhouse gases into the atmosphere. Nitrous oxide, the chlorofluorocarbons—which are notorious for their ability to destroy the planet's ozone layer—and several more all trap warmth with greater efficiency than carbon dioxide. Methane and the rest of these gases, even though their concentrations are small, will together account for fifty per cent of the projected greenhouse warming. They are as much of a problem as carbon dioxide. And as all these compounds warm the atmosphere it will be able to hold more water vapor—itself a potent greenhouse gas. The British Meteorological Office calculates that this extra water vapor will warm the earth two-thirds as much as the carbon dioxide alone.

Most discussion of the greenhouse gases rushes immediately to their future consequences, without pausing to let the simple fact of what has already happened sink in: the air around us—even where it's clean, and smells like spring, and is filled with birds—is significantly changed. We have substantially altered the earth's atmosphere.

That said, the question of what this new atmosphere means must arise. The direct effects are unnoticeable. Anyone who lives indoors breathes carbon dioxide at a level several times the atmospheric concentration without suffering any harm; the federal government limits industrial workers to a chronic exposure of five thousand parts per million, or almost fifteen times the current atmospheric level. A hundred years from now, a child at recess will still breathe far less carbon dioxide than a child in a classroom. This, however, is only mildly good news. Changes in the atmosphere will change the weather, and *that* will change recess. The weather—the temperature, the amount of rainfall, the speed of the wind—will change. The chemistry of the atmosphere may seem an abstraction, a text written in a foreign language. But its translation into the weather of New York and Cincinnati and San Francisco will change the life of each of us.

Theories about the effects all begin with an estimate of expected warming. The wave of concern that began with Revelle and Suess's article and Keeling's Mauna Loa and South Pole data has led to the development of complex computer models of the entire globe. The models agree that when, as has been predicted, carbon dioxide (or the equivalent combination of carbon dioxide and other greenhouse gases) doubles from the pre-Industrial Revolution level, the average global temperature will increase, and that the increase will be one and a half to five and a half degrees Celsius, or three to ten degrees Fahrenheit. Perhaps the most famous of these computer models has been constructed by James Hansen and his colleagues at the National

Aeronautics and Space Administration's Goddard Institute for Space Studies. Even though it remains a rough simulation of the real world, they have improved it to the point where they are willing to forecast not just the effects of a doubling of carbon dioxide but the incremental effects along the way—that is, not just the forecast for 2050 but the one for 2000.

Take Dallas, for instance. According to Hansen's calculations, the doubled level of gases would increase the annual number of days with temperatures above 100°F. from nineteen to seventy-eight. On sixty-eight days, as opposed to the current four, the nighttime temperature wouldn't fall below 80°F. A hundred and sixty-two days a year—half the year, essentially—the temperature would top 90°F. New York City would have forty-eight days a year above the ninety-degree mark, up from fifteen at present. And so on. This would clearly change the world as we know it. One of Hansen's colleagues told reporters, "It reaches a hundred and twenty degrees in Phoenix now. Will people still live there if it's a hundred and thirty degrees? A hundred and forty?" (And such heat waves are possible even if the average global increase, figured over a year, is only a couple of degrees, since any average conceals huge swings.) These changes, Hansen and his colleagues said in a paper published last fall in the *Journal of Geophysical Research*, should begin to be obvious to the man in the street by the early nineteen-nineties; that is, the odds of a very hot summer will, thanks to the greenhouse effect, become better than even beginning now.

In recent years, there have, of course, been any number of doom-laden prophecies that haven't come true—oil is selling at eighteen dollars a barrel, half its price just a few years ago. Is the warming theory valid? The obvious way to check is to measure the temperature and see if it's going up. But this is easier said than done. In the first place, the warming doesn't show up immediately. The oceans can hold a lot of heat; the warming so far may be stored there, ready to re-radiate out to the atmosphere, the way the sun's heat is held through the night by a rock. This "thermal lag" may be as little as ten years, as much as a hundred. And when you check the thermometers it won't do to measure only a few places for only a few years, because climate is "noisy"—full of random fluctuations. (If you had spent this summer in Tucson, for example, you would have been sure that something was happening: the city set forty-seven high-temperature records. New York, by contrast, has had fairly normal summer temperatures.) To find what climatologists call the "warming signal" through the static of naturally cold and hot years requires an enormous effort. Two such studies have been done—one by Hansen and his NASA colleagues, the other at the University of East Anglia. The studies reach back to 1880, when scientists first began systematic weather observations. To find truly global averages, they include readings from thousands of land-based and shipboard monitoring stations. Both studies conclude that the earth's temperature increased a little more than a degree Fahrenheit from 1880 to 1980. This is consistent with what most of the greenhouse models indicate. Updates of both studies show that the four warmest years on record occurred in the nineteen-eighties; the rise is accelerating as more gases enter the air, just as the models indicate. The British study lists the six warmest years on record as (in descending order) 1988, 1987, 1983, 1981, 1980, and 1986.

In 1988, the American drought hit the heart of the Grain Belt, where most of the nation's and much of the world's food is grown. It followed a dry fall and winter, so its effects were quickly evident; the Mississippi River, for example, sank to its lowest level since 1872, when the Navy began taking measurements. And just about the time that the pictures on television began to grab everyone's attention it got very, very hot in the urban East, where those in the government and the media establishment, among others, have their homes. It happened that in late June, as the anxiety intensified—newscasters telling us that the next two weeks were crucial for corn pollination, meteorologists issuing pessimistic sixty-day forecasts—the Senate Committee on Energy and Natural Resources held a hearing on the greenhouse effect. It was actually the second part of the hearing. Part I had been held the previous November, when, according to the Louisiana Democrat J. Bennett Johnston, the senators listened with "concern" as they were told that one expected result of the greenhouse effect would be a drying of the Midwest and the Southeast. But now, "as we experience a-hundred-and-one-degree temperatures in Washington, D.C., and [reduced] soil moisture across the Midwest is ruining the soybean crops, the corn crops, the cotton crops," Senator Johnston said, concern was giving way to "alarm." Several of the senators said that they had already read the report of Dr. Hansen, the chief witness, and predicted that it would startle listeners. Hansen's report, Dale Bumpers, of Arkansas, said, should be "cause for headlines in every newspaper in America tomorrow morning." As it turned out, he was not exaggerating. Hansen testified that he was ready to state that the warming signal was beginning to emerge above the noise of normal weather, that there was only a one per cent chance that the temperature increases seen in the last few years were accidental, and that we now lived in the greenhouse world.

It was a claim no other established scientist had made—certainly not one on a government payroll. The reaction was much as the senators had expected. The next day's *Times*, for instance, ran a story at the top of the front page under the headline "GLOBAL WARMING HAS BEGUN, EXPERT TELLS SENATE." The message was finally getting across, nearly a century after Arrhenius and three decades after Revelle and Suess. But the heat of the day may have been a mixed blessing; though it focussed everyone's attention on the issue, it also led most people to think that what Hansen had said was that the heat and drought of 1988 were greenhouse-related. Strictly speaking, that is not what he had testified to. "It is not possible to blame a specific heat wave or drought on the greenhouse effect," he said—and, indeed, some experts think that the drought and heat of 1988 were mainly the result of a fluctuation of tropical ocean currents which steered the North American jet stream, with its cargo of rainstorms, north of the Great Plains.

What we *can* blame the carbon dioxide and the methane for is a longer-range pattern. Even if the summer of 1988 had been cool and damp, even if there had been mushrooms growing in the wheat fields of Kansas,

Hansen would have said the same thing. What had convinced him was not the devastation in the Midwest or the misery in the Eastern cities but the numbers that his computer kept spitting at him. "There are two time scales to consider," he explained, some months after giving his testimony. "One is the last three complete decades, for which the natural variability in temperature has been calculated—it is about point thirteen degrees Celsius. This coincides roughly with the thirty years for which we have precise measurements of carbon dioxide and other gases. And our readings show that the global mean temperature has risen about point four degrees in the three-decade period. The other is the larger record—the observations back to the eighteen-eighties. Over that period, there's been about a point-six-degree-Celsius rise. Now, over a longer period there's also more natural variability—sources like fluctuations in solar activity, deep ocean circulation, and so forth." The standard deviation over the longer period, he noted, was about .2°C. So in both cases Hansen's observed rise was almost exactly three times the standard deviation. "There's no magic point where you pick out the signal," he said. "But when it gets to three sigma, when it gets to three times the standard deviation—you're getting to a level where it's unlikely to be an accidental warming."

Some recent studies tend to agree with Hansen's conclusion that the warming has already begun: precipitation appears to have increased above 35 degrees north latitude and decreased below it since the early nineteen-fifties, for instance—a result anticipated in the greenhouse models. And some investigators have found a "variable but widespread" warming of the Alaskan permafrost, which changes temperature much more slowly than the air and thus may provide a better record.

But not all scientists—not even all those committed to the greenhouse theory—believe that the warming has already begun. Hansen, though well respected, is out on a limb, if a fairly stout one. Some have taken issue with his use of statistics, and others with his outspokenness. At a workshop on global warming at Amherst College this spring, the assembled climatologists concluded that while it was "tempting to attribute" the recent warm years to the greenhouse effect "such an attribution cannot now be made with any degree of confidence." Stephen Schneider, a senior scientist at the National Center for Atmospheric Research and a longtime proponent of the greenhouse theory, offers a gambler's analogy: the warm years of the nineteen-eighties, he says, are not "proof" of a warming any more than a dealer' s drawing four aces "proves" that he's dealing from the bottom of the deck. "Different tastes cause some people to accept the reality of a hypothesized climatic change at a low signal-to-noise ratio, whereas others might not believe in the reality of the change until a large signal has persisted for a very long time," Schneider told the Senate two months after Hansen testified. "Quite simply, accepting any particular signal-to-noise ratio as 'proof' of global warming reflects the personal judgment of the investigator."

Kenneth Watt, a professor of zoology and environmental studies at the University of California at Davis, says that studies such as Hansen's fail to correct enough for the "urban-heat island effect"—a phenomenon well known to meteorologists, in which, as cities grow up around thermometers, concrete and exhaust skew readings. There's also no guarantee that other factors—solar flares, perhaps, which coincide with both warming and cooling trends, or the strong El Niño current of recent years—aren't skewing the readings. Last January, Tim Barnett, a climatologist at Scripps, correctly forecast much cooler low-latitude temperatures for the first part of this year as a result of "La Niña," a tropical "cold event" that is the opposite of El Niño. During the summer of 1988, in some parts of the ocean off equatorial South America the water temperature dropped 7°F. Hansen saw the dip in his computer data, and he agrees that it may make this year's overall readings go down. "But such things are bumps," he says.

But few of the objections are to the theory as a whole. Everyone in the scientific community agrees that carbon dioxide is on the rise, and almost everyone believes that the rise cannot help having some effect. An occasional scientist says that the onset of the effect may be delayed as much as forty years, but this is considerably different from dismissing it. Last May, Hansen returned to Capitol Hill to tell the Senate's Science, Technology, and Space Subcommittee that his studies showed a definite danger of future drought. The White House tried to alter his testimony, arguing that, in the words of the Presidential press secretary, Marlin Fitzwater, "there are many points of view on the global warming issue." But Fitzwater didn't cite any studies undercutting Hansen's, and the same day Stephen Schneider assured the subcommittee that "there is virtually no scientific controversy" over the contention that more carbon dioxide in the atmosphere will produce higher temperatures. "That's not a speculative theory," he said.

There is debate, though, over the question of what will happen as the heating begins. A large-scale change in the climate will set off a series of other changes, and while some of these would make the problem worse, others might lessen it. Skeptics are inclined to argue that the warming will trigger some natural compensatory brake. S. Fred Singer, a professor of environmental sciences at the University of Virginia, has assumed a part-time role as greenhouse curmudgeon, expressing his doubts to reporters and on various Op-Ed pages. He grants that the earth's temperature should increase "provided that all other factors remain the same." But, he says, they won't. "For example, as oceans warm and more water vapor enters the atmosphere, the greenhouse effect will increase somewhat, but so should cloudiness—which can keep out incoming solar radiation and thereby reduce the warming." There are other possibilities. "The feedbacks are enormously complicated," Michael MacCracken, of the Lawrence Livermore National Laboratory, in California, told *Time* in 1987. "It's like a Rube Goldberg machine in the sense of the number of things that interact in order to tip the world into fire or ice."

The computer models have tried to incorporate such factors. In some cases, Hansen admits, we simply don't have enough knowledge to make more than educated guesses; the behavior of the oceans is something of a wild card, and so are the clouds. (The difficulty of estimating cloud feedback is a major reason that most warming predictions are expressed as a range of temperatures, and not as a single number.) But almost every doubt is double-edged. Low-level stratocumulus clouds reflect a lot of solar radiation and might tend to cool the earth. Monsoon clouds, on the other hand, are long and thin, and let in the sun's heat while preventing its escape. Hansen's work suggests that the overall effect of clouds will be to increase the warming.

A variety of other feedback effects have also been identified and tallied up. For instance, every surface has an albedo—a degree to which it reflects light. A polar ice cap, or a white shirt, has a high albedo—a large proportion of the sun's rays are reflected back into space. If the ice is replaced by dark blue ocean, more heat will be absorbed. Tropical rain forests absorb a lot of heat now; if they turn to deserts, these deserts will reflect heat. The feedbacks are products of the warming signal, and are distinct from phenomena that always have affected and always will affect temperature—volcanoes, say, which can throw up so much dust that it acts as a veil, or EI Niños, or solar flares. In any event, the warming estimates provided by the computer models are not worst-case scenarios. They are the middle ground. Stephen Schneider told the Senate energy committee last year that it was "equally likely" that the warming forecasts were too low as that they were too high.

Some of the potential feedbacks are so enormous that they may someday make us almost forget what originally caused the greenhouse warming. Twenty thousand years ago, the land that surrounds my house in the Adirondacks was covered by glaciers that had spread slowly down from Canada, and eventually retreated there. As the ice disappeared, "the fierce ruthlessness of nature gave way to a benevolent mood," in the words of a local writer. "Rains came over the years to chasten the harshness of the landscape. The startling gaping holes in the earth were filled with crystal-clear water. Soft green foliage came to clothe the naked rock-hewn slopes." This was a slow process, and is even now incomplete. Some plant and animal species are still migrating up here. Great forests rose on the glacial till and soon created more soil for greater forests, and so on—a process that was first interrupted a couple of hundred years ago, when men began cutting down most of the Adirondack woods. But this interruption was only temporary; just before the turn of this century, New York State, in an early burst of environmental consciousness, began buying huge tracts of land in the Adirondacks and stipulating that they be "forever kept as wild"—off limits to loggers and real-estate developers alike. As a result, this area, though still threatened, is a happy exception—a reforested, replenished zone, a second-chance wilderness.

But the trees that live here don't do so because of the laws; they do so because of the climate. They have slowly marched north as the climate warmed since the end of the last ice age, and if it continued slowly warming they would slowly keep marching; the convoy of pines might march right out of here, and the mass of hardwoods found in lower Appalachian latitudes might march in to replace them. But before we get too used to this marching metaphor it is worth recalling that trees are rooted in the ground; forests move only by the slow growth of new trees along their edges. In a year, a forest moves, naturally, a half mile at most. Which is fine, if that's how slowly the climate is changing. The computer models, however, project an increase in average global temperature as high as one degree Fahrenheit per decade. An increase of one degree in average global temperature moves the climatic zone some thirty-five to fifty miles north. So if the temperature increases one degree per decade the forest surrounding my home would be due at the Canadian border by 2020, which is just about the time that we'd be expecting the trees from a hundred miles south to start arriving. They won't—half a mile a year is as fast as forests move. The trees outside my window will still be there, but they'll be dead or dying.

Eventually, perhaps within a few decades, forests—or, at least, scrub better adapted to the new conditions—will replace the forests that expired. But in the meantime those dead forests will release tremendous amounts of carbon to the atmosphere. Last year's Yellowstone fires released carbon amounting to 2.8 per cent of this country's annual emissions from fossil fuels; that is, in a dozen weeks, on only about a million and a half acres, the fires released as much carbon as ten days' worth of driving, home heating, factory production, motor-boating, and so on. The world's forests, plants, and soil (which gives up its carbon much more rapidly as trees die) contain more than two trillion tons of carbon, probably more than a third of it in the middle and high latitudes. By contrast, the atmosphere at present contains only about seven hundred and fifty billion tons. So even a fairly small change in the forests could substantially increase the amount of carbon dioxide in the atmosphere, intensifying the warming.

This vast decline, this forest "dieback," is not some distant proposition. A 1988 study issued by the World Meteorological Organization and the United Nations Environmental Program found that, given a fairly rapid warming, "reproductive failure and forest dieback is estimated to begin between 2000 and 2050." A University of Virginia study predicts what Michael

Oppenheimer, of the Environmental Defense Fund, calls "biomass crashes" in the pine forests of the southeastern United States over the next forty years if the warming continues. Last September, James Hansen told reporters that the birch trees and many of the evergreens of the Northeast "may have a hard time surviving, even in the next ten to twenty years." There are signs—frightening signs—that some of the feedback loops are starting to kick in. In May, George Woodwell, a biologist at the Woods Hole Research Center, told the Senate's science-and-technology subcommittee that the annual one-and-a-half-parts-per-million increase in atmospheric carbon dioxide seemed to have surged upward in the last eighteen months to two and a half parts per million. "I'm suggesting that the warming of the earth is increasing the decay of organic matter," he said, adding that such an event had not been worked into the computer climate models—in other words, their estimates of future warming might well be too low.

For the moment, though, forget about the higher temperatures and the dead trees and the other effects. The physical consequences of increasing the level of carbon dioxide will be staggering, but no more staggering than the simple fact of what we have already done. Carbon-dioxide levels have gone up significantly, and globally. Elevated levels can be measured far from industry and miles above the ground. And the changes are irrevocable. They are not possibilities. They cannot be wished away, and they cannot be legislated away. To prevent them, we would have had to clean up our collective act many decades ago. We have done this ourselves—by driving our cars, running our factories, clearing our forests, growing our rice, turning on our air-conditioners. In the years since the Civil War, and especially in the years since the Second World War, we have changed the atmosphere—changed it enough so that the climate will change dramatically. Most of the major events of human history gradually lose their meaning: wars that seemed at the time all-important are now a series of dates that schoolchildren don't even try to remember; great feats of engineering crumble in the desert. But now the way of life of one part of the world in one half century is altering every inch and every hour of the planet.

[. . .]

The single most talked-about consequence of a global warming is probably the expected rise in sea level as a result of polar melting. For the last several thousand years, sea level has been rising, but so slowly that it has almost been a constant. In consequence, people have extensively developed the coastlines. But a hundred and twenty thousand years ago, during the previous interglacial period, sea level was twenty feet above the current level; at the height of the last ice age, when much of the world's water was frozen at the poles, sea level was three hundred feet below what it is now. Scientists estimate that the world's remaining ice cover contains enough water so that if it should all melt it would raise sea level more than two hundred and fifty feet. This potential inundation is stored in the Greenland Ice Sheet (if it melts, it will raise the world's oceans twenty-three feet), the West Antarctic Ice Sheet (another twenty-three feet), and East Antarctica (more than two hundred feet), with a smaller amount—perhaps half a metre—in the planet's alpine glaciers. (Melting the ice currently over water, such as the sea ice of the Arctic Ocean, won't raise sea level, any more than a melting ice cube overflows a gin-and-tonic.) The East Antarctic is relatively safe; the direst fears of a rising sea came as the result of a 1968 study concluding that the Ross and Filchner-Ronne ice shelves, which support the West Antarctic Ice Sheet, could disintegrate within forty years. Subsequent investigations, however, seem to have demonstrated that such a disintegration would take at least two centuries, and probably more like five (though several investigators have speculated that it might become irreversible within the next century).

But the salvation of the West Antarctic does not mean the salvation of Bangladesh, or even of East Hampton. A number of other factors may raise sea level significantly. Glaciers bordering the Gulf of Alaska, for example, have been melting for decades, and constitute a source of fresh water about the size of the entire Mississippi River system. And even if nothing at all melted, the increased heat would raise sea level considerably. Warm water takes up more space than cold water; this thermal expansion, given a global temperature increase of between one and a half and five and a half degrees Celsius, should raise sea level a foot, according to James Hansen. It is by now widely accepted that sea level will rise significantly over the next decades. The E.P.A. has estimated that it will rise between five and seven feet by 2100, and speculated about worst-case scenarios that might lead to an eleven-foot rise; the National Academy of Sciences has been more conservative; other researchers have turned in even scarier numbers. Suffice it to say that included in the range of guesses of almost every panel and scientist studying the problem is an increase in global sea level of better than three feet over the next century.

That may not sound like very much, but it means that the sea would reach a height unprecedented in the history of civilization. The immediate effects of the swollen sea would be seen in a place like the Maldives. By most accounts, this archipelago of eleven hundred and ninety small islands about four hundred miles southwest of Sri Lanka is fairly paradisal. Its residents had never heard a gun fired in anger until last year, when a short-lived coup attempt was mounted by foreign mercenaries. They survived the downturn in the coir business (coir is an elastic fibre made from coconut husks); breadfruit and citron trees are abundant. But most of this happy nation rises only two metres above the Indian Ocean. If sea level were to rise one metre, storm surges would become an enormous, crippling danger; were it to rise two metres—well within the range of possibilities predicted by many studies—the country would all but disappear. In October of 1987, the Maldivian President, Maumoon Abdul Gayoom, went before the United Nations General Assembly. He described his country as "an endangered nation." The Maldivians, he pointed out, "did not contribute to the impending catastrophe . . . and alone we cannot save ourselves." A map drawn a hundred years from now may not show the Maldives at all, except as a danger to mariners.

Other nations, though not extinguished, would be very badly hurt. A two-metre rise in sea level would flood twenty per cent of the land in Bangladesh, much of which is built on the floodplains at the mouth of the Brahmaputra. In Egypt, such a rise would inundate less than one per cent of the land, but that area constitutes much of the Nile Delta, where most of the population lives. Nor is the danger only to the Third World. Several years ago, the E.P.A. distributed a worksheet to allow local governments to calculate their future position vis-à-vis the salt water. In Sandy Hook, New Jersey, for instance, add thirteen inches to the projected increase in sea level to account for local geologic subsidence, for a net ocean rise, in the next hundred years, of four feet one inch. In Massachusetts, between three thousand and ten thousand acres of oceanfront land worth between three billion and ten billion dollars might disappear by 2025, and that figure does not include land lost to encroaching ponds and bogs as the rising sea lifts the water table. But storm surges would do the most dramatic damage: in Galveston, Texas, ninety-four per cent of the land is within the plain that would be flooded by the worst storms. Such surges are the reason that Holland built many of its protective dikes. The most extensive barriers went up after the winter of 1953, when a surge breached the existing dikes in eighty-nine places along the central delta, killing nearly two thousand people and tens of thousands of cattle. Afterward, the Dutch decided to spend more than three billion dollars building new defenses.

As the Dutch effort indicates, much can be done to defend against increases in sea level. The literature abounds with studies of how much it would cost to protect coastal areas. The trouble is, spending the money to protect the shoreline would lead to ecological costs harder to calculate but easy to understand. Coastal marshes or wetlands exist in a nearly unbroken chain along the Gulf and Atlantic Coasts of the United States. Protected from the waves of the ocean by barrier islands or peninsulas, they are part land and part water, and are home to an abundance of plants and animals. They are more biologically productive than either the ocean or the dry land, in part because tidal flows spread food and flush out waste; it is a cycle that encourages quick growth and rapid decay. These communities support an immense variety of birds, fish, shellfish, and plants. Early settlers (with noble exceptions, like William Bartram) thought coastal marshes "miasmal," and drained or filled many of them. In recent years, federal and state authorities have grudgingly begun to protect them. As King Canute demonstrated, however, the ocean disregards governments, and as its level rises the area of the wetlands will dwindle. This is not axiomatic: if the marsh has room and time enough to back up, it will, and the drowned wetland will be replaced by a new one. But, as another recent E.P.A. report pointed out, "in most areas . . . the slope above the marsh is steeper than the marsh; so a rise in sea level causes a net loss of marsh acreage." That is, in many cases the marsh will run into a cliff it can't climb. In a number of places, the cliffs will be man-made. If I have a house on Cape Cod, and my choice is to build a wall in front of it or let a marsh come in and colonize my basement, I will probably build the wall.

Should the ocean go up a metre, at least half the nation's coastal wetlands will be lost one way or another. "Most of today's wetland shorelines still would have wetlands," according to the E.P.A. report to Congress. "The strip would simply be narrower. By contrast, protecting all mainland areas would generally replace natural shorelines with bulkheads and levees." The relentlessly practical authors add that "this distinction is important because for many species of fish, the length of a wetland shoreline is more critical than the total area." It's also important if you are used to the idea of the ocean meeting the land with ease and grace instead of bumping into an endless concrete wall.

There are other reasons to fear a sea-level rise. In normal times, the water pouring out of a river pushes the ocean back. But in a drought the reduced flow creates a vacuum that the sea oozes in to fill. The "salt front" advances. In the drought of the nineteen-sixties, it nearly reached the point in the Delaware River where Philadelphia's water intake is located. During a drought, New York City must release vast quantities of water from its reservoirs on the upper Delaware to keep the salt front from creeping upriver. New Yorkers, however, continue to take showers and wash their hands. In the summer of 1985, city officials made up for the diminished flow from reservoirs by pumping water straight from the Hudson. This worked well—the water turned out to be considerably cleaner than many had expected—except that as the flow of the Hudson was reduced the salt front began to move up that river, and the town fathers of Poughkeepsie grew worried about *their* supply's getting salty. As the greenhouse warming kicks in, increased evaporation could steal from ten to twenty-four per cent of the water in New York's reservoirs, the E.P.A.'s 1988 report continues. In addition, a one-metre sea-level rise could push the salt front up past the city's water intake on the Hudson. In all, the report observes, "doubled carbon dioxide could produce a shortfall equal to twenty-eight to forty-two per cent of planned supply in the Hudson River Basin."

The expected effects of sea-level rise typify the many other consequences of a global warming. On the one hand, they are of such magnitude that we can't grasp them. If there is significant polar melting, the earth's center of gravity will shift, tipping the globe in such a way that sea level might actually drop at Cape Horn and along the coast of Iceland. (I read this in the E.P.A. report and found that I didn't really know what it meant to tip the earth, though I was awed by the idea.) On the other hand, the changes ultimately acquire a quite personal dimension: Should I put a wall in front of my house? Does this taste salty to you? What's more, many of the various effects of the warming compound one another. If the weather grows hotter and I take more showers, more water must be diverted from the river, and the salt front moves upstream, and so on. The complications multiply almost endlessly (more air-conditioning means more power generated means more water sucked from the rivers to cool the generators means less water flowing downstream, et cetera ad infinitum). These aren't the simple complexities of, say, last summer, when everyone on the East Coast rushed to the beaches to escape the hot weather, only to discover a tide of syringes and fecal matter. These complexities are the result of throwing every single natural system into an uproar at the same time, so that none of nature's reliable compensations can be counted on. For example, at the same time that sea level is increasing, and the warmer air is gathering up more water vapor and presumably increasing the overall precipitation, the temperature is continuing to go up. The result, the computer modellers say, will be greatly increased levels of evaporation; in many parts of the world, there will be a drier interior to complement the sodden coasts.

It's not simply a matter of heat. If the temperature rises, the number of days with snow cover will likely fall. When the snow-melting season ends, more of the sun's energy is absorbed by the ground instead of being reflected back to space, and, as a result, the soil begins to dry out. In the greenhouse world, this seasonal change will begin earlier, because the snow will melt faster. In some areas, other weather changes may offset the evaporation. Roger Revelle, of the Scripps Institution, has estimated that flows in the Niger, the Senegal, the Volta, the Blue Nile, the Mekong, and the Brahmaputra would increase—probably with disastrous results in the last two cases—whereas flows might diminish in the Hwang Ho, in China; the Amu Darya and the Syr

Darya, which run through the Soviet Union's principal agricultural areas; the Tigris-Euphrates system; and the Zambezi. The United States, as usual, has been most closely studied. America is blessed with ample water; on an average day, four trillion two hundred billion gallons of precipitation fall on the lower forty-eight states. Most of that water evaporates, leaving only about one trillion four hundred and thirty-five billion gallons a day, of which in 1985 only about three hundred and forty billion gallons a day were withdrawn for human use. It seems like more than enough. However, as anyone who has ever flown across the nation (and looked out the window) can attest, the water is not spread evenly. In its report to Congress, the E.P.A. notes that total water use exceeds average stream flow in twenty-four of the fifty-three Western water-resource regions, a difference made up by "mining" groundwater stocks and importing water. Much of the Colorado River's flow, for example, is dammed, diverted, and consumed by irrigation projects and by the millions upon millions of people living in places that would otherwise be too dry.

And matters may get worse. After studying the temperature and streamflow records, Revelle and the climatologist Paul Waggoner concluded that if a "conservative," two-degree Celsius increase in temperature occurs, the virgin flow of the Colorado could fall by nearly a third; the same study predicts that if, as some of the computer models suggest, this temperature rise is accompanied by a ten-per-cent decrease in precipitation in the Southwest because of new weather patterns, runoff into the upper Colorado could fall by forty per cent. Even if rainfall went up ten per cent, the runoff would still drop by nearly twenty per cent. Across the West, the picture is similar: in the Missouri, Arkansas, lower-Colorado, and Rio Grande irrigation regions, supply could fall by more than half. In the Missouri, Rio Grande, and Colorado basins, the estimated water needs in the year 2000 could not be met by stream flows after the expected climatic changes. One model predicts a twenty-five-per-cent increase in the demand for irrigation water from the Ogallala Aquifer, the subterranean lake that irrigates the Great Plains and is already badly depleted.

A compelling question is what all this means for agriculture. The answer comes on several levels, the first being that of the individual plant. Quite apart from heat and drought, the simple increase of atmospheric carbon dioxide affects plants. Ninety per cent of the dry weight of a plant comes from the conversion of carbon dioxide into carbohydrates by photosynthesis. If nothing else limits a plant's growth—if it has plenty of sunshine, water, and nutrients—then increased carbon dioxide should increase the yield. And in ideal laboratory conditions this is what happens; as a result, some journalists have rhapsodized about "supercucumbers" and found other green linings to the cloud of greenhouse gases. But there are drawbacks. If some crops grow more quickly, farmers may need to buy more fertilizer, since leaves may become richer in carbon but poorer in nitrogen, reducing food quality not only for human beings but for nitrogen-craving insects, who may eat more leaf to get their fix. In the best case, direct effects of increased carbon dioxide on yield are expected to be small; the annual harvest of well-tended crops might rise about five per cent when the carbon-dioxide level reaches four hundred parts per million, all other things being equal.

But all other things won't be equal. All other things—moisture, temperature, growing season—will be different. It is an obvious point, but worth repeating: most of what we eat spends its growing life in the open air, "exposed," in the words of Paul Waggoner, "to the annual lottery of the weather." About fifty million acres of America's cropland and rangeland are irrigated, but even those fields depend on the weather over any long stretch. And we can't just stick the wheat crop under glass.

It is a tricky business trying to predict what changes in the weather will do to crops. A longer growing season—the period between killing frosts—surely helps; a lack of moisture surely hurts. If temperatures stay warm, plants grow nicely. If temperatures get really hot, they wither. (A long stretch above ninety-five degrees Fahrenheit, for instance, can sterilize corn.) The climate models are too crude to project with any precision what will happen in a given area, and too many variables make even the broadest predictions difficult. The severe droughts of the Dust Bowl years provide scant guidance: on the one hand, the technological revolution in agriculture has tripled yields since then, but on the other, as a government report notes, "the economic robustness associated with general multiple-enterprise farms has long since passed from the scene on any significant scale," and therefore "the current vulnerability of our agricultural system to climate change may be greater in some ways than in the past." Most of the experts have simply thrown up their hands. The best guesses seem to be that the northern reaches of the Soviet Union and Canada will be able to grow more food and the Great Plains of the United States less—not so little that America couldn't feed itself but enough below present production so that United States food exports, which earn the country some forty billion dollars in a good year, might fall by seventy per cent. "It has been suggested," Stephen Schneider told the Senate energy committee last year, "that a future with soil-moisture change . . . would translate to a loss of comparative advantage of United States agricultural products on the world market"—a sentence to make an economist shiver on an August day.

This sounds like somewhat comforting news—as if we would still have enough to eat—but when computers are modelling something as complex as all of agriculture the potential for error is enormous (or the potential for accuracy is small). The effect of the heat and drought of 1988 made liars of most of the computer models in just a few weeks. They had concluded that the expected doubling of carbon dioxide in several decades might make the weather hot and dry enough to cut American corn and soybean yields as much as twenty-seven per cent, but in the summer of 1988, when the rains held off, the American corn crop fell thirty per cent—down by about two and a half billion bushels.

Even if, as seems likely, that heat wave had little to do with the greenhouse effect, we now have some idea what it will feel like once it is here. As of late August, the grain stored around the world amounted to only about two hundred and eighteen million metric tons—enough to last forty-seven days, and the lowest level since 1973. Worldwide consumption of grain outpaced worldwide production by sixty million metric tons last year. You can live with a budget deficit for quite a while, but when the food runs out there's no central bank to mint some more.

The thing to remember is that all these changes may be happening at once. It's hotter, and it's drier, and sea level is rising as fast as food prices, and hurricanes are strengthening, and so on. And not least is the simple fact of daily life in a hotter climate. The American summer of 1988, when no one talked about anything but the heat and how soon it would end, was only a degree or two warmer, on average, than what we were used to. But the models predict that summers could eventually be five or six or seven degrees warmer than the old "normal." Science has yet to devise a way of determining what percentage of people feel like human beings on any given August afternoon, or the number of work hours lost to the third cold bath of the day—or, for that matter, the loss of wit and civility in a population concerned mainly with keeping its shirts dry. These are important matters, and a future full of summers like that one is a grim prospect. Summer will come to mean something different—not the carefree season anymore but a time to grit one's teeth and get through. To anyone who lived through the 1988 heat it seems unlikely that people will simply get used to it.

A certain number of people who didn't get used to the heat died of it. Public health researchers have correlated mortality and temperature tables. When the weather gets hot, they find, preterm births and perinatal deaths both rise. Mortality from heart disease goes up during heat waves, and emphysema gets worse. If, the E.P.A. notes, in its 1988 report to Congress, "climate change encourages a transition from forest to grassland in some areas, grass pollens could increase," worsening hay fever and asthma. "A variety of other U.S. diseases indicate a sensitivity to changes in weather," the report continues. "Higher humidity may increase the incidence and severity of fungal skin diseases (such as ringworm and athlete's foot), and yeast infections (candidiasis). Studies on soldiers stationed in Vietnam during the war indicated that outpatient visits for skin diseases (the largest single cause of outpatient visits) were directly correlated to increases in humidity."

That last sentence suggests that a useful place to look for information about American weather might be Vietnam. There is nothing wrong with the Vietnamese climate—it is not "better" or "worse" than the various American climates, or the weather in Britain, or the cold of Canada. And people have been able to move back and forth between all these zones, adapting to conditions. In fact, we often want to—a change of climate is perhaps the single biggest inducement to travel. But now the climate is travelling. A recent United Nations study estimates that sometime in the next century the climate of Finland will have become similar to that of northern Germany, that of southern Saskatchewan to northern Nebraska, of the Leningrad region to the western Ukraine, of the central Urals to central Norway, of Hokkaido to northern Honshu, and of Iceland to northeast Scotland. If we felt like keeping the weather we're accustomed to, it's we who would have to move, travelling north ahead of the heat.

The list of miscellaneous circumstances that might result from changes in the atmosphere looks to be infinite. In New York City, the heat of the summer of 1988 softened asphalt and caused thousands of "hummocks"—potholes in reverse—in the streets. "When it's over ninety degrees for a prolonged period of time, the problem is virtually out of control," Lucius Riccio, of the New York City Bureau of Highway Operations, told the *Times.* Steel expansion joints buckled along Interstate 66 around Washington, D.C., during the heat wave, and a hundred and sixty people were injured when a train derailed in Montana, apparently because the heat warped the rails. Coupled with the physical predictions are endless political and financial conjectures. Francis Bretherton, of the National Center for Atmospheric Research, told *Time* that if the Great Plains became a dust bowl and people followed the seasonable temperatures north, Canada might replace the United States as the Western superpower.

This game swings from the specific to the wildly speculative. There is no easy way to say that something can't happen or is unlikely to happen; forecasts have to be based on the past, and there is no longer a relevant past. Jesse Ausubel, the director of programs at the National Academy of Engineers, told *Fortune* that it "may become difficult to find a site for a dam or an airport or a public transportation system or anything designed to last thirty to forty years," and asked, "What do you do when the past is no longer a guide to the future?" We are left with a vast collection of "mights," and only one certainty: we have changed the world, and therefore some of the "mights" are inevitable. I find myself thinking often of some purple-martin chicks that Penny Moser found "cooked to death" near her Illinois farm in 1988's heat. This was an actual event and also a metaphor. The heat will cook the eggs of birds, and that destruction—and the hurricanes and the rising sea and the dying forests will rob us of our sense of security. That the temperature had never reached a hundred degrees at the airport in Glens Falls, the city nearest my home, made it a decent bet that it never would. And then, in July of 1988, it did. There is no good reason anymore to say that it won't reach a hundred and ten degrees. The old planet is a different planet. There is no reason to feel secure, because there is no reason to be secure.

Last summer, I paddled across a northern Adirondack lake with a state biologist to visit an eagle's nest. Thirty years before, in an effort to curb black flies, communities in this area put big blocks of DDT in the streams. The black flies survived (they hung in clouds around us the whole morning), but the eagles, among others, didn't. The chemical thinned the shells of their eggs; when the mother eagles sat on the eggs, the shells collapsed. Finally, last year, three pairs of eagles returned to the Adirondacks and built nests. We sat in the canoe and watched a big eagle circle above us with patient irritation, head ruffled. His mate was on the nest, and we were too close. He swooped nearby; we backed off; he rose with a beat or two of his six-foot wingspan and flew for the nest. When he got there, he flared his wings, stalled, and dropped softly down.

Had Rachel Carson not written when she did about the dangers of DDT, it might well have been too late by the time anyone cared about what was happening. She pointed out the problem; she offered a solution;

the world shifted course. That is how this discussion should end, too. At this writing, the greenhouse effect shows every sign of becoming an important political issue. President George Bush has called for an international scientific workshop on the subject; there is talk of drawing up an international treaty on climate change modelled on the recent international accords to phase out production of chlorofluorocarbons. It all sounds promisingly rational. We ought to come up with a good practical response, a plan, a series of steps, a seven-point proposal to offset the greenhouse effect. That is our reflex. The minute the scientists at the June, 1988, congressional hearings finished explaining that we were heating up the earth, senators began talking about nuclear power; it was literally their first reaction. Senator Frank Murkowski, of Alaska, asked, "Is it indeed a reality that we must look more aggressively to nuclear as a release? Because I don't see the public demanding any reduction in the power requirements that our air-conditioners run off of, everything else that we enjoy." Not even the senator from Alaska can imagine life without air-conditioning, so we must come up with some solution, and fast. But is nuclear power a solution? Lay aside the questions of whether it's safe and what we will do with the resulting waste (though it is a sign of the depth of our addiction that we would be willing to lay aside such considerations). Nuclear energy is, at the moment and for the foreseeable future, useful for generating electricity but not for, say, powering my Honda. We may well need to swallow our fears about safety and build more reactors, but doing so won't make everything all right. Returning to the same mix of natural gas and coal that America used in 1973 could save as much carbon dioxide as expanding nuclear power fifty per cent. And we have no spare decades in which to build more Shorehams and Seabrooks; putting off the solution twenty or thirty or forty years would give us thirty or forty or sixty more parts per million of carbon dioxide.

But what about increasing efficiency—what about conservation? There is—no question—waste, even sixteen years after the energy crisis. For example, most of the electricity consumed by industry is used to drive motors; companies, anticipating expansion, tend to buy larger motors than they need; however, large motors are inefficient when they run at less than full speed. The latest edition of the *World Resources* yearbook estimates that if every industrial motor in the United States were to be equipped with available speed-control technology. America's total electricity consumption would fall seven per cent. We must end waste, the sooner the better. But will this kind of action *solve* the problem? Consider a few numbers supplied by Irving Mintzer, of the World Resources Institute. He describes a "base case" scenario that "reflects conventional wisdom in its assumptions about technological change, economic growth, and the evolution of the global energy system." In this model, nations do not enact policies to slow carbon-dioxide emissions, nor do they provide more than minimal support for increased energy efficiency and solar research and development, though they do slow the rate of chlorofluorocarbon production. The result is an average global warming of up to 4.7°F. by the year 2000, and of up to 8.5°F. by 2030. This, Mintzer says, "is by no means the worst possible outcome." If the use of coal and synthetic fuels is encouraged, and tropical deforestation continues to increase, the planet would be doomed to an increase of up to 12.6°F. by 2030, and, by 2075, to a nearly thirty-degree jump—a level with implications too sci-fi for us to imagine. The good news, such as it is, concerns Mintzer's "slow buildup scenario." In this one, strong international efforts to reduce greenhouse-gas emissions "eventually stabilize the atmosphere's composition." Coal, gas, and oil prices are markedly increased, per-capita energy use declines in industrialized countries, and governments actively pursue the development of solar energy. The world embarks on "massive" reforestation efforts. And so on. If all these heroic efforts had begun in 1980, by 2075 we would experience a warming of between 2.5°F. and 7.6°F., which is still "greater than any experienced during recorded human history."

Carbon dioxide and other greenhouse gases come from everywhere, so the situation they create can be fixed only by fixing everything. Small substitutions and quick fixes are not the answer. One common suggestion is to replace much of the coal and oil we burn with methane, since it produces considerably less carbon dioxide. But, as I have noted, any methane that escapes unburned into the atmosphere traps solar radiation twenty times as efficiently as carbon dioxide does. And methane does leak—from wells, from pipelines, from appliances; some estimates suggest that as much as three per cent of the natural gas tapped in this country escapes unburned. So converting from oil to natural gas might make the situation worse. The size and complexity of the industrial system we have built makes even small course corrections physically difficult.

Not only is that system huge but the trend toward growth is incredibly powerful. At the simplest level—population—the increase continues, if not unabated, then only slightly abated. In some of the developing countries, thirty-seven per cent of the population is under fifteen years of age; in Africa, the figure is forty-five per cent. Without a static population, even the most immediate and obvious goals, like slowing deforestation or reducing fossil-fuel use, seem far-fetched. Over the last century, a human life has become a machine for burning petroleum. At least in the West, the system that produces excess carbon dioxide is not only huge and growing but also psychologically all-encompassing. It makes no sense to talk about cars and power plants and so on as if they were something apart from our lives—they *are* our lives. Moreover, for any program to be a success we must act not only as individuals and as nations but as a community of nations. The trouble is, though, that some countries may perceive themselves to be potential winners in a climatic change. The Russians may decide that the chance of increased harvests from a longer growing season is worth the risk of global warming. And the United States, the Soviet Union, and China own about two-thirds of the world's coal reserves, so any one of them can scuttle progress. The possibilities of other divisions—rich nations versus poor nations, say—are large. Every country has its own forms of despoliation to protect; the Canadians, for instance, who complain loudly about their position as helpless victims of American acid rain, are cutting down the virgin forests of British Columbia at an almost

Brazilian pace. And the fact that decisions must be made now for the decades ahead means that, in the words of Richard Benedick, our Deputy Assistant Secretary of State for Environment, Health, and Natural Resources, "somehow, political leaders and government processes and budget-makers must accustom themselves to a new way of thinking." Of all the quixotic ideas discussed here, that may top the list.

The greenhouse effect is often compared to the destruction of the ozone layer, another example of atmospheric pollution with global implications. But the destruction of the ozone layer can and likely will be solved by our ceasing to produce the chemicals currently destroying it. Though this step won't end the problem overnight, it will take care of it eventually. And, though the necessary international negotiations may be complex, steps like this are easy enough so that they will certainly be taken. Essentially, it's like controlling DDT. The problem of global warming, however, does not yield to the same sort of solution. With aggressive action—as Mintzer's numbers indicate—we can "stabilize" the situation at a level that is only mildly horrific, but we cannot solve it.

This is not to say that we should not act. We must act, and in every way possible, and immediately. We stand at the end of an era—the hundred years' binge of oil, gas, and coal which has given us both the comforts and the predicament of the moment. Even those countries which wouldn't object to a degree or two of warming for a longer growing season can't endure endless heating. The choice of doing nothing—of continuing to burn ever more oil and gas and coal—is not a choice. It will lead us, if not straight to hell, then straight to a place with a comparable temperature. But even the scientists calling most vociferously for controls on emissions say they are doing so in order to slow down the warming so that we can adapt to it. That adaptation is all that remains to be discussed.

Adjustment to the greenhouse world will not be easy; our addiction to oil is deep. Our every comfort—especially the freedom from hard labor, for those of us who enjoy such freedom—depends on fossil fuels. They allowed us to dominate the earth, instead of letting the earth dominate us. Our impulse will be to adapt not ourselves but the earth—to figure out a new way to continue our domination, and hence our accustomed life styles, our hopes for our children. This defiance is our reflex. Our impulse will be to defy the doomsayers and press ahead into a new world.

The futurist Julian Simon has infuriated environmentalists by predicting that before we ran out of anything essential, scientists would discover new ways to produce it; if we started to run out of copper, say, we would find out how to make it from other metals. In a 1981 book called "The Ultimate Resource," he writes that with knowledge, imagination, and enterprise "we can manipulate the elements in such fashion that we can have all the mineral raw materials that we need and desire at prices ever smaller relative to other prices and to our total incomes. In short, our cornucopia is the human mind and heart." This is not a scientific treatise—Simon has not discovered how to produce copper from other metals. It is, despite its reliance on "long-run economic indicators" and such, a religious argument, an article of faith. "The main fuel to speed our progress is our stock of knowledge, and the brake is our lack of imagination," he writes. "To have more children grow up is also to have more people who can find ways to avert catastrophe." The religiosity of this view can be seen as well in books like "The Hopeful Future," of 1983, whose author, G. Harry Stine, argues that to make predictions based on current rates of growth and progress is absurd. Even a curve that shows the rate of human progress as increasing from its present mind-boggling pace is too conservative. Only "Curve E," a "cubic curve that continues to turn upward ever more steeply with no limit in sight," makes sense. "It means that we can expect eight times as much progress in the next fifty years as we have seen in the past fifty," he says. This is not, strictly speaking, blind faith; the optimists can explain their reasoning. But it is faith, and it comes with other religious trappings—a dark view of people who think differently, for instance. ("Some of the 'futurists' making 'downside' forecasts don't like people. That means they don't like themselves either," Stine chides.) And there is a vision of a not too distant utopia. In the twenty-first century, Stine writes, enormous orbiting satellites will beam down "enough energy for everybody to do everything."

I am not dismissing the futurists. On the contrary, I think it possible that they are right: we can keep progressing, even in the teeth of the greenhouse effect. We will invent new tools, new technologies, to keep ourselves alive on the planet. We will figure out ways to extend our control so far that not even the rogue nature we have inadvertently created in our last century of progress will escape our domination. I can imagine scenarios—a nuclear war, for instance—that would cancel this future. But my guess is that the defiant optimists are likely correct in their assertion that we can have a "macromanaged" world—one that may well allow us to continue our ways of life even in the face of the coming heat. People with sincere and "progressive" ideas about man's future profess their hope for such a world. Buckminster Fuller is probably the great example. He was not an enemy of the environment; his geodesic domes, for instance, are as stable as conventional buildings, at about three per cent of the weight. Were we all to live in them, there would be a lot more forests standing.

He was not an enemy of the environment, but he was a champion of man. "We have to deal with our spaceship, Earth, as a machine, which is what it is," he wrote in "Approaching the Benign Environment." I doubt whether Fuller would have viewed the end of nature with much trepidation, for he never believed that we would or should stay long in the surroundings we had grown accustomed to. Instead, we were like a chick in a shell. This shell had just enough food in it—enough coal and oil and oxygen and whatever—to allow us to develop to a certain point. "But then, by design, the nutriment is exhausted just at the time when the chick is large enough to be able to locomote on its own legs," he wrote. "And so as the chick pecks at the shell seeking more nutriment it inadvertently breaks open the shell." The analogy is somewhat selfish—that there are other species in the shell with us seems not to have crossed his mind—but it may well be correct.

As a mild example of our hubris, consider a recent book: "Gaia: An Atlas of Planet Management." Despite this title, it does not, I think, reflect fully the Gaia hypothesis, which was first outlined by James Lovelock in the nineteen-seventies; namely, that the earth is a self-sustaining, self-regulating organism. Instead, it argues that man should take ever more control of the planet. Its editor, Norman Myers, seems almost thrilled by the current state of affairs. The approaching crises represent "our final evolutionary examination." We must rise to the occasion, pass the test. And we will: "We are grown up. We have acquired the power of life and death for our planet and most of its inhabitants. . . . Our 'satellite vision' means that all the planet's resources—soils, forests, rivers, oceans, minerals—can be not only mapped in fine detail, but vetted for pollution, erosion, or drought; for changes in albedo or humidity . . . for movements of shoaling fish and migratory creatures." We can process this data at high speed in our computers; we can communicate it around the world instantly. And we can act on it." With the power of life in our hands, we could, for instance, make forests spring up on bare lands, safeguard species against the pressures for extinction." It is time for us, "as incipient planet managers," to "use this power and use it well," Myers goes on. "The ancient Greeks, the Renaissance communities, the founders of America, the Victorians, enjoyed no such challenge as this. What a time to be alive!" The physician Lewis Thomas is quoted as saying that if we succeed "we could become a sort of collective mind for the earth."

This is the defiant reflex, cloaked in a filmy veil of New Age ecological thinking. Many of the proposals of the planet managers are sound—the usual suggestions of environmentalists. In the world we have created, they may offer us our only chance. But the planet managers have respect mostly for man: they understand that the current methods of domination will overheat the planet, but they have new and improved methods. In their forests of the future, cloned Douglas firs and American sycamores will "sprout like mushrooms," growing straighter, producing "denser wood." Almost all wildlife can be "harvested" from preserves, so that "conservation and profit can go hand-in-hand." Even at its most far-reaching, though, "macromanagement" remains a fairly crude enterprise. You may be able to keep track of fish by satellite, but they are still wild creatures, growing at their own pace. The next step—the step we stand about to take—is much more radical.

The first time I gave much thought to biotechnology, I was a young reporter covering the weekly meetings of the city council of Cambridge, Massachusetts. For several years, the councillors debated how to regulate the genetic-engineering work then under way at Harvard and M.I.T. Week after week, Nobel Prize winners and brilliant young researchers would arrive at the meetings to answer questions; their biggest doubter was Alfred E. Vellucci, the councillor from Italian and Portuguese East Cambridge, who would long ago have won a Nobel himself if only they were awarded for local politics. Gifted with a strong imagination, Vellucci conjured up countless possible ways for "these bugs," the reprogrammed organisms, to be accidentally released. Could they escape through the sewers? The air-conditioning? On the soles of people's shoes? Eventually, and over the protests of the universities, the city enacted fairly strict regulations governing "containment"—the thickness of laboratory doors, and so on. I remember thinking that gene-splicing was something like nuclear power—potentially useful, albeit risky. It didn't occur to me then to think much more deeply about it.

But genetic engineering is the first way to create new life. It is a staggering idea—"the second big bang," as one biologist put it. Just in time—just as the clouds of carbon dioxide threaten to heat the atmosphere—we are figuring out a new method of domination, a method more thorough, and therefore more promising, than burning coal and oil and natural gas. It is the method that offers us the most hope of continuing our way of life, our economic growth. It promises crops that need little water and can survive the heat; it promises cures for the new ailments we are creating as well as the old ones we have yet to defeat; it promises a way to survive in almost any environment we may create.

And for this reason it is without a doubt the most important scientific advance ever, in conceptual and moral terms. When I say "moral," I am not thinking primarily of the uses to which such technology might be put—eugenics, for instance. I am thinking of the very fact of the technology. The environmental lobbyist Jeremy Rifkin, who has emerged as one of the few vigorous opponents of genetic engineering, says that for thousands of years human beings have lived "pyrotechnically," burning, melting, mixing inanimate materials—coal, say, or iron. We have worked from the outside in, to alter our environment. Now we are starting to work from the inside out, and that changes everything. Everything except the driving force, the endless desire to master our planet. The British writer Brian Stableford declares in his celebratory book "Future Man" that genetic engineering "will eventually enable us to turn the working of all living things on earth—the entire biosphere—to the particular advantage of our own species." No clearer and crisper definition exists of what I have been calling "defiance."

Watson and Crick described the double helix of DNA in 1953. Just twenty years later, Stanley Cohen, of Stanford, and Herbert Boyer, of the University of California at San Francisco, took two unrelated organisms and cut out a piece of DNA from each. They knit the pieces together, and when they were done they had a new form of life, a kind of life that had not existed five minutes before. In 1981, scientists from the University of Ohio and from Jackson Laboratory, in Bar Harbor, Maine, transferred a gene that controlled the manufacture of part of the hemoglobin in rabbits to a mouse embryo, which they brought to term. The mouse was not exactly a mouse; it had a functioning rabbit gene, which it passed on to subsequent generations. This proof of the possibility of blends between unrelated species was soon followed by others. English researchers crossed a goat and a sheep, two animals that wouldn't dream of mating in the barnyard (or, if they did—for dreams are widespread—nothing would come of it). At the University of Pennsylvania, biologists managed to insert human growth-hormone genes in the fetus of a mouse. After it was born, the mouse grew twice as fast as other mice and to twice their size. Having passed the gene on to its offspring, it made forever moot the question "Are you a man or a mouse?" These mice are both, and neither.

By the end of 1988, according to a tally in the *Times*, there were more than a thousand different strains of such "transgenic" mice, as well as twelve breeds of pig and several varieties of rabbits and fish. In the spring of that year, two Harvard researchers announced the creation of a mouse that was genetically altered to develop cancer, so that oncologists could use it for studying new treatments. Unlike the earlier inventions, this mouse had commercial possibilities and was awarded the nation's first animal patent. The patent was licensed to Du Pont, and the mice will go on sale this fall, for fifty dollars apiece.

Even these mice, though, will be confined to laboratories (until they escape). A bigger barrier probably fell in April of 1987, when workers from a company called Advanced Genetic Sciences released the first genetically altered bacteria to the great outdoors—in a strawberry field in Brentwood, California. Trademarked Frostban, the bacteria—*Pseudomonas syringae* and *Pseudomonas fluorescens*—lacked an "ice-nucleating gene" in their DNA and were designed to prevent crop losses from frost damage. Environmental activists had ripped up many of the strawberry plants in an attempt to delay the test, but it was an empty gesture. A few days later, Steven Lindow, the man who discovered the operative gene, sprayed Frostban bacteria on a field of potato plants in Tule Lake, California, without any interference.

The pace of this revolution keeps speeding up. Genetically "improved" trees, for instance, already exist. A Seattle company selects "élite" redwoods from its wild stands, on the basis of such qualities as trunk straightness, height, "specific gravity" of the wood, and "proper branch drop." Then it clones the trees and plants the seedlings; eventually, gnarly, crooked trees will be gone from its stands. Classical methods of improving seeds do not "adequately satisfy the criteria of the rapid availability of trees of superior quality," according to a 1982 report by Congress's Office of Technology Assessment. Christmas-tree growers are now attempting to clone trees with branches that rise at the proper forty-five-degree angle and carry thick needles that "do not fall off to litter the living room floor." A company called Calgene has isolated a gene that gives tobacco plants some resistance to the herbicide glyphosate. The herbicide works by blocking a pathway in plants that synthesizes aromatic amino acids; once the tobacco plants have been genetically retuned, however, you can spray the herbicide on the surrounding weeds without hurting the tobacco.

The future—the fairly near future—holds much more, at least according to the most fanciful accounts. In his book, Brian Stableford promises that the "battery chickens" of tomorrow will look very different from the birds of the moment, and in fact the accompanying illustration shows them looking rather like hunks of flesh. This is because we may be able to design chickens without the unnecessary heads, wings, and tails. "Nutrients would be pumped in and wastes pumped out through tubes connected to the body," Stableford says. Perhaps we could "grow" lamb chops on an "infinite production line, with red meat and fat attached to an ever-elongating spine of bone." Eventually, all plants might "become unnecessary," having been replaced by artificial leaves that would waste none of the sunlight they received on such luxuries as roots but instead would employ "the energy they trap to make things for us to use."

And then there's us. What about night vision, or sonar (although, Stableford says, "this would involve whole new anatomical structures being added to the head"), or double-glazed eyes for living in space, or the "very minor modification" that would allow us to digest cellulose? These developments, though in the future (and farther in the future, I would guess, than these authors predict), are not conceptually different from what we have begun to do in the last twenty years, and what we have started to do in a large way in the last two years. The line is not in the distance; the line is here and now, and we shall very soon be on the other side, if we're not there already. And on the other side of the line is the second end of nature.

Some people tend not to worry very much about genetic engineering or other such developments, because they think of them as extensions of traditional practices—selective breeding, for example. But nature put definite limits on such activities. Mendel could cross two peas, but he couldn't cross a pea with a pine, much less with a pig, much less with a person. We could pen up chickens in batteries, but they still had heads. Our understanding of the natural limits helped define nature in our minds. Such notions will quickly become quaint. The idea that nature—that *anything*—could be defined will soon be outdated. Because anything can be changed. A rabbit may be a rabbit for the moment, but tomorrow "rabbit" will have no meaning. "Rabbit" will be a few strands of genetic code, no more important than a set of plans for a 1940 Ford. Why not make a rabbit more like a dog, or a duck? Whatever suits us. In such a world, nothing will be impossible—including, perhaps, immortality. Why die? (Why age?) Whether eternal life will have any meaning is another matter. "Eventually," Stableford says, "there may well be a complete breakdown in the distinction between the living and the non-living: the boundary between the two will be blurred and filled in by systems which involve both the machinery of life and the machinery of metal, plastic, and glass."

All this is speculation, certainly. No one can say with any exactness what will result from a development as awesome as the cracking of the gene. But if that technology falters some other may emerge. It is the logical outcome of our belief that we must forever dominate the world to our advantage. The problem, in other words, is not simply that burning oil releases carbon dioxide, which happens, by virtue of its molecular structure, to trap the sun's heat. The problem is that nature, the independent force that has surrounded us since our earliest days, cannot coexist with our numbers and our habits. We may well be able to create a world that can support our numbers and our habits, but it will be an artificial world—a space station.

Or, just possibly, we could change our habits.

One very small example of an idea so large as to be unwieldy: To cope with the greenhouse world, people in the developed countries will probably begin to install much more energy efficient washing machines. That would reduce somewhat the amount of carbon dioxide each of us puffs into the atmosphere. But what if, instead, people got together with their neighbors and agreed to buy a single washing machine for the entire block (not such a novel concept to people in big-city apartment houses)? And what if they also decided that instead of continually buying fashionable clothes they would reduce their wardrobes to a comfortable, or even uncomfortable, minimum? What if, in other words, we began to reject a pervasive individual consumerism, and began to alter a basic way we look at ourselves? Mightn't such a path, broadened to include other facets of daily life, offer the best way not only to avoid overheating the planet but also to keep from transforming it in the other sad ways I have discussed?

As long as the desire for endless material advancement drives us, there is no way to set limits. We are unlikely to develop genetic engineering to eradicate disease and then not use it to manufacture perfectly efficient chickens; there is nothing in the logic of our beliefs that would lead us to draw that line. If there is one item that virtually all successful politicians on earth—socialist and fascist and capitalist—agree on, it is that "economic growth" is good, necessary, the proper end of organized human activity.

Our present environmental troubles, though, just might give us the chance to change the way we think. Spurred by the realization of what we have done, we might begin to think and then behave more humbly. As the effects of man's domination have become clearer in recent years, a new idea has begun to spread, both in America and abroad. Some environmentalists have begun to talk of two approaches to the world: the traditional anthropocentric view, and the biocentric vision of mankind as just another part of the world. This concept is foreign to most of us. My first sense of what it might mean came a couple of summers ago in Idaho, when I was camped next to a man who hikes almost every year from Mexico to Canada. A dozen times, he told me, he had met grizzly bears, the grandest mammals left on the continent: "The last one, he stood on his hind legs, clicked his jaws, woofed three times. I was too close to him, and he was just letting me know. Another one, he circled me about forty feet away and wouldn't look me in the eye. When you get that close, you realize you're part of the food chain."

The idea that man doesn't necessarily belong at the top—that the hierarchy we've spent many thousands of years establishing is dangerous to other species and also to ourselves—is a strange and powerful idea. The few philosophers and environmentalists interested in such a Copernican shift have taken to calling this alternative path "deep ecology"—as opposed to the "shallow ecology" of conventional environmentalism, which seeks merely to turn mankind into better stewards. Deep ecology suggests that instead of just giving better orders we learn to give fewer and fewer orders—to sink back into the natural world. Deep ecologists question the industrial basis of our civilization, the need to forever grow in wealth and numbers, the entire way we live. We should, they say, work toward a smaller world population—half the current one, maybe, or even less. And we should lay aside our desire for material advancement in favor of "doing with enough."

Such ideas are not blueprints; they aren't even outlines. But they are at least a starting point for those who seek to save a world fast vanishing. They are radical ideas, but we live at a radical moment. We live at the end of nature, the instant when the essential character of the world is changing. If our way of life is ending nature, it is not radical to talk about transforming our way of life. When I climb the hill out back, I often pause on a ledge from which I can see my house—the car in the driveway, the chimney above the stove. I love the life that that house represents, love it very much. But I love the hemlocks around me on the hill, too, and the coyotes, and the deer. And it seems that either that life down there must change or the life up here around me will change—the trees will wilt in the sun or else sprout in perfect, heat-tolerant, genetically improved rows.

Exactly what a humbler world would look like I cannot say. We are used to planning utopias, worlds engineered for human happiness. But this would be something different—an "atopia," perhaps, where the integrity of the planet, and not our desires, would be the engine. If our thinking changed, the details would follow of their own accord. Perhaps we might all begin to use the "appropriate technology" of "sustainable development" which we urge on Third World peasants—solar cookstoves, or bicycle-powered pumps. Probably many more of us would be growing our own food. Such solutions are not beyond our imagination. When we decided that accumulation and growth were our economic ideals, we invented wills and lending at interest and puritanism and supersonic aircraft. Why would we come up with ideas less powerful in an all-out race to do with less?

The difficulty in accomplishing this transformation is almost certainly more psychological than intellectual—less that we can't figure out major alterations in our way of life than that we don't want to. The people whose lives may point the way—Thoreau, say, or Gandhi—we dismiss as exceptional, a polite way of saying that there is no reason we should be expected to go where they pointed. The challenge they presented with the example of their lives is much more subversive than anything they wrote or said, for if they could live that simply it's no use saying we couldn't. And maybe now we should—not just for moral or aesthetic reasons but for reasons of chemistry and physics.

Such a change would obviously be colossally difficult. For one thing, while we as individuals would have to change our habits, it would mean very little—save as a good gesture—for any one of us to, say, drive less. *Most* people have to be persuaded to drive less, and persuaded quickly; this is the first environmental crisis one can't escape by heading for the woods. It's also difficult for us to turn our backs on the idea of economic growth, because it has been sold as the answer to the poverty that afflicts most of the planet. For example,

S. Fred Singer, the greenhouse skeptic, writes, "Drastically limiting the emission of carbon dioxide means cutting deeply into global energy use. But limiting economic growth condemns the poor, especially in the Third World, to continued poverty, if not outright starvation." I am sometimes dubious about the actual depth of feeling for the Third World such arguments imply; they mesh too conveniently with our desires. An overheated, ozone-depleted world would probably be crueller to the poor than to the rich, and if our desire is to alleviate poverty, limiting our standard of living and sharing our surplus would likely work as well But I have no doubt about the power of arguments like Singer's to stall effective action of any sort if we are reluctant to take such action in the first place.

Still, problems like the inertia of affluence, the push of poverty, and the soaring population are traditional problems. We can think about them, deal with them, perhaps overcome them. In my lifetime, in this country, we have gone from Jim Crow to affirmative action, and there is no saying we can't do something similar with regard to the planet.

I fear that we won't, though, and for an entirely different set of reasons—reasons intimately linked to the unique and depressing moment in which we find ourselves. As we have seen, nature is already ending. And not only does its passing prevent us from returning to the world we previously knew but also, for a couple of powerful reasons, it makes any of the fundamental changes I've discussed even more unlikely than they might be in easier times.

In the first place, the end of nature is a plunge into the unknown, fearful as much *because* it is unknown as because the world may become hot or dry or whipped by hurricanes. But the type of shift in attitudes I've been describing—the deep-ecology alternative, for instance—would make life even more unpredictable. One would have to begin to forgo the traditional methods of securing one's future—children, possessions, and so on. As the familiar world around us starts to change, every threatened instinct will have us scrambling to preserve at least our familiar style of life. We can—we may well—make the adjustments necessary for our survival. For instance, some of the early work in agricultural biotechnology has focussed on inventing plants able to survive heat and drought. It seems the sensible thing to do—the way to keep life as "normal" as possible in the face of change. It leads, though, as I have said, to the second end of nature: the imposition of our artificial world in place of the broken natural one.

I got a glimpse of this particular future a few years ago, when I spent some time along the La Grande River in sub-Arctic Quebec. It is barren land but beautiful—a taiga of tiny ponds and hummocks stretching to the horizon, carpeted in light-green caribou moss. There are trees—almost all black spruce, and all spindly, sparse. No one lived there save a small number of Indians and Eskimos—about the number the area could support. A decade or so ago, Hydro-Québec, the provincial utility, decided to exploit the power of the La Grande by building three huge dams along a three-hundred-and-fifty-mile stretch of the river. The largest is the size of fifty-four thousand two-story houses, a HydroQuébec spokesman told me. Its spillway could carry the combined flow of all the rivers of Europe. Erecting it was a Bunyanesque task: eighteen thousand men carved the roads north through the taiga and poured the concrete. (Photographs show the cooks stirring spaghetti sauce with canoe paddles.) This is the perfect example of "environmentally sound" energy generation; the dams produce a tremendous amount of power without giving off any greenhouse gas. They are the sort of structure we will be clamoring to build as the warming progresses.

But environmentally sound is not the same as natural. The dams have altered an area larger than Switzerland.

The flow of the Caniapiscau River has been partly reversed to provide more water for the turbines. In September of 1984, at least ten thousand caribou drowned trying to cross the river during their annual migration. They were crossing at their usual spot, but the river was not its usual size; it was so swollen that many of the animals were swept forty-five miles downstream. Every good argument—the argument that fossil fuels cause the greenhouse effect, the argument that in a drier, hotter world we will need more water, the argument that as our margin of security dwindles we must act to restore it—will lead us to more La Grande projects, more dams on the Colorado, more "management." Every argument that the warmer weather and increased ultraviolet are killing plants and causing cancer will have us looking to genetic engineering for salvation. And with each such step we will move farther from nature.

And as that happens the counter-argument—the argument for nature—will grow ever fainter. Wendell Berry once argued that in the absence of a "fascination" with the wonder of the natural world "the energy needed for its preservation will never be developed"—that "there must be a mystique of the rain if we are ever to restore the purity of the rainfall." This makes sense when the problem is transitory—sulfur-dioxide emissions drifting over the Adirondacks. But how can there be a mystique of the rain, now that every drop—even the drops that fall as snow on the Arctic, even the drops that fall deep in the remaining forest primeval—bears the permanent stamp of man? Having lost its separateness, nature loses its special power. Instead of being a category like God—something beyond our control—it is now a category like the defense budget or the minimum wage, a problem we must work out. This alone changes its meaning completely, and changes our reaction to it. The end of nature probably also makes us reluctant to attach ourselves to its remnants, for the same reason that we usually don't choose new friends from among the terminally ill. I love the mountain outside my back door—the stream that runs along its flank, and the stream that slides down a quarter-mile mossy chute, and the place where the slope flattens into an open plain of birch and oak. But I know that in some way I resist getting to know it better—for fear, weak-kneed as it sounds, of getting hurt. I fear that if I knew as well as a forester what sick trees look like I would see them everywhere. I find now that I like the woods best in winter, when it is harder to tell what might be dying, but I try not to love even winter too much, because of the January perhaps not so distant when the snow will fall as warm rain. There is no future in loving nature.

And there may not even be much past. Though Thoreau's writings grew in value and importance the closer we drew to the end of nature, the time fast approaches when he will be inexplicable, his notions less comprehensible to future men than cave paintings are to us. Thoreau writes of the land around Katahdin that it "was vast, Titanic, and such as man never inhabits. Some part of the beholder, even some vital part, seems to escape through the loose grating of his ribs. . . . Nature has got him at a disadvantage, caught him alone, and pilfers him of some of his divine faculty. She does not smile on him as in the plains. She seems to say sternly, Why came ye here before your time. This ground is not prepared for you." That sentiment describes perfectly the last stage of the relationship of man to nature; though we had subdued her in the low places, the peaks, the poles, the jungles still rang with her pure message. But what will this passage mean in the years to come, when Katahdin, the "cloud factory," is ringed by clouds that are the work of man? When the great pines around its base have been genetically improved for straightness of trunk and "proper branch drop," or, more likely, have sprung from the cones of genetically improved trees that began a few miles and a few generations distant on some timber plantation? When the moose that ambles by is part of a herd whose rancher is committed to the enlightened notion that "conservation and profit can go hand-in-hand"? Soon Thoreau will make no sense. And when that happens the end of nature, which began with our alteration of the atmosphere and continued with the responses of the planetary managers and the genetic engineers, will be final. The loss of memory will be the eternal loss of meaning.

I understand perfectly well that defiance may bring prosperity, and a sort of security—that more hydropower will mean less carbon dioxide, and that genetic engineering will help the sick, and that much progress can still be made against human misery. And I have no plans to live in a cave, or even in an unheated cabin. If it took twelve thousand years to get where we are, it will take a few generations to climb back down. But this could be the epoch in which people decide at least to go no farther along the path we have been following—when we make not only the necessary technological adjustments to preserve the world from overheating but also the necessary mental adjustments to insure that we will never again put our good ahead of everything else's. This is the path I choose, for it offers at least a shred of hope for a living, eternal, meaningful world.

As birds have flight, our special gift is reason. Part of that reason drives the intelligence that allows us to master DNA or build big power plants. But our reason could also keep us from following blindly the biological imperatives toward endless growth in numbers and territory. Our reason allows us to conceive of our species as a species, and to recognize the danger that our growth poses to it, and to feel something for the other species we threaten. Should we so choose, we could exercise our reason to do what no other animal can do: we could limit ourselves voluntarily, choose to remain God's creatures instead of making ourselves gods. What a towering achievement that would be, so much more impressive than the largest dam—beavers can build dams—because so much harder. Such restraint, not genetic engineering or planetary management, is the real challenge. If we now, today, began to limit our numbers and our desires and our ambitions, perhaps nature could someday resume its independent working. Perhaps the temperature could someday adjust itself down to its own setting, and the rain fall of its own accord.

The Climate of Man

Elizabeth Kolbert

May 2, 2005

The world's first empire was established forty-three hundred years ago, between the Tigris and Euphrates Rivers. The details of its founding, by Sargon of Akkad, have come down to us in a form somewhere between history and myth. Sargon—Sharru-kin, in the language of Akkadian—means "true king"; almost certainly, though, he was a usurper. As a baby, Sargon was said to have been discovered, Moses-like, floating in a basket. Later, he became cupbearer to the ruler of Kish, one of ancient Babylonia's most powerful cities. Sargon dreamed that his master, Ur-Zababa, was about to be drowned by the goddess Inanna in a river of blood. Hearing about the dream, Ur-Zababa decided to have Sargon eliminated. How this plan failed is unknown; no text relating the end of the story has ever been found.

Until Sargon's reign, Babylonian cities like Kish, and also Ur and Uruk and Umma, functioned as independent city-states. Sometimes they formed brief alliances—cuneiform tablets attest to strategic marriages celebrated and diplomatic gifts exchanged—but mostly they seem to have been at war with one another. Sargon first subdued Babylonia's fractious cities, then went on to conquer, or at least sack, lands like Elam, in present-day Iran. He presided over his empire from the city of Akkad, the ruins of which are believed to lie south of Baghdad. It was written that "daily five thousand four hundred men ate at his presence," meaning, presumably, that he maintained a huge standing army. Eventually, Akkadian hegemony extended as far as the Khabur plains, in northeastern Syria, an area prized for its grain production. Sargon came to be known as "king of the world"; later, one of his descendants enlarged this title to "king of the four corners of the universe."

Akkadian rule was highly centralized, and in this way anticipated the administrative logic of empires to come. The Akkadians levied taxes, then used the proceeds to support a vast network of local bureaucrats.

They introduced standardized weights and measures—the gur equalled roughly three hundred litres—and imposed a uniform dating system, under which each year was assigned the name of a major event that had recently occurred: for instance, "the year that Sargon destroyed the city of Mari." Such was the level of systematization that even the shape and the layout of accounting tablets were imperially prescribed. Akkad's wealth was reflected in, among other things, its art work, the refinement and naturalism of which were unprecedented.

Sargon ruled, supposedly, for fifty-six years. He was succeeded by his two sons, who reigned for a total of twenty-four years, and then by a grandson, Naram-sin, who declared himself a god. Naram-sin was, in turn, succeeded by his son. Then, suddenly, Akkad collapsed. During one three-year period, four men each, briefly, claimed the throne. "Who was king? Who was not king?" the register known as the Sumerian King List asks, in what may be the first recorded instance of political irony.

The lamentation "The Curse of Akkad" was written within a century of the empire's fall. It attributes Akkad's demise to an outrage against the gods. Angered by a pair of inauspicious oracles, Naram-sin plunders the temple of Enlil, the god of wind and storms, who, in retaliation, decides to destroy both him and his people:

> For the first time since cities were built and
> founded,
> The great agricultural tracts produced no grain,
> The inundated tracts produced no fish,
> The irrigated orchards produced neither syrup nor
> wine,
> The gathered clouds did not rain, the masgurum did
> not grow.
> At that time, one shekel's worth of oil was only
> one-half quart,
> One shekel's worth of grain was only one-half
> quart. . . .
> These sold at such prices in the markets of all the
> cities!
> He who slept on the roof, died on the roof,
> He who slept in the house, had no burial,
> People were flailing at themselves from hunger.

For many years, the events described in "The Curse of Akkad" were thought, like the details of Sargon's birth, to be purely fictional.

In 1978, after scanning a set of maps at Yale's Sterling Memorial Library, a university archeologist named Harvey Weiss spotted a promising-looking mound at the confluence of two dry riverbeds in the Khabur plains, near the Iraqi border. He approached the Syrian government for permission to excavate the mound, and, somewhat to his surprise, it was almost immediately granted. Soon, he had uncovered a lost city, which in ancient times was known as Shekhna and today is called Tell Leilan.

Over the next ten years, Weiss, working with a team of students and local laborers, proceeded to uncover an acropolis, a crowded residential neighborhood reached by a paved road, and a large block of grain-storage rooms. He found that the residents of Tell Leilan had raised barley and several varieties of wheat, that they had used carts to transport their crops, and that in their writing they had imitated the style of their more sophisticated neighbors to the south. Like most cities in the region at the time, Tell Leilan had a rigidly organized, state-run economy: people received rations—so many litres of barley and so many of oil—based on how old they were and what kind of work they performed. From the time of the Akkadian empire, thousands of similar potsherds were discovered, indicating that residents had received their rations in mass-produced, one-litre vessels. After examining these and other artifacts, Weiss constructed a time line of the city's history, from its origins as a small farming village (around 5000 B.C.), to its growth into an independent city of some thirty thousand people (2600 B.C.), and on to its reorganization under imperial rule (2300 B.C.).

Wherever Weiss and his team dug, they also encountered a layer of dirt that contained no signs of human habitation. This layer, which was more than three feet deep, corresponded to the years 2200 to 1900 B.C., and it indicated that, around the time of Akkad's fall, Tell Leilan had been completely abandoned. In 1991, Weiss sent soil samples from Tell Leilan to a lab for analysis. The results showed that, around the year 2200 B.C., even the city's earthworms had died out. Eventually, Weiss came to believe that the lifeless soil of Tell Leilan and the end of the Akkadian empire were products of the same phenomenon—a drought so prolonged and so severe that, in his words, it represented an example of "climate change."

Weiss first published his theory, in the journal *Science*, in August, 1993. Since then, the list of cultures whose demise has been linked to climate change has continued to grow. They include the Classic Mayan civilization, which collapsed at the height of its development, around 800 A.D.; the Tiwanaku civilization, which thrived near Lake Titicaca, in the Andes, for more than a millennium, then disintegrated around 1100 A.D.; and the Old Kingdom of Egypt, which collapsed around the same time as the Akkadian empire. (In an account eerily reminiscent of "The Curse of Akkad," the Egyptian sage Ipuwer described the anguish of the period: "Lo, the desert claims the land. Towns are ravaged. . . . Food is lacking. . . . Ladies suffer like maidservants. Lo, those who were entombed are cast on high grounds.") In each of these cases, what began as a provocative hypothesis has, as new information has emerged, come to seem more and more compelling. For example, the notion that Mayan civilization had been undermined by climate change was first proposed in the late nineteen-eighties, at which point there was little climatological evidence to support it. Then, in the mid-nineteen-nineties, American scientists studying sediment cores from Lake Chichancanab, in north-central Yucatán, reported that precipitation patterns in the region had indeed shifted during the ninth and tenth centuries, and that this shift had led to periods of prolonged drought. More recently, a group of researchers examining ocean-sediment cores collected off the coast of Venezuela produced an even more detailed record of rainfall in the area. They found that the region experienced a series of severe, "multi-year drought events" beginning around 750 A.D. The collapse of the Classic Mayan civilization, which has been described as "a demographic disaster as profound as any other in human history," is thought to have cost millions of lives.

The climate shifts that affected past cultures predate industrialization by hundreds—or, in the case of the Akkadians, thousands—of years. They reflect the climate system's innate variability and were caused by forces that, at this point, can only be guessed at. By contrast, the climate shifts predicted for the coming century are attributable to forces that are now well known. Exactly how big these shifts will be is a matter of both intense scientific interest and the greatest possible historical significance. In this context, the discovery that large and sophisticated cultures have already been undone by climate change presents what can only be called an uncomfortable precedent.

The Goddard Institute for Space Studies, or GISS, is situated just south of Columbia University's main campus, at the corner of Broadway and West 112th Street. The institute is not well marked, but most New Yorkers would probably recognize the building: its ground floor is home to Tom's Restaurant, the coffee shop made famous by "Seinfeld."

GISS, an outpost of NASA, started out, forty-four years ago, as a planetary-research center; today, its major function is making forecasts about climate change. GISS employs about a hundred and fifty people, many of whom spend their days working on calculations that may—or may not—end up being incorporated in the institute's climate model. Some work on algorithms that describe the behavior of the atmosphere, some on the behavior of the oceans, some on vegetation, some on clouds, and some on making sure that all these algorithms, when they are combined, produce results that seem consistent with the real world. (Once, when some refinements were made to the model, rain nearly stopped falling over the rain forest.) The latest version of the GISS model, called ModelE, consists of a hundred and twenty-five thousand lines of computer code.

GISS's director, James Hansen, occupies a spacious, almost comically cluttered office on the institute's seventh floor. (I must have expressed some uneasiness the first time I visited him, because the following day I received an e-mail assuring me that the office was "a lot better organized than it used to be.") Hansen, who is sixty-three, is a spare man with a lean face and a fringe of brown hair. Although he has probably done as much to publicize the dangers of global warming as any other scientist, in person he is reticent almost to the point of shyness. When I asked him how he had come to play such a prominent role, he just shrugged. "Circumstances," he said.

Hansen first became interested in climate change in the mid-nineteen-seventies. Under the direction of James Van Allen (for whom the Van Allen radiation belts are named), he had written his doctoral dissertation on the climate of Venus. In it, he had proposed that the planet, which has an average surface temperature of eight hundred and sixty-seven degrees Fahrenheit, was kept warm by a smoggy haze; soon afterward, a space probe showed that Venus was actually insulated by an atmosphere that consists of ninety-six per cent carbon dioxide. When solid data began to show what was happening to greenhouse-gas levels on earth, Hansen became, in his words, "captivated." He decided that a planet whose atmosphere could change in the course of a human lifetime was more interesting than one that was going to continue, for all intents and purposes, to broil away forever. A group of scientists at NASA had put together a computer program to try to improve weather forecasting using satellite data. Hansen and a team of half a dozen other researchers set out to modify it, in order to make longer-range forecasts about what would happen to global temperatures as greenhouse gases continued to accumulate. The project, which resulted in the first version of the GISS climate model, took nearly seven years to complete.

At that time, there was little empirical evidence to support the notion that the earth was warming. Instrumental temperature records go back, in a consistent fashion, only to the mid-nineteenth century. They show that average global temperatures rose through the first half of the twentieth century, then dipped in the nineteen-fifties and sixties. Nevertheless, by the early nineteen-eighties Hansen had gained enough confidence in his model to begin to make a series of increasingly audacious predictions. In 1981, he forecast that "carbon dioxide warming should emerge from the noise of natural climate variability" around the year 2000. During the exceptionally hot summer of 1988, he appeared before a Senate subcommittee and announced that he was "ninety-nine per cent" sure that "global warming is affecting our planet now." And in the summer of 1990 he offered to bet a roomful of fellow-scientists a hundred dollars that either that year or one of the following two years would be the warmest on record. To qualify, the year would have to set a record not only for land temperatures but also for sea-surface temperatures and for temperatures in the lower atmosphere. Hansen won the bet in six months.

Like all climate models, GISS's divides the world into a series of boxes. Thirty-three hundred and twelve boxes cover the earth's surface, and this pattern is repeated twenty times moving up through the atmosphere, so that the whole arrangement might be thought of as a set of enormous checkerboards stacked on top of one another. Each box represents an area of four degrees latitude by five degrees longitude. (The height of the box varies depending on altitude.) In the real world, of course, such a large area would have an incalculable number of features; in the world of the model, features such as lakes and forests and, indeed, whole mountain ranges are reduced to a limited set of properties, which are then expressed as numerical approximations. Time in this grid world moves ahead for the most part in discrete, half-hour intervals, meaning that a new set of calculations is performed for each box for every thirty minutes that is supposed to have elapsed in actuality. Depending on what part of the globe a box represents, these calculations may involve dozens of different algorithms, so that a model run that is supposed to simulate climate conditions over the next hundred years involves more than a quadrillion separate operations. A single run of the GISS model, done on a supercomputer, usually takes about a month.

Very broadly speaking, there are two types of equations that go into a climate model. The first group expresses fundamental physical principles, like the conservation of energy and the law of gravity. The second group describes—the term of art is "parameterize"—patterns and interactions that have been observed in nature but may be only partly understood, or processes that occur on a small scale, and have to be averaged out over huge spaces. Here, for example, is a tiny piece of ModelE, written in the computer language FORTRAN, which deals with the formation of clouds:

```
C**** COMPUTE THE AUTOCONVERSION RATE OF
CLOUD WATER TO PRECIPITATION
  RHO=1.E5*PL(L)/(RGAS*TL(L))
  TEM=RHO*WMX(L)/(WCONST*FCLD+ 1.E-20)
  IF(LHX.EQ.LHS) TEM=RHO*WMX(L)/
  (WMUI*FCLD+1.E-20)
  TEM=TEM*TEM
  IF(TEM.GT.10.) TEM=10.
  CM1=CM0
  IF(BANDF) CM1=CM0CBF*
  IF(LHX.EQ.LHS) CM1=CM0
  CM=CM1*(1.-1./EXP(TEM*TEM))+1.
  *100.*(PREBAR(L+1)+
  * PRECNVL(L+1)*BYDTsrc*)
  IF(CM.GT.BYDTsrc*) CM=BYDTsrc
  PREP(L)=WMX(L)CM
  END IF
C**** FORM CLOUDS ONLY IF RH GT RH00 219
IF(RH1(L).LT.RH00(L)) GO TO 220.
```

All climate models treat the laws of physics in the same way, but, since they parameterize phenomena like cloud formation differently, they come up with different results. (At this point, there are some fifteen major climate models in operation around the globe.) Also, because the real-world forces influencing the climate are so numerous, different models tend, like medical students, to specialize in different processes. GISS's model, for example, specializes in the behavior of the atmosphere, other models in the behavior of the oceans, and still others in the behavior of land surfaces and ice sheets.

Last fall, I attended a meeting at GISS which brought together members of the institute's modelling team. When I arrived, about twenty men and five women were sitting in battered chairs in a conference room across from Hansen's office. At that particular moment, the institute was performing a series of runs for the U.N. Intergovernmental Panel on Climate Change. The runs were overdue, and apparently the I.P.C.C. was getting impatient. Hansen flashed a series of charts on a screen on the wall summarizing some of the results obtained so far.

The obvious difficulty in verifying any particular climate model or climate-model run is the prospective nature of the results. For this reason, models are often run into the past, to see how well they reproduce trends that have already been observed. Hansen told the group that he was pleased with how ModelE had reproduced the aftermath of the eruption of Mt. Pinatubo, in the Philippines, which took place in June of 1991. Volcanic eruptions release huge quantities of sulfur dioxide—Pinatubo produced some twenty million tons of the gas—which, once in the stratosphere, condenses into tiny sulfate droplets. These droplets, or aerosols, tend to cool the earth by reflecting sunlight back into space. (Man-made aerosols, produced by burning coal, oil, and biomass, also reflect sunlight and are a countervailing force to greenhouse warming, albeit one with serious health consequences of its own.) This cooling effect lasts as long as the aerosols remain suspended in the atmosphere. In 1992, global temperatures, which had been rising sharply, fell by half of a degree. Then they began to climb again. ModelE had succeeded in simulating this effect to within nine-hundredths of a degree. "That's a pretty nice test," Hansen observed laconically.

One day, when I was talking to Hansen in his office, he pulled a pair of photographs out of his briefcase. The first showed a chubby-faced five-year-old girl holding some miniature Christmas-tree lights in front of an even chubbier-faced five-month-old baby. The girl, Hansen told me, was his granddaughter Sophie and the boy was his new grandson, Connor. The caption on the first picture read, "Sophie explains greenhouse warming." The caption on the second photograph, which showed the baby smiling gleefully, read, "Connor gets it."

When modellers talk about what drives the climate, they focus on what they call "forcings." A forcing is any ongoing process or discrete event that alters the energy of the system. Examples of natural forcings include, in addition to volcanic eruptions, periodic shifts in the earth's orbit and changes in the sun's output, like those linked to sunspots. Many climate shifts of the past have no known forcing associated with them; for instance, no one is certain what brought about the so-called Little Ice Age, which began in Europe some five hundred years ago. A very large forcing, meanwhile, should produce a commensurately large—and obvious—effect. One GISS scientist put it to me this way: "If the sun went supernova, there's no question that we could model what would happen."

Adding carbon dioxide, or any other greenhouse gas, to the atmosphere by, say, burning fossil fuels or levelling forests is, in the language of climate science, an anthropogenic forcing. Since pre-industrial times, the concentration of CO_2 in the earth's atmosphere has risen by roughly a third, from 280 parts per million to 378 p.p.m. During the same period, concentrations of methane, an even more powerful (but more short-lived) greenhouse gas, have more than doubled, from .78 p.p.m. to 1.76 p.p.m. Scientists measure forcings in terms of watts per square metre, or w/m2, by which they mean that a certain number of watts of energy have been added (or, in the case of a negative forcing, subtracted) for every single square metre of the earth's surface. The size of the greenhouse forcing is estimated, at this point, to be 2.5 w/m2. A miniature Christmas light gives off about four tenths of a watt of energy, mostly in the form of heat, so that, in effect (as Sophie supposedly explained to Connor), we have covered the earth with tiny bulbs, six for every square metre. These bulbs are burning twenty-four hours a day, seven days a week, year in and year out.

If greenhouse gases were held constant at today's levels, it is estimated that it would take several decades for the full impact of the forcing that is already in place to be felt. This is because raising the earth's temperature involves not only warming the air and the surface of the land but also melting sea ice, liquefying glaciers, and, most significant, heating the oceans—all processes that require tremendous amounts of energy. (Imagine trying to thaw a gallon of ice cream or warm a pot of water using an Easy-Bake oven.) It could be argued that the delay that is built into the system is socially useful, because it enables us—with the help of climate models—to prepare for what lies ahead, or that it is socially disastrous, because it allows us to keep adding CO_2 to the atmosphere while fobbing the impacts off on our children and grandchildren. Either way, if current trends continue, which is to say, if steps are not taken to reduce emissions, carbon-dioxide levels will probably reach 500 parts per million—nearly double pre-industrial levels—sometime around the middle of the century. By that point, of course, the forcing associated with greenhouse gases will also have increased, to four watts per square metre and possibly more. For comparison's sake, it is worth keeping in mind that the total forcing that ended the last ice age—a forcing that was eventually sufficient to melt mile-thick ice sheets and raise global sea levels by four hundred feet—is estimated to have been just six and a half watts per square metre.

There are two ways to operate a climate model. In the first, which is known as a transient run, greenhouse gases are slowly added to the simulated atmosphere—just as they would be to the real atmosphere—and the model forecasts what the effect of these additions will be at any given moment. In the second, greenhouse gases are added to the atmosphere all at once, and the model is run at these new levels until the climate has fully adjusted to the forcing by reaching a new equilibrium. Not surprisingly, this is known as an equilibrium run. For doubled CO_2, equilibrium runs of the GISS model predict that average global temperatures will rise by 4.9 degrees Fahrenheit. Only about a third of this increase is directly attributable to more greenhouse gases; the rest is a result of indirect effects, the most important among them being the so-called "water-vapor feedback." (Since warmer air holds more moisture, higher temperatures are expected to produce an atmosphere containing more water vapor, which is itself a greenhouse gas.) GISS's forecast is on the low end of the most recent projections; the Hadley Centre model, which is run by the British Met Office, predicts that for doubled CO_2 the eventual temperature rise will be 6.3 degrees Fahrenheit, while Japan's National Institute for Environmental Studies predicts 7.7 degrees.

In the context of ordinary life, a warming of 4.9, or even of 7.7, degrees may not seem like much to worry about; in the course of a normal summer's day, after all, air temperatures routinely rise by twenty degrees or more. Average global temperatures, however, have practically nothing to do with ordinary life. In the middle of the last glaciation, Manhattan, Boston, and Chicago were deep under ice, and sea levels were so low that Siberia and Alaska were connected by a land bridge nearly a thousand miles wide. At that point, average global temperatures were roughly ten degrees colder than they are today. Conversely, since our species evolved, average temperatures have never been much more than two or three degrees higher than they are right now.

This last point is one that climatologists find particularly significant. By studying Antarctic ice cores, researchers have been able to piece together a record both of the earth's temperature and of the composition of its atmosphere going back four full glacial cycles. (Temperature data can be extracted from the isotopic composition of the ice, and the makeup of the atmosphere can be reconstructed by analyzing tiny bubbles of trapped air.) What this record shows is that the planet is now nearly as warm as it has been at any point in the last four hundred and twenty thousand years. A possible consequence of even a four- or five-degree temperature rise—on the low end of projections for doubled CO_2—is that the world will enter a completely new climate regime, one with which modern humans have no prior experience. Meanwhile, at 378 p.p.m., CO_2 levels are significantly higher today than they have been at any other point in the Antarctic record. It is believed that the last time carbon-dioxide levels were in this range was three and a half million years ago, during what is known as the mid-Pliocene warm period, and they likely have not been much above it for tens of millions of years. A scientist with the National Oceanic and Atmospheric Administration (NOAA) put it to me—only half-jokingly—this way: "It's true that we've had higher CO_2 levels before. But, then, of course, we also had dinosaurs."

David Rind is a climate scientist who has worked at GISS since 1978. Rind acts as a trouble-shooter for the institute's model, scanning reams of numbers known as diagnostics, trying to catch problems, and he also works with GISS's Climate Impacts Group. (His office, like Hansen's, is filled with dusty piles of computer printouts.) Although higher temperatures are the most obvious and predictable result of increased CO_2, other, second-order consequences—rising sea levels, changes in vegetation, loss of snow cover—are likely to be just as significant. Rind's particular interest is how CO_2 levels will affect water supplies, because, as he put it to me, "you can't have a plastic version of water."

One afternoon, when I was talking to Rind in his office, he mentioned a visit that President Bush's science adviser, John Marburger, had paid to GISS a few years earlier. "He said, 'We're really interested in adaptation to climate change,'" Rind recalled. "Well, what does 'adaptation' mean?" He rummaged through one of his many file cabinets and finally pulled out a paper that he had published in the *Journal of Geophysical Research* entitled "Potential Evapotranspiration and the Likelihood of Future Drought." In much the same way that wind velocity is measured using the Beaufort scale, water availability is measured using what's known as the Palmer Drought Severity Index. Different climate models offer very different predictions about future water availability; in the paper, Rind applied the criteria used in the Palmer index to GISS's model and also to a model operated by NOAA's Geophysical Fluid Dynamics Laboratory. He found that as carbon-dioxide levels rose the world began to experience more and more serious water shortages, starting near the equator and then spreading toward the poles. When he applied the index to the GISS model for doubled CO_2, it showed most of the continental United States to be suffering under severe drought conditions. When he applied the index to the G.F.D.L. model, the results were even more dire. Rind created two maps to illustrate these findings. Yellow represented a forty-to-sixty-per-cent chance of summertime drought, ochre a sixty-to-eighty-per-cent chance, and brown an eighty-to-a-hundred-per-cent chance. In the first map, showing the GISS results, the Northeast was yellow, the Midwest was ochre, and the Rocky Mountain states and California were brown. In the second, showing the G.F.D.L. results, brown covered practically the entire country.

"I gave a talk based on these drought indices out in California to water-resource managers," Rind told me. "And they said, 'Well, if that happens, forget it.' There's just no way they could deal with that."

He went on, "Obviously, if you get drought indices like these, there's no adaptation that's possible. But let's say it's not that severe. What adaptation are we talking about? Adaptation in 2020? Adaptation in 2040? Adaptation in 2060? Because the way the models project this, as global warming gets going, once you've adapted to one decade you're going to have to change everything the next decade.

"We may say that we're more technologically able than earlier societies. But one thing about climate change is it's potentially geopolitically destabilizing. And we're not only more technologically able; we're more technologically able destructively as well. I think it's impossible to predict what will happen. I guess—though I won't be around to see it—I wouldn't be shocked to find out that by 2100 most things were destroyed." He paused. "That's sort of an extreme view."

On the other side of the Hudson River and slightly to the north of GISS, the Lamont-Doherty Earth Observatory occupies what was once a weekend estate in the town of Palisades, New York. The observatory is an outpost of Columbia University, and it houses, among its collections of natural artifacts, the world's largest assembly of ocean-sediment cores—more than thirteen thousand in all. The cores are kept in steel compartments that look like drawers from a filing cabinet, only longer and much skinnier. Some of the cores are chalky, some are clayey, and some are made up almost entirely of gravel. All can be coaxed to yield up—in one way or another—information about past climates.

Peter deMenocal is a paleoclimatologist who has worked at Lamont-Doherty for fifteen years. He is an expert on ocean cores, and also on the climate of the Pliocene, which lasted from roughly five million to two million years ago. Around two and a half million years ago, the earth, which had been warm and relatively ice-free, started to cool down until it entered an era—the Pleistocene—of recurring glaciations. DeMenocal has argued that this transition was a key event in human evolution: right around the time that it occurred, at least two types of hominids—one of which would eventually give rise to us—branched off from a single ancestral line. Until quite recently, paleoclimatologists like deMenocal rarely bothered with anything much closer to the present day; the current interglacial—the Holocene—which began some ten thousand years ago, was believed to be, climatically speaking, too stable to warrant much study. In the mid-nineties, though, deMenocal, motivated by a growing concern over global warming—and a concomitant shift in government research funds—decided to look in detail at some Holocene cores. What he learned, as he put it to me when I visited him at Lamont-Doherty last fall, was "less boring than we had thought."

One way to extract climate data from ocean sediments is to examine the remains of what lived or, perhaps more pertinently, what died and was buried there. The oceans are rich with microscopic creatures known as foraminifera. There are about thirty planktonic species in all, and each thrives at a different temperature, so that by counting a species' prevalence in a given sample it is possible to estimate the ocean temperatures at the time the sediment was formed. When deMenocal used this technique to analyze cores that had been collected off the coast of Mauritania, he found that they contained evidence of recurring cool periods; every fifteen hundred years or so, water temperatures dropped for a few centuries before climbing back up again. (The most recent cool period corresponds to the Little Ice Age, which ended about a century and a half ago.) Also, perhaps even more significant, the cores showed profound changes in precipitation. Until about six thousand years ago, northern Africa was relatively wet—dotted with small lakes. Then it became dry, as it is today. DeMenocal traced the shift to periodic variations in the earth's orbit, which, in a generic sense, are the same forces that trigger ice ages. But orbital changes occur gradually, over thousands of years, and northern Africa appears to have switched from wet to dry all of a sudden. Although no one knows exactly how this happened, it seems, like so many climate events, to have been a function of feedbacks—the less rain the continent got, the less vegetation there was to retain water, and so on until, finally, the system just flipped. The process provides yet more evidence of how a very small forcing sustained over time can produce dramatic results.

"We were kind of surprised by what we found," deMenocal told me about his work on the supposedly stable Holocene. "Actually, more than surprised. It was one of these things where, you know, in life you take certain things for granted, like your neighbor's not going to be an axe murderer. And then you discover your neighbor *is* an axe murderer."

Not long after deMenocal began to think about the Holocene, a brief mention of his work on the climate of Africa appeared in a book produced by *National Geographic.* On the facing page, there was a piece on Harvey Weiss and his work at Tell Leilan. DeMenocal vividly remembers his reaction. "I thought, Holy cow, that's just amazing!" he told me. "It was one of these cases where I lost sleep that night, I just thought it was such a cool idea."

DeMenocal also recalls his subsequent dismay when he went to learn more. "It struck me that they were calling on this climate-change argument, and I wondered how come I didn't know about it," he said. He looked at the *Science* paper in which Weiss had originally laid out his theory. "First of all, I scanned the list of authors and there was no paleoclimatologist on there," deMenocal said. "So then I started reading through the paper and there basically was no paleoclimatology in it." (The main piece of evidence Weiss adduced for a drought was that Tell Leilan had filled with dust.) The more deMenocal thought about it, the more unconvincing he found the data, on the one hand, and the more compelling he found the underlying idea, on the other. "I just couldn't leave it alone," he told me. In the summer of 1995, he went with Weiss to Syria to visit Tell Leilan. Subsequently, he decided to do his own study to prove—or disprove—Weiss's theory.

Instead of looking in, or even near, the ruined city, deMenocal focussed on the Gulf of Oman, nearly a thousand miles downwind. Dust from the Mesopotamian floodplains, just north of Tell Leilan, contains heavy concentrations of the mineral dolomite, and since arid soil produces more wind-borne dust, deMenocal figured that if there had been a drought of any magnitude it would show up in gulf sediments. "In a wet period, you'd be getting none or very, very low amounts of dolomite, and during a dry period you'd be getting a lot," he explained. He and a graduate student named Heidi Cullen developed a highly sensitive test to detect dolomite, and then Cullen assayed, centimetre by centimetre, a sediment core that had been extracted near where the Gulf of Oman meets the Arabian Sea.

"She started going up through the core," DeMenocal told me. "It was like nothing, nothing, nothing, nothing, nothing. Then one day, I think it was a Friday afternoon, she goes, 'Oh, my God.' It was really classic." DeMenocal had thought that the dolomite level, if it were elevated at all, would be modestly higher; instead, it went up by four hundred per cent. Still, he wasn't satisfied. He decided to have the core re-analyzed using a different marker: the ratio of strontium 86 and strontium 87 isotopes. The same spike showed up. When deMenocal had the core carbon-dated, it turned out that the spike lined up exactly with the period of Tell Leilan's abandonment.

Tell Leilan was never an easy place to live. Much like, say, western Kansas today, the Khabur plains received enough annual rainfall—about seventeen inches—to support cereal crops, but not enough to grow much else. "Year-to-year variations were a real threat, and so they obviously needed to have grain storage and to have ways to buffer themselves," deMenocal observed. "One generation would tell the next, 'Look, there are these things that happen that you've got to be prepared for.' And they were good at that. They could manage that. They were there for hundreds of years."

He went on, "The thing they couldn't prepare for was the same thing that we won't prepare for, because in their case they didn't know about it and because in our case the political system can't listen to it. And that is that the climate system has much greater things in store for us than we think."

Shortly before Christmas, Harvey Weiss gave a lunchtime lecture at Yale's Institute for Biospheric Studies. The title was "What Happened in the Holocene," which, as Weiss explained, was an allusion to a famous archeology text by V. Gordon Childe, entitled "What Happened in History." The talk brought together archeological and paleoclimatic records from the Near East over the last ten thousand years.

Weiss, who is sixty years old, has thinning gray hair, wire-rimmed glasses, and an excitable manner. He had prepared for the audience—mostly Yale professors and graduate students—a handout with a time line of Mesopotamian history. Key cultural events appeared in black ink, key climatological ones in red. The two alternated in a rhythmic cycle of disaster and innovation. Around 6200 B.C., a severe global cold snap—red ink—produced aridity in the Near East. (The cause of the cold snap is believed to have been a catastrophic flood that emptied an enormous glacial lake—called Lake Agassiz—into the North Atlantic.) Right around the same time—black ink—farming villages in northern Mesopotamia were abandoned, while in central and southern Mesopotamia the art of irrigation was invented. Three thousand years later, there was another cold snap, after which settlements in northern Mesopotamia once again were deserted. The most recent red event, in 2200 B.C., was followed by the dissolution of the Old Kingdom in Egypt, the abandonment of villages in ancient Palestine, and the fall of Akkad. Toward the end of his talk, Weiss, using a PowerPoint program, displayed some photographs from the excavation at Tell Leilan. One showed the wall of a building—probably intended for administrative offices—that had been under construction when the rain stopped. The wall was made from blocks of basalt topped by rows of mud bricks. The bricks gave out abruptly, as if construction had ceased from one day to the next.

The monochromatic sort of history that most of us grew up with did not allow for events like the drought that destroyed Tell Leilan. Civilizations fell, we were taught, because of wars or barbarian invasions or political unrest. (Another famous text by Childe bears the exemplary title "Man Makes Himself.") Adding red to the time line points up the deep contingency of the whole enterprise. Civilization goes back, at the most, ten thousand years, even though, evolutionarily speaking, modern man has been around for at least ten times that long. The climate of the Holocene was not boring, but at least it was dull enough to allow people to sit still. It is only after the immense climatic shifts of the glacial epoch had run their course that writing and agriculture finally emerged.

Nowhere else does the archeological record go back so far or in such detail as in the Near East. But similar red-and-black chronologies can now be drawn up for many other parts of the world: the Indus Valley, where, some four thousand years ago, the Harappan civilization suffered a decline after a change in monsoon patterns; the Andes, where, fourteen hundred years ago, the Moche abandoned their cities in a period of diminished rainfall; and even the United States, where the arrival of the English colonists on Roanoke Island, in 1587, coincided with a severe regional drought. (By the time English ships returned to resupply the colonists, three years later, no one was left.) At the height of the Mayan civilization, population density was five hundred per square mile, higher than it is in most parts of the U. S. today. Two hundred years later, much of the territory occupied by the Mayans had been completely depopulated. You can argue that man through culture creates stability, or you can argue, just as plausibly, that stability is for culture an essential precondition.

After the lecture, I walked with Weiss back to his office, which is near the center of the Yale campus, in the Hall of Graduate Studies. This past year, Weiss decided to suspend excavation at Tell Leilan. The site lies only fifty miles from the Iraqi border, and, owing to the uncertainties of the war, it seemed like the wrong sort of place to bring graduate students. When I visited, Weiss had just returned from a trip to Damascus, where he had gone to pay the guards who watch over the site when he isn't there. While he was away from his office, its contents had been piled up in a corner by repairmen who had come to fix some pipes. Weiss considered the piles disconsolately, then unlocked a door at the back of the room.

The door led to a second room, much larger than the first. It was set up like a library, except that instead of books the shelves were stacked with hundreds of cardboard boxes. Each box contained fragments of broken pottery from Tell Leilan. Some were painted, others were incised with intricate designs, and still others were barely distinguishable from pebbles. Every fragment had been inscribed with a number, indicating its provenance.

I asked what he thought life in Tell Leilan had been like. Weiss told me that that was a "corny question," so I asked him about the city's abandonment. "Nothing allows you to go beyond the third or fourth year of a drought, and by the fifth or sixth year you're probably gone," he observed. "You've given up hope for the rain, which is exactly what they wrote in 'The Curse of Akkad.'" I asked to see something that might have been used in Tell Leilan's last days. Swearing softly, Weiss searched through the rows until he finally found one particular box. It held several potsherds that appeared to have come from identical bowls. They were made from a greenish-colored clay, had been thrown on a wheel, and had no decoration. Intact, the bowls had held about a litre, and Weiss explained that they had been used to mete out rations—probably wheat or barley—to the workers of Tell Leilan. He passed me one of the fragments. I held it in my hand for a moment and tried to imagine the last Akkadian who had touched it. Then I passed it back.

The Darkening Sea

Elizabeth Kolbert

November 12, 2006

Pteropods are tiny marine organisms that belong to the very broad class known as zooplankton. Related to snails, they swim by means of a pair of winglike gelatinous flaps and feed by entrapping even tinier marine creatures in a bubble of mucus. Many pteropod species—there are nearly a hundred in all—produce shells, apparently for protection; some of their predators, meanwhile, have evolved specialized tentacles that they employ much as diners use forks to spear escargot. Pteropods are first male, but as they grow older they become female.

Victoria Fabry, an oceanographer at California State University at San Marcos, is one of the world's leading experts on pteropods. She is slight and soft-spoken, with wavy black hair and blue-green eyes. Fabry fell in love with the ocean as a teen-ager after visiting the Outer Banks, off North Carolina, and took up pteropods when she was in graduate school, in the early nineteen-eighties. At that point, most basic questions about the animals had yet to be answered, and, for her dissertation, Fabry decided to study their shell growth. Her plan was to raise pteropods in tanks, but she ran into trouble immediately. When disturbed, pteropods tend not to produce the mucus bubbles, and slowly starve. Fabry tried using bigger tanks for her pteropods, but the only correlation, she recalled recently, was that the more time she spent improving the tanks "the quicker they died." After a while, she resigned herself to constantly collecting new specimens. This, in turn, meant going out on just about any research ship that would have her.

Fabry developed a simple, if brutal, protocol that could be completed at sea. She would catch some pteropods, either by trawling with a net or by scuba diving, and place them in one-litre bottles filled with seawater, to which she had added a small amount of radioactive calcium 45. Forty-eight hours later, she would remove the pteropods from the bottles, dunk them in warm ethanol, and pull their bodies out with a pair of tweezers. Back on land, she would measure how much calcium 45 their shells had taken up during their two days of captivity.

In the summer of 1985, Fabry got a berth on a research vessel sailing from Honolulu to Kodiak Island. Late in the trip, near a spot in the Gulf of Alaska known as Station Papa, she came upon a profusion of *Clio pyramidata*, a half-inch-long pteropod with a shell the shape of an unfurled umbrella. In her enthusiasm, Fabry collected too many specimens; instead of putting two or three in a bottle, she had to cram in a dozen. The next day, she noticed that something had gone wrong. "Normally, their shells are transparent," she said. "They look like little gems, little jewels. They're just beautiful. But I could see that, along the edge, they were becoming opaque, chalky."

Like other animals, pteropods take in oxygen and give off carbon dioxide as a waste product. In the open sea, the CO_2 they produce has no effect. Seal them in a small container, however, and the CO_2 starts to build up, changing the water's chemistry. By overcrowding her *Clio pyramidata*, Fabry had demonstrated that the organisms were highly sensitive to such changes. Instead of growing, their shells were dissolving. It stood to reason that other kinds of pteropods—and, indeed, perhaps any number of shell-building species—were similarly vulnerable. This should have represented a major discovery, and a cause for alarm. But, as is so often the case with inadvertent breakthroughs, it went unremarked upon. No one on the boat, including Fabry, appreciated what the pteropods were telling them, because no one, at that point, could imagine the chemistry of an entire ocean changing.

Since the start of the industrial revolution, humans have burned enough coal, oil, and natural gas to produce some two hundred and fifty billion metric tons of carbon. The result, as is well known, has been a transformation of the earth's atmosphere. The concentration of CO_2 in the air today—three hundred and eighty parts per million—is higher than it has been at any point in the past six hundred and fifty thousand years, and probably much longer. At the current rate of emissions growth, CO_2 concentration will top five hundred parts per million—roughly double preindustrial levels—by the middle of this century. It is expected that such an increase will produce an eventual global temperature rise of between three and a half and seven degrees Fahrenheit, and that this, in turn, will prompt a string of disasters, including fiercer hurricanes, more deadly droughts, the disappearance of most remaining glaciers, the melting of the Arctic ice cap, and the inundation of many of the world's major coastal cities. But this is only half the story.

Ocean covers seventy per cent of the earth's surface, and everywhere that water and air come into contact there is an exchange. Gases from the atmosphere get absorbed by the ocean and gases dissolved in the water are released into the atmosphere. When the two are in equilibrium, roughly the same quantities are being dissolved as are getting released. But change the composition of the atmosphere, as we have done, and the exchange becomes lopsided: more CO_2 from the air enters the water than comes back out. In the nineteen-nineties, researchers from seven countries conducted nearly a hundred cruises, and collected more than seventy thousand seawater samples from different depths and locations. The analysis of these samples, which was completed in 2004, showed that nearly half of all the carbon dioxide that humans have emitted since the start of the nineteenth century has been absorbed by the sea.

When CO_2 dissolves, it produces carbonic acid, which has the chemical formula H_2CO_3. As acids go, H_2CO_3 is relatively innocuous—we drink it all the time in Coke and other carbonated beverages—but in sufficient quantities it can change the water's pH. Already, humans have pumped enough carbon into the oceans—some hundred and twenty billion tons—to produce a .1 decline in surface pH. Since pH, like the Richter scale, is a logarithmic measure, a .1 drop represents a rise in acidity of about thirty per cent. The process is generally referred to as "ocean acidification," though it might more accurately be described as a decline in ocean alkalinity. This year alone, the seas will absorb an additional two billion tons of carbon, and next year it is expected that they will absorb another two billion tons. Every day, every American, in effect, adds forty pounds of carbon dioxide to the oceans.

Because of the slow pace of deep-ocean circulation and the long life of carbon dioxide in the atmosphere, it is impossible to reverse the acidification that has already taken place. Nor is it possible to prevent still more from occurring. Even if there were some way to halt the emission of CO_2 tomorrow, the oceans would continue to take up carbon until they reached a new equilibrium with the air. As Britain's Royal Society noted in a recent report, it will take "tens of thousands of years for ocean chemistry to return to a condition similar to that occurring at pre-industrial times."

Humans have, in this way, set in motion change on a geologic scale. The question that remains is how marine life will respond. Though oceanographers are just beginning to address the question, their discoveries, at this early stage, are disturbing. A few years ago, Fabry finally pulled her cloudy shells out of storage to examine them with a scanning electron microscope. She found that their surfaces were riddled with pits. In some cases, the pits had grown into gashes, and the upper layer had started to pull away, exposing the layer underneath.

The term "ocean acidification" was coined in 2003 by two climate scientists, Ken Caldeira and Michael Wickett, who were working at the Lawrence Livermore National Laboratory, in Northern California. Caldeira has since moved to the Carnegie Institution, on the campus of Stanford University, and during the summer I went to visit him at his office, which is housed in a "green" building that looks like a barn that has been taken apart and reassembled at odd angles. The building has no air-conditioning; temperature control is provided by a shower of mist that rains down into a tiled chamber in the lobby. At the time of my visit, California was in the midst of a record-breaking heat wave; the system worked well enough that Caldeira's office, if not exactly cool, was at least moderately comfortable.

Caldeira is a trim man with wiry brown hair and a boyish sort of smile. In the nineteen-eighties, he worked as a software developer on Wall Street, and one of his clients was the New York Stock Exchange, for whom he designed computer programs to help detect insider trading. The programs functioned as they were supposed to, but after a while Caldeira came to the conclusion that the N.Y.S.E. wasn't actually interested in catching insider traders, and he decided to switch professions. He went back to school, at N.Y.U., and ended up becoming a climate modeller.

Unlike most modellers, who focus on one particular aspect of the climate system, Caldeira is, at any given moment, working on four or five disparate projects. He particularly likes computations of a provocative or surprising nature; for example, not long ago he calculated that cutting down all the world's forests and replacing them with grasslands would have a slight cooling effect. (Grasslands, which are lighter in color than forests, absorb less sunlight.) Other recent calculations that Caldeira has made show that to keep pace with the present rate of temperature change plants and animals would have to migrate poleward by thirty feet a day, and that a molecule of CO_2 generated by burning fossil fuels will, in the course of its lifetime in the atmosphere, trap a hundred thousand times more heat than was released in producing it.

Caldeira began to model the effects of carbon dioxide on the oceans in 1999, when he did some work for the Department of Energy. The department wanted to know what the environmental consequences would be of capturing CO_2 from smokestacks and injecting it deep into the sea. Caldeira set about calculating how the ocean's pH would change as a result of deep-sea injection, and then compared that result with the current practice of pouring carbon dioxide into the atmosphere and allowing it to be taken up by surface waters. In 2003, he submitted his work to *Nature*. The journal's editors advised him to drop the discussion of deep-ocean injection, he recalled, because the calculations concerning the effects of ordinary atmospheric release were so startling. Caldeira published the first part of his paper under the subheading "The coming centuries may see more ocean acidification than the past 300 million years."

Caldeira told me that he had chosen the term "ocean acidification" quite deliberately, for its shock value. Seawater is naturally alkaline, with a pH ranging from 7.8 to 8.5—a pH of 7 is neutral—which means that, for now, at least, the oceans are still a long way from actually turning acidic. Meanwhile, from the perspective of marine life, the drop in pH matters less than the string of chemical reactions that follow.

The main building block of shells is calcium carbonate—$CaCO_3$. (The White Cliffs of Dover are a huge $CaCO_3$ deposit, the remains of countless tiny sea creatures that piled up during the Cretaceous—or "chalky"—period.) Calcium carbonate produced by marine organisms comes in two principal forms, aragonite and calcite, which have slightly different crystal structures. How, exactly, different organisms form calcium carbonate remains something of a mystery. Ordinarily in seawater, $CaCO_3$ does not precipitate out as a solid. To build their shells, calcifying organisms must, in effect, assemble it. Adding carbonic acid to the water complicates their efforts, because it reduces the number of carbonate ions in circulation. In scientific terms, this is referred to as "lowering the water's saturation state with respect to calcium carbonate." Practically, it means shrinking the supply of material available for shell formation. (Imagine trying to build a house when someone keeps stealing your bricks.) Once the carbonate concentration gets pushed low enough, even existing shells, like those of Fabry's pteropods, begin to dissolve.

To illustrate, in mathematical terms, what the seas of the future will look like, Caldeira pulled out a set of graphs. Plotted on one axis was aragonite saturation levels; on the other, latitude. (Ocean latitude is significant because saturation levels tend naturally to decline toward the poles.) Different colors of lines represented different emissions scenarios. Some scenarios project that the world's economy will continue to grow rapidly and that this growth will be fuelled mostly by oil and coal. Others assume that the economy will grow more slowly, and still others that the energy mix will shift away from fossil fuels. Caldeira considered four much studied scenarios, ranging from one of the most optimistic, known by the shorthand B1, to one of the most pessimistic, A2. The original point of the graphs was to show that each scenario would produce a different ocean. But they turned out to be more similar than Caldeira had expected.

Under all four scenarios, by the end of this century the waters around Antarctica will become undersaturated with respect to aragonite—the form of calcium carbonate produced by pteropods and corals. (When water becomes undersaturated, it is corrosive to shells.) Meanwhile, surface pH will drop by another .2, bringing acidity to roughly double what it was in pre-industrial times. To look still further out into the future, Caldeira modelled what would happen if humans burned through all the world's remaining fossil-fuel resources, a process that would release some eighteen thousand gigatons of carbon dioxide. He found that by 2300 the oceans would become undersaturated from the poles to the equator. Then he modelled what would happen if we pushed still further and burned through unconventional fuels, like low-grade shales. In that case, we would drive the pH down so low that the seas would come very close to being acidic.

"I used to think of B1 as a good scenario, and I used to think of A2 as a terrible scenario," Caldeira told me. "Now I look at them as different flavors of bad scenarios."

He went on, "I think there's a whole category of organisms that have been around for hundreds of millions of years which are at risk of extinction—namely, things that build calcium-carbonate shells or skeletons. To a first approximation, if we cut our emissions in half it will take us twice as long to create the damage. But we'll get to more or less the same place. We really need an order-of-magnitude reduction in order to avoid it."

Caldeira said that he had recently gone to Washington to brief some members of Congress. "I was asked, 'What is the appropriate stabilization target for atmospheric CO_2?'" he recalled. "And I said, 'Well, I think it's inappropriate to think in terms of stabilization targets. I think we should think in terms of emissions targets.' And they said, 'O.K., what's the appropriate emissions target?' And I said, 'Zero.'

"If you're talking about mugging little old ladies, you don't say, 'What's our target for the rate of mugging little old ladies?' You say, 'Mugging little old ladies is bad, and we're going to try to eliminate it.' You recognize you might not be a hundred per cent successful, but your goal is to eliminate the mugging of little old ladies. And I think we need to eventually come around to looking at carbon-dioxide emissions the same way."

Coral reefs grow in a great swath that stretches like a belt around the belly of the earth, from thirty degrees north to thirty degrees south latitude. The world's largest reef is the Great Barrier, off the coast of northeastern Australia, and the second largest is off the coast of Belize. There are extensive coral reefs in the tropical Pacific, in the Indian Ocean, and in the Red Sea, and many smaller ones in the Caribbean. These reefs, home to an estimated twenty-five per cent of all marine fish species, represent some of the most diverse ecosystems on the planet.

Much of what is known about coral reefs and ocean acidification was originally discovered, improbably enough, in Arizona, in the self-enclosed, supposedly self-sufficient world known as Biosphere 2. A three-acre glassed-in structure shaped like a ziggurat, Biosphere 2 was built in the late nineteen-eighties by a private group—a majority of the funding came from the billionaire Edward Bass—and was intended to demonstrate how life on earth (Biosphere 1) could be re-created on, say, Mars. The building contained an artificial "ocean," a "rain forest," a "desert," and an "agricultural zone." The first group of Biosphereans—four men and four women—managed to remain, sealed inside, for two years. They produced all their own food and, for a long stretch, breathed only recycled air, but the project was widely considered a failure. The Biosphereans spent much of the time hungry, and, even more ominously, they lost control of their artificial atmosphere. In the various "ecosystems," decomposition, which takes up oxygen and gives off CO_2, was supposed to be balanced by photosynthesis, which does the reverse. But, for reasons mainly having to do with the richness of the soil that had been used in the "agricultural zone," decomposition won out. Oxygen levels inside the building kept falling, and the Biosphereans developed what amounted to altitude sickness. Carbon-dioxide levels soared, at one point reaching three thousand parts per million, or roughly eight times the levels outside.

When Biosphere 2 officially collapsed, in 1995, Columbia University took over the management of the building. The university's plan was to transform it into a teaching and research facility, and it fell to a scientist named Chris Langdon to figure out something pedagogically useful to do with the "ocean," a tank the size of an Olympic swimming pool. Langdon's specialty was measuring photosynthesis, and he had recently finished a project, financed by the Navy, that involved trying to figure out whether blooms of bioluminescent algae could be used to track enemy submarines. (The answer was no.) Langdon was looking for a new project, but he wasn't sure what the "ocean" was good for. He began by testing various properties of the water. As would be expected in such a high-CO_2 environment, he found that the pH was low.

"The very first thing I did was try to establish normal chemistry," he recalled recently. "So I added chemicals—essentially baking soda and baking powder—to the water to bring the pH back up." Within a week, the alkalinity had dropped again, and he had to add more chemicals. The same thing happened. "Every single time I did it, it went back down, and the rate at which it went down was proportional to the concentration. So, if I added more, it went down faster. So I started thinking, What's going on here? And then it dawned on me."

Langdon left Columbia in 2004 and now works at the Rosenstiel School of Marine and Atmospheric Science, at the University of Miami. He is fifty-two, with a high forehead, deep-set blue eyes, and a square chin.

When I went to visit him, not long ago, he took me to see his coral samples, which were growing in a sort of aquatic nursery across the street from his office. On the way, we had to pass through a room filled with tanks of purple sea slugs, which were being raised for medical research. In the front row, the youngest sea slugs, about half an inch long, were floating gracefully, as if suspended in gelatine. Toward the back were slugs that had been fed for several months on a lavish experimental diet. These were the size of my forearm and seemed barely able to lift their knobby, purplish heads.

Langdon's corals were attached to tiles arranged at the bottom of long, sinklike tanks. There were hundreds of them, grouped by species: *Acropora cervicornis*, a type of staghorn coral that grows in a classic antler shape; *Montastrea cavernosa*, a coral that looks like a seafaring cactus; and *Porites divaricata*, a branching coral made up of lumpy, putty-colored protuberances. Water was streaming into the tanks, but when Langdon put his hand in front of the faucet to stop the flow, I could see that every lobe of *Porites divaricata* was covered with tiny pink arms and that every arm ended in soft, fingerlike tentacles. The arms were waving in what looked to be a frenzy either of joy or of supplication.

Langdon explained that the arms belonged to separate coral polyps, and that a reef consisted of thousands upon thousands of polyps spread, like a coating of plaster, over a dead calcareous skeleton. Each coral polyp is a distinct individual, with its own tentacles and its own digestive system, and houses its own collection of symbiotic algae, known as zooxanthellae, which provide it with most of its nutrition. At the same time, each polyp is joined to its neighbors through a thin layer of connecting tissue, and all are attached to the colony's collective skeleton. Individual polyps constantly add to the group skeleton by combining calcium and carbonate ions in a medium known as the extracytoplasmic calcifying fluid. Meanwhile, other organisms, like parrot fish and sponges, are constantly eating away at the reef in search of food or protection. If a reef were ever to stop calcifying, it would start to shrink and eventually would disappear.

"It's just like a tree with bugs," Langdon explained. "It needs to grow pretty quickly just to stay even."

As Langdon struggled, unsuccessfully, to control the pH in the Biosphere "ocean," he started to wonder whether the corals in the tank might be to blame. The Biosphereans had raised twenty different species of coral, and while many of the other creatures, including nearly all the vertebrates selected for the project, had died out, the corals had survived. Langdon wondered whether the chemicals he was adding to raise the pH were, by increasing the saturation state, stimulating their growth. At the time, it seemed an unlikely hypothesis, because the prevailing view among marine biologists was that corals weren't sensitive to changes in saturation. (In many textbooks, the formula for coral calcification is still given incorrectly, which helps explain the prevalence of this view.) Just about everyone, including Langdon's own postdoc, a young woman named Francesca Marubini, thought that his theory was wrong. "It was a total pain in the ass," Langdon recalled.

To test his hypothesis, Langdon employed a straightforward but time-consuming procedure. Conditions in the "ocean" would be systematically varied, and the growth of the coral monitored. The experiment took more than three years to complete, produced more than a thousand measurements, and, in the end, confirmed Langdon's hypothesis. It revealed a more or less linear relationship between how fast the coral grew and how highly saturated the water was. By proving that increased saturation spurs coral growth, Langdon also, of course, demonstrated the reverse: when saturation drops, coral growth slows. In the artificial world of Biosphere 2, the implications of this discovery were interesting; in the real world they were rather more grim. Any drop in the ocean's saturation levels, it seemed, would make coral more vulnerable.

Langdon and Marubini published their findings in the journal *Global Biogeochemical Cycles* in the summer of 2000. Still, many marine biologists remained skeptical, in no small part, it seems, because of the study's association with the discredited Biosphere project. In 2001, Langdon sold his house in New York and moved to Arizona. He spent another two years redoing the experiments, with even stricter controls. The results were essentially identical. In the meantime, other researchers launched similar experiments on different coral species. Their findings were also the same, which, as Langdon put it to me, "is the best way to make believers out of people."

Coral reefs are under threat for a host of reasons: bottom trawling, dynamite fishing, coastal erosion, agricultural runoff, and, nowadays, global warming. When water temperatures rise too high, corals lose—or perhaps expel, no one is quite sure—the algae that nourish them. (The process is called "bleaching," because without their zooxanthellae corals appear white.) For a particular reef, any one of these threats could potentially be fatal. Ocean acidification poses a different kind of threat, one that could preclude the very possibility of a reef.

Saturation levels are determined using a complicated formula that involves multiplying the calcium and carbonate ion concentrations, and then dividing the result by a figure called the stoichiometric solubility product. Prior to the industrial revolution, the world's major reefs were all growing in water whose aragonite saturation level stood between 4 and 5. Today, there is not a single remaining region in the oceans where the saturation level is above 4.5, and there are only a handful of spots—off the northeastern coast of Australia, in the Philippine Sea, and near the Maldives—where it is above 4. Since the takeup of CO_2 by the oceans is a highly predictable physical process, it is possible to map the saturation levels of the future with great precision. Assuming that current emissions trends continue, by 2060 there will be no regions left with a level above 3.5. By 2100, none will remain above 3.

As saturation levels decline, the rate at which reefs add aragonite through calcification and the rate at which they lose it through bioerosion will start to approach each other. At a certain point, the two will cross, and reefs will begin to disappear. Precisely where that point lies is difficult to say, because erosion may well accelerate as ocean pH declines. Langdon estimates that the crossing point will be reached when atmospheric CO_2 levels exceed six hundred and fifty parts per million, which, under a "business as usual" emissions scenario, will occur sometime around 2075.

"I think that this is just an absolute limit, something they can't cope with," he told me. Other researchers put the limit somewhat higher, and others somewhat lower.

Meanwhile, as global temperatures climb, bleaching events are likely to become more common. A major worldwide bleaching event occurred in 1998, and many Caribbean reefs suffered from bleaching again during the summer of 2005. Current conditions in the equatorial Pacific suggest that 2007 is apt to be another bleaching year. Taken together, acidification and rising ocean temperatures represent a kind of double bind for reefs: regions that remain hospitable in terms of temperature are becoming increasingly inhospitable in terms of saturation, and vice versa.

"While one, bleaching, is an acute stress that's killing them off, the other, acidification, is a chronic stress that's preventing them from recovering," Joanie Kleypas, a reef scientist at the National Center for Atmospheric Research, in Boulder, Colorado, told me. Kleypas said she thought that some corals would be able to migrate to higher latitudes as the oceans warm, but that, because of the lower saturation levels, as well as the difference in light regimes, the size of these migrants would be severely limited. "There's a point where you're going to have coral but no reefs," she said.

The tropical oceans are, as a rule, nutrient-poor; they are sometimes called liquid deserts. Reefs are so dense with life that they are often compared to rain forests. This rain-forest-in-the-desert effect is believed to be a function of a highly efficient recycling system, through which nutrients are, in effect, passed from one reef-dwelling organism to another. It is estimated that at least a million, and perhaps as many as nine million, distinct species live on or near reefs.

"Being conservative, let's say it's a million species that live in and around coral," Ove Hoegh-Guldberg, an expert on coral reefs at the University of Queensland, in Australia, told me. "Some of these species that hang around coral reefs can sometimes be found living without coral. But most species are completely dependent on coral—they literally live in, eat, and breed around coral. And, when we see coral get destroyed during bleaching events, those species disappear. The key question is how vulnerable all these various species are. That's a very important question, but at the moment you'd have to say that a million different species are under threat."

He went on, "This is a matter of the utmost importance. I can't really stress it in words strong enough. It's a do-or-die situation."

Around the same time that Langdon was performing his coral experiments at the Biosphere, a German marine biologist named Ulf Riebesell decided to look into the behavior of a class of phytoplankton known as coccolithophores. Coccolithophores build plates of calcite—coccoliths—that they arrange around themselves, like armor, in structures known as coccospheres. (Viewed under an electron microscope, they look like balls that have been covered with buttons.) Coccolithophores are very tiny—only a few microns in diameter—and also very common. One of the species that Riebesell studied, *Emiliani huxleyi*, produces blooms that can cover forty thousand square miles, turning vast sections of the ocean an eerie, milky blue.

In his experiments, Riebesell bubbled CO_2 into tanks of coccolithophores to mimic the effects of rising atmospheric concentrations. Both of the species he was studying—*Emiliani huxleyi* and *Gephyrocapsa oceanica*—showed a clear response to the variations. As CO_2 levels rose, not only did the organisms' rate of calcification slow; they also started to produce deformed coccoliths and ill-shaped coccospheres.

"To me, it says that we will have massive changes," Riebesell, who works at the Leibniz Institute of Marine Sciences, in Kiel, told me. "If a whole group of calcifiers drops out, are there other organisms taking their place? What is the rate of evolution to fill those spaces? That's awfully difficult to address in experimental work. These organisms have never, ever seen this in their entire evolutionary history. And if they've never seen it they probably will find it difficult to deal with."

Calcifying organisms come in a fantastic array of shapes, sizes, and taxonomic groups. Echinoderms like starfish are calcifiers. So are mollusks like clams and oysters, and crustaceans like barnacles, and many species of bryozoans, or sea mats, and tiny protists known as foraminifera—the list goes on and on. Without experimental data, it's impossible to know which species will prove to be particularly vulnerable to declining pH and which will not. In the natural world, the pH of the water changes by season, and even time of day, and many species may be able to adapt to new conditions, at least within certain bounds. Obviously, though, it's impractical to run experiments on tens of thousands of different species. (Only a few dozen have been tested so far.) Meanwhile, as the example of coral reefs makes clear, what's more important than how acidification will affect any particular organism is how it will affect entire marine ecosystems—a question that can't be answered by even the most ambitious experimental protocol. The recent report on acidification by Britain's Royal Society noted that it was "not possible to predict" how whole communities would respond, but went on to observe that "without significant action to reduce CO_2 emissions" there may be "no place in the future oceans for many of the species and ecosystems we know today."

Carol Turley is a senior scientist at Plymouth Marine Laboratory, in Plymouth, England, and one of the authors of the Royal Society report. She observed that pH is a critical variable not just in calcification but in other vital marine processes, like the cycling of nutrients.

"It looks like we'll be changing lots of levels in the food chain," Turley told me. "So we may be affecting the primary producers. We may be affecting larvae of zooplankton and so on. What I think might happen, and it's pure speculation, is that you may get a shortening of the food chain so that only one or two species comes out on top—for instance, we may see massive blooms of jellyfish and things like that, and that's a very short food chain."

Thomas Lovejoy, who coined the term "biological diversity" in 1980, compared the effects of ocean acidification to "running the course of evolution in reverse."

"For an organism that lives on land, the two most important factors are temperature and moisture," Lovejoy, who is now the president of the Heinz Center for Science, Economics, and the Environment, in Washington, D.C., told me. "And for an organism that lives in the water the two most important factors are temperature and acidity. So this is just a profound, profound change. It is going to send all kinds of ripples through marine ecosystems, because of the importance of calcium carbonate for so many organisms in the oceans, including those at the base of the food chain. If you back off and look at it, it's as if you or I went to our annual physical and the body chemistry came back and the doctor looked really, really worried. It's a systemic change. You could have food chains collapse, and fisheries ultimately with them, because most of the fish we get from the ocean are at the end of long food chains. You probably will see shifts in favor of invertebrates, or the reign of jellyfish."

Riebesell put it this way: "The risk is that at the end we will have the rise of slime."

Paleooceanographers study the oceans of the geologic past. For the most part, they rely on sediments pulled up from the bottom of the sea, which contain what might be thought of as a vast library written in code. By analyzing the oxygen isotopes of ancient shells, paleooceanographers can, for example, infer the temperature of the oceans going back at least a hundred million years, and also determine how much—or how little—of the planet was covered by ice. By analyzing mineral grains and deposits of "microfossils," they can map archaic currents and wind patterns, and by examining the remains of foraminifera they can recreate the history of ocean pH.

In September, two dozen paleooceanographers met with a roughly equal number of marine biologists at a conference hosted by Columbia University's Lamont-Doherty Earth Observatory. The point of the conference, which was titled "Ocean Acidification—Modern Observations and Past Experiences," was to use the methods of paleooceanography to look into the future. (The ocean-acidification community is still a relatively small one, and at the conference I ran into half the people I had spoken to about the subject, including Victoria Fabry, Ken Caldeira, and Chris Langdon.) Most of the meeting's first day was devoted to a discussion of an ecological crisis known as the Paleocene-Eocene Thermal Maximum, or P.E.T.M.

The P.E.T.M. took place fifty-five million years ago, at the border marking the end of the Paleocene epoch and the beginning of the Eocene, when there was a sudden, enormous release of carbon into the atmosphere. After the release, temperatures around the world soared; the Arctic, for instance, warmed by ten degrees Fahrenheit, and Antarctica became temperate. Presumably because of this, vertebrate evolution veered off in a new direction. Many of the so-called archaic mammals became extinct, and were replaced by entirely new orders: the ancestors of today's deer, horses, and primates all appeared right around the time of the P.E.T.M. The members of these new orders were curiously undersized—the earliest horse was no bigger than a poodle—a function, it is believed, of hot, dry conditions that favored smallness.

In the oceans, temperatures rose dramatically and, because of all the carbon, the water became increasingly acidic. Marine sediments show that many calcifying organisms vanished—more than fifty species of foraminifera, for example, died out—while others that were once rare became dominant. On the seafloor, the usual buildup of empty shells from dead calcifiers ceased. In ocean cores, the P.E.T.M. shows up vividly as a band of reddish clay sandwiched between thick layers of calcium carbonate.

No one is sure exactly where the carbon of the P.E.T.M. came from or what triggered its release. (Deposits of natural gas known as methane hydrates, which sit, frozen, underneath the ocean floor, are one possible source.) In all, the release amounted to about two trillion metric tons, or eight times as much carbon as humans have added to the atmosphere since industrialization began. This is obviously a significant difference in scale, but the consensus at the conference was that if there was any disparity between then and now it was that the impact of the P.E.T.M. was not drastic enough.

The seas have a built-in buffering capacity: if the water's pH starts to drop, shells and shell fragments that have been deposited on the ocean floor begin to dissolve, pushing the pH back up again. This buffering mechanism is highly effective, provided that acidification takes place on the same timescale as deep-ocean circulation. (One complete exchange of surface and bottom water takes thousands of years.) Paleoceanographers estimate that the release of carbon during the P.E.T.M. took between one and ten thousand years—the record is not detailed enough to be more exact—and thus occurred too rapidly to be completely buffered. Currently, CO_2 is being released into the air at least three times and perhaps as much as thirty times as quickly as during the P.E.T.M. This is so fast that buffering by ocean sediments is not even a factor.

"In our case, the surface layer is bearing all the burden," James Zachos, a paleooceanographer at the University of California at Santa Cruz, told me. "If anything, you can look at the P.E.T.M. as a best-case scenario." Ken Caldeira said that he thought a better analogy for the future would be the so-called K-T, or Cretaceous-Tertiary, boundary event, which occurred sixty-five million years ago, when an asteroid six miles wide hit the earth. In addition to dust storms, fires, and tidal waves, the impact is believed to have generated huge quantities of sulfuric acid.

"The K-T boundary event was more extreme but shorter-lived than what we could do in the coming centuries," Caldeira said. "But by the time we've burned conventional fossil-fuel resources what we've done will be comparable in extremeness, except that it will last millennia instead of years." More than a third of all marine genera disappeared at the K-T boundary. Half of all coral species became extinct, and it took the other half more than two million years to recover.

Ultimately, the seas will absorb most of the CO_2 that humans emit. (Over the very long term, the figure will approach ninety per cent.) From a certain vantage point, this is a lucky break. Were the oceans not providing a vast carbon sink, almost all of the CO_2 that humans have emitted would still be in the air. Atmospheric concentrations would now be nearing five hundred parts per million, and the disasters predicted for the end of the century would already be upon us. That there is still a chance to do something to avert the worst consequences of global warming is thanks largely to the oceans.

But this sort of accounting may be misleading. As the process of ocean acidification demonstrates, life on land and life in the seas can affect each other in unexpected ways. Actions that might appear utterly unrelated—say, driving a car down the New Jersey Turnpike and secreting a shell in the South Pacific—turn out to be connected. To alter the chemistry of the seas is to take a very large risk, and not just with the oceans.

Writers in the Storm

Kathryn Schulz

November 23, 2015

"No weather will be found in this book," Mark Twain declares in the opening pages of his 1892 novel "The American Claimant." He has determined to do without it, he explains, on the ground that it usually just gets in the way of the story. "Many a reader who wanted to read a tale through was not able to do it," he writes, "because of delays on account of the weather."

Twain was not alone in mistrusting meteorological activity in fiction. As literary subjects go, weather has a terrible reputation. More precisely, it has two terrible reputations that do not get along. On the one hand, weather is widely regarded as the most banal topic in the world—in print as in conversation, the one we resort to when we have nothing else to say. On the other hand, it stands perpetually accused of melodrama. "It was a dark and stormy night," begins Edward Bulwer-Lytton's 1830 novel "Paul Clifford," which goes on to invoke torrential rain, gusting wind, guttering lamplight, and rattling rooftops: weather as plot, setting, star, and supporting cast of what is, by broad consensus, the worst sentence in the history of English literature.

Melodramatic or banal prose mostly gets blamed on the author, reasonably enough. But melodrama and banality are aesthetic judgments, and, as such, they are sometimes also products of their context. Twain was writing in the late nineteenth century, a time when the field of meteorology was belatedly coming into its own. With that scientific model of weather in ascendance, the literary models came to seem suspect. Weather facts served to make weather fictions seem overwrought, while the newly empirical understanding of the atmosphere—and, more staggering at the time, the ability to predict its behavior—made weather itself seem suddenly more prosaic.

That was the context in which Twain joked about eradicating weather from his work. But even he conceded that "weather is necessary to a narrative of human experience." Through the ages, we have used weather in our stories to illuminate the workings of our universe, our culture, our politics, our relationships, and ourselves. Before "The American Claimant" was published, sans weather, you might as easily have searched the canon for a novel without adverbs. Twain was likely correct when he called his weatherless book "the first attempt of the kind in fictitious literature."

Twain died in 1910, too soon to discover that his joke turned out to be borderline prophetic. After maintaining its centrality in Western literature for millennia, weather, while by no means vanishing entirely, faded in importance in the twentieth century. Only in our own time are we seeing it return in significant ways to our stories—thanks, as it happens, to the same forces that drove it away in the first place.

Storms sent to punish, lightning to frighten, thunder to humble, floods to obliterate: across nearly all cultures, the first stories that we told about weather were efforts to explain it, and the explanations invariably came down to divine agency. From the bag of winds gifted to Aeolus to the Biblical drought visited on Jerusalem, meteorological phenomena first appear in the narrative record as tools used by deities to battle one another and to help or hinder humans.

Early religions distributed those tools profligately. In Greek mythology, the wind alone was apportioned among more than a dozen gods, goddesses, nymphs, and demons—to say nothing of Zeus, who ruled the sky, and Poseidon, who could stir up storms. But, with the rise of monotheism, dominion over the elements was consolidated into a single God, and bad weather, like suffering and death, became one of those things which we brought down on ourselves through sin. In Eden, the climate was perfectly temperate. Only after the banishment of Adam and Eve did God—in the words of Milton, in "Paradise Lost"—"Affect the Earth with cold and heat / Scarce tolerable," and summon "ice / And snow and haile and stormie gust."

Meteorology would never entirely shed these religious undertones; even the eminently dry and secular field of contract law continues to call an unexpected weather event an "act of God." But by the time that Milton was writing, in the mid-seventeenth century, the role of weather in literature was shifting. While our earliest weather stories tried to explain meteorological phenomena, subsequent ones used meteorological phenomena to explain ourselves. Weather, in other words, went from being mythical to being metaphorical. In a symbolic system that is now so familiar as to be intuitive, atmospheric conditions came to stand in for the human condition.

That symbolic use of weather is the subject of Alexandra Harris's "Weatherland," a forthcoming history of weather in English literature. "My subject is not the weather itself," she writes, "but the weather as it is daily recreated in the human imagination." Her survey begins with an astute observation: weather works so well as a symbol partly because its literal manifestation is oddly slippery. "Meteorological phenomena are serially elusive," she writes. "Winds and air-fronts reveal their characters only in the effects they have on other things." A breeze sends smoke drifting northward from a chimney; a thermal betrays itself in the effortless upward trajectory of a hawk; low temperatures make themselves visible as our breath hanging in the air. Weather, one of the most potent forces in our lives, is often imperceptible, perpetually changing, and frequently mysterious.

As Harris points out, all of this makes it a convenient substitute for another "serially elusive" phenomenon: the self. King Lear, Shakespeare tells us, was "minded like the weather"—as charged and turbulent as the storm that raged around him on the heath. In a way, we have all been minded like the weather ever since, so accustomed have we become to using meteorology to describe mental activity. Minds are foggy (unless they are experiencing a brainstorm), temperaments sunny, attitudes chilly; moods blow in and out. Wordsworth wandered lonely as a cloud; Robert Frost, in "Tree at My Window," explicitly compared outer and inner weather. Harris draws particular attention to the association between minds and clouds, from the cumulus shape of the cartoon thought bubble to the early Christian belief that Adam's mind was made from a pound of clouds. She might also have cited Sartre, who memorably described consciousness as "a wind blowing from nowhere toward the world."

As a set of symbols, weather also blows toward the world; we use it to describe not only ourselves but our private relationships and our societies as a whole. Nabokov characterized his marriage to Véra Slonim with a one-word emotional-weather report: "cloudless." Emily Brontë conjured the opposite kind of relationship in "Wuthering Heights." When we first meet Catherine Earnshaw, she is a ghostly hand rapping on a window in a storm—which is to say, she is essentially the storm itself, rattling the glass panes of her former home. At every point thereafter, emotional drama and atmospheric drama are one. If Lear is minded like the weather, Catherine and Heathcliff are bodied like it—together, the most famous storm ever to strike the Yorkshire moors.

Six years later and two hundred miles to the southeast, Dickens summoned vastly drearier conditions for "Bleak House"—which, outside of the Book of Revelation, might have the most consistently dreadful weather of any work of Western literature. "It rains for the first twelve chapters," Harris notes, "before pausing and raining again." The skies are further blackened by soot and smoke—in Dickens's words, "gone into mourning, one might imagine, for the death of the sun." Fog smothers the city. The mud is so abundant that it is "as if the waters had but newly retired from the face of the earth"; in what may be the only dinosaur cameo in Victorian literature, Dickens imagines a forty-foot Megalosaurus slogging through it up Holborn Hill.

Much of this (though not the dinosaur) reflected a reality of contemporaneous London, where clouds mixed with soot from the unregulated chimneys of the early industrial era to darken clothes, lungs, and skies. Yet the weather in "Bleak House" is unmistakably symbolic: the mud is that of a hopelessly sullied culture, the fog that of an opaque and unnavigable legal system. As in earlier, religious stories, meteorology here is morality, and the prevailing conditions leave everything hidden, murky, and stained. Lest anyone miss the point, Dickens names his saintly heroine Esther Summerson.

This kind of heavy-handed meteorological symbolism was not to everyone's liking. To be specific, it was not to the liking of John Ruskin, the most influential critic in the nineteenth century. In 1856, in the third volume of "Modern Painters," Ruskin criticized writers for attributing human emotions to the natural world, a tendency that he famously termed the pathetic fallacy. ("Pathetic," in this context, refers to pathos, and the fallacy to something sham; the phrase might best be translated from the Victorian as "emotional falseness.") The sun does not shine mercilessly, Ruskin insisted, and the skies have never once wept, and, Dickens notwithstanding, fog cannot be found "cruelly pinching the toes and fingers" of a little apprentice boy. "It is one of the signs of the highest power in a writer," Ruskin argued, "to check all such habits of thought, and to keep his eyes fixed firmly on the *pure fact*"—on the "ordinary, proper, and true appearance of things."

Ruskin was reacting in part to sentimental literature and gothic novels, in which every dewdrop and tree limb was apt to quiver with human emotion. But he was also motivated by his own unusual attentiveness to the natural world—and, in particular, to its weather. "Modern Painters" includes scores of chapters on rain, mist, clouds, lightning, sunlight, and storms, and it dwells at length on the fidelity, or lack thereof, with which artists render the sky. Ruskin's own commitment to fidelity was impressive: he once stood outside on a winter morning and counted the cirrus clouds above him—all fifty thousand of them. His first public talk, given when he was eighteen, was on the color and formation of alpine clouds. He delivered another speech two years later, in 1839, on the "Present State of Meteorological Science" to the Meteorological Society of London.

That society, founded sixteen years earlier, was the first of its kind in the world. Ruskin was resisting the personification of weather and insisting on the "pure facts" of it just at the moment in history when those facts were becoming known. In his speech, he called on those who loved meteorology to "zealously come forward to deprecate the apathy with which it has long been regarded." The society of meteorologists, he continued, "wishes its influence and its power to be omnipotent over the globe, so that it may be able to know, at any given instant, the state of the atmosphere at every point on its surface."

It would take the better part of a century, but that vision eventually became a reality. What Ruskin did not predict, however (though it might have pleased him), was that the rise of an empirical model of weather would occasion the decline of the symbolic one—and, with it, the overall decline of weather in literature.

Ruskin was right to note that meteorology lagged far behind other sciences, though he might have gone on to observe that there were good reasons for the delay. It is harder to study things in the air than things on the ground, harder to study things that change rapidly than things that change slowly, if at all, and nearly impossible to study a global system such as weather in the absence of any kind of real-time global communications. As a result, weather science got almost nowhere in the two thousand years between Aristotle's mostly erroneous Meteorologica, written sometime around 350 B.C., and the development of the telegraph, in the eighteen-forties.

At the dawn of the nineteenth century, then, nearly everything about weather remained a mystery. No one understood the wind. No one knew why temperatures dropped as you climbed closer to the sun. No one could explain how clouds, with their countless tons of rainwater, somehow remained suspended in midair. No one knew what caused lightning, or why it tended to strike the tallest thing around—a problem for Christian meteorology, since it appeared that God had a special propensity for destroying church steeples. No one even knew what the sky was made of. Above all, no one knew what it was likely to do next. In 1854, when the Irish barrister John Ball suggested to the House of Commons that one day "we might know in this metropolis the condition of the weather twenty-four hours beforehand," he drew incredulous laughter.

That anecdote appears in Peter Moore's "The Weather Experiment," an account of the rise of modern meteorology. A huge cast of characters brought that field into being, but Moore, while giving them their due, focusses chiefly on Robert FitzRoy, a British naval officer and a towering figure in the history of meteorology. Among other achievements, FitzRoy improved the barometer, pioneered the use of statistics to track the weather, created Britain's storm-warning system, and established the government bureau that would later become the Met Office, the British equivalent of the National Weather Service. He is best known, however, for something he did in his capacity as a ship's captain: in 1831, while casting about for someone to keep him company on an upcoming voyage to South America, he met and invited along a young naturalist named Charles Darwin.

Moore observes a nice parallel: just as Darwin sought to explain the past, FitzRoy sought to explain the future—or, anyway, the portion of it that pertained to the weather. Given the prevailing belief that God reigned over the earth and the sky, both lines of inquiry were unpopular. Even meteorologists struggled to reconcile their profession with their faith. In 1838, William Reid, a British engineer who devoted himself to the study of hurricanes after witnessing the destruction they wrought in the Caribbean, felt compelled to publicly affirm his belief that the laws of nature were "designed by incomprehensible wisdom, arranged by supreme power, and tending to the most benevolent ends." FitzRoy, a devout Christian, ultimately rejected Darwin's work, at the expense of the friendship. Yet his own study contributed as much as anyone's to the forging of a new narrative about the weather—one as different from earlier accounts as Darwinism is from the creation story. Thanks in no small part to FitzRoy's influence, religious explanations of weather gave way to empirical ones, and "the heavens" gradually turned into "the atmosphere": a place that could be subjected, like an island full of finches, to scientific inquiry.

That shift in terminology was telling. Early meteorologists not only developed an entirely new story about the weather; they developed a new language to describe it. Prior to the nineteenth century, Moore writes, those trying to make sense of the weather "had no linguistic framework to scientifically explain what they saw." He quotes a Worcestershire diarist, writing in 1703: "Our Language is exceeding scanty & barren of words to use & express ye various notions I have of Weather &c. I tire myself with Pumping for apt terms & similes to illustrate my Thoughts."

The real problem, though, was not too few words but too many. One could describe the weather in any number of ways (that diarist characterized skies as, among other things, "loaded, "varnished," "bloated," "pendulous," and "like a tall fresco ceiling"), but the terms had no consistent and universal meaning. The problem had been identified as early as 1663, when the British polymath Robert Hooke, who later coined the word "cell" (in its biological sense), proposed a uniform vocabulary for describing clouds. His terms, unsurprisingly, did not stick. "Let Water'd, signify a Sky that has many high thin & small clouds looking almost like waterd tabby, calld in some places a maccarell sky from the Resemblance it has to the spots on the Backs of those fishes," Hooke suggested. He also recommended categorizing certain clouds as "hairy."

It took a hundred and forty years, the influence of Linnaeus, and at least one other rival plan (by the misguided evolutionary theorist Jean-Baptiste Lamarck) for a universal taxonomy of weather to catch on. In

1803, the British pharmacist Luke Howard suggested that clouds be described as *cirrus*, *stratus*, *cumulus*, and *nimbus*—the Latin words for "curl," "layer," "mass," and "rain." Two years later, Francis Beaufort, a British naval officer frustrated by the idiosyncratic weather descriptions recorded at sea, proposed twelve standardized gradations of wind strength, from "calm" to "hurricane": the Beaufort scale. FitzRoy himself contributed perhaps the most significant weather term of all: "forecast."

It is difficult, in our era of tornado watches and storm warnings, to appreciate how catastrophic the weather could be before we had any ability to forecast it. More than eight thousand people died in Britain's Great Storm of 1703, as did fifteen thousand grazing sheep when the storm surge hit the River Severn. (We know as much as we do about the calamity thanks to Daniel Defoe, who chronicled it in scrupulous detail in his 1704 book, "The Storm." Widely credited as the father of the modern novel, Defoe also pioneered the genre of the modern disaster narrative.) Things were no better a century and a half later and across the pond; in 1869, there were 1,914 shipwrecks in the Great Lakes alone. Not coincidentally, the ship-salvage industry was instrumental in lobbying against early weather forecasting. Partly owing to its influence, the British government effectively eliminated FitzRoy's position at the Meteorological Department shortly after his death, and suspended his two major innovations—weather forecasts and storm warnings—until scientific and public outcry sufficed to get them reinstated. Weather still wreaks havoc, but the rise of forecasting has saved untold numbers of lives, to say nothing of ships, crops, money, picnics, horse races, and weddings.

Quite aside from its practical value, the advent of forecasting indicated that meteorology had finally matured. As anyone in a long-term relationship knows, the more thoroughly you understand a system, the better you can predict how it is likely to behave. Mythological and religious explanations of weather not only failed at prediction but excluded it as a possibility; you cannot accurately forecast the caprice of Zeus, or the will of an omnipotent God. By contrast, in the nineteenth and early twentieth centuries, every accurate forecast served as evidence in favor of the new model of weather. (It is easy, these days, to kvetch about the inaccuracy of forecasts, but such complaints are relative. Weather remains imperfectly predictable, and probably always will be—meteorology is the field that gave us chaos theory—but we take for granted just how good prediction has become. It is one thing not to know where Hurricane Joaquin will make landfall, and something else entirely not to know that it exists until it strikes.) By the beginning of the twentieth century, forecasting was commonplace, meteorologists had cracked most major atmospheric phenomena, and the empirical model of weather had become, as Ruskin had hoped, "omnipotent over the globe."

In the visual arts, the rise of this new model occasioned a revolution in the representation of weather. For centuries, the sky in paintings was heavenly (azure, angel-stuffed) or else was rendered unobtrusively, as a backdrop for the presumptively more important activities on the ground. That changed in the early nineteenth century, thanks largely to the British artist John Constable. Keenly interested in contemporary meteorology, Constable monitored the latest developments in the field, painted outside in all weather, and, on the backs of more than a hundred studies of the sky, recorded the precise climatic conditions under which he painted them. The resulting landscapes featured such realistic weather that one critic, the Swiss painter Henry Fuseli, said that Constable's work "makes me call for my great-coat and umbrella."

Fuseli did not mean this as a compliment. Initially, Constable's meteorological accuracy met with widespread resistance. The first person to purchase one of his landscapes, outside his own circle of acquaintances, had another artist paint over the sky with a more tepid version. Eventually, though, both the critics and the public came around. Together with his colleague J. M. W. Turner (whose realistic skies Ruskin vigorously defended in "Modern Painters"), Constable paved the way for Delacroix, then Whistler, then Winslow Homer, until, in the visual arts, weather as iconography gave way to weather as weather: a natural phenomenon whose force and majesty were immense and sufficient in their own right.

A commensurate shift notably failed to take place in literature. Meteorology had constructed a new story about weather, down to the vocabulary used to tell it, yet writers seemed unable or unwilling to make use of it, even as their traditional strategies were becoming less viable. With the rise of a scientific understanding of weather, both its mythological and metaphorical clout diminished. Storms seem less like the verdict of God when you can track them by satellite two weeks out, and lightning loses some of its gothic thrill when you know that it is merely electrostatic discharge. A forecast, meanwhile, is a kind of anti-pathetic fallacy: it insists that the weather is the product of natural forces, utterly unrelated to the goings on in our culture, our relationships, and our soul.

While meteorology was advancing, then, the role of weather in literature began to decline. At the same time, the role of weather in real life was declining as well. As Western nations shifted from largely rural to largely urban economies, fewer people worked the kind of jobs that kept them exposed to the elements. As more automobiles hit the road, and more of those roads were paved, it became less of an ordeal to get from A to B in mud and sleet and snow. And, as indoor heating and cooling systems became common, more people were insulated from the vagaries of the weather.

In response to these changes, fiction, too, became climate-controlled: in the modern novel, as in modern housing, outside conditions seldom intruded. It is easy enough to find a rainstorm or a humid afternoon in twentieth-century prose, of course. But, with some notable exceptions (John Steinbeck's "The Grapes of Wrath," Willa Cather's prairie trilogy), weather dwindled as a pervasive and determinative force in fiction. It mattered in the burgeoning field of nature writing, but it lingered elsewhere mainly in poetry (though much less so than in earlier eras) and in children's books, with their tendency toward anachronism and nostalgia. Already robbed of most of its mythological weight, weather gradually lost the rest of its literary status, too. Only in the past few decades, as the facts about weather have become more and more pressing, is the subject beginning to reassert itself in fiction.

A hundred and sixty years after Dickens filled his skies with soot, a hundred and seventy-five years after Ruskin yearned for omnipotence over the globe, four hundred years after Shakespeare made a reckless ruler pull down his kingdom on his head, a hundred and twenty excess parts per million of carbon dioxide in our atmosphere: that is where we stand today. Unlike in Mark Twain's time, there is nothing remotely banal about the weather. If anything, we are in mourning for that banality. What used to be idle chitchat about the unusually warm day or last weekend's storm has become both premonitory and polarizing. Nor is there any innate melodrama left in meteorology. Weather is, instead, at the heart of the great drama of our time. Accordingly, the comedy has leached from Twain's line. "No weather will be found in this book" now reads either as denialist—a refusal to face climatic reality—or, very simply, as sad.

But we do not need that line anymore. After a long wait, quite a lot of weather can suddenly be found in our books again. We owe that revival to the same thing that first led to the decline of weather in literature: developments in the field of meteorology. It is not just that the facts about climate change have become clear; it is that, in establishing those facts, the scientific model of weather, which eclipsed the symbolic one in the nineteenth century, is now colliding with it. These days, the atmosphere really does reflect human activity, and, as in our most ancient stories, our own behavior really is bringing disastrous weather down on our heads. Meteorological activity, so long yoked to morality, finally has genuine ethical stakes.

That shift began to be reflected in literature in the later decades of the twentieth century, with the emergence of the genre now known as cli-fi—short for climate fiction, and formed by analogy to "sci-fi." As that suggests, novels about the weather have tended to congregate in genre fiction. The dystopian novelist J. G. Ballard wrote about climate change before the climate was known to be changing; later, Kim Stanley Robinson, Margaret Atwood, and many others used the conventions of science fiction to create worlds in which the climate is in crisis. More recently, though, books about weather are displaying a distinct migratory pattern—farther from genre fiction and closer to realism; backward in time from the future and ever closer to the present. See, among others, Ian McEwan's "Solar," Barbara Kingsolver's "Flight Behavior," Nathaniel Rich's "Odds Against Tomorrow,"

Karen Walker's "The Age of Miracles," Jesmyn Ward's "Salvage the Bones," and Dave Eggers's "Zeitoun." (Weather is on the rise in nonfiction, too. In addition to "Weatherland" and "The Weather Experiment," recent or forthcoming titles include Tim Flannery's "Atmosphere of Hope," Christine Corton's "London Fog," Lauren Redniss's "Thunder & Lightning," and Cynthia Barnett's "Rain.")

The emergent canon of weather-related fiction got an excellent addition this fall in Claire Vaye Watkins's début novel, "Gold Fame Citrus." It is set in the future, at a time when extended drought and rapid desertification have turned much of the American West into one "mega-dune," known as the Amargosa, after the first mountain range it consumed. Watkins's title refers to the fantasies that once made people head west; now almost everyone in the region is desperate to move away. Most seek refuge in other states, where they are pejoratively referred to as Mojavs—Okies in reverse, discriminated against and, increasingly, turned away at the borders. Some stay in the desiccated and dangerous remnants of Los Angeles. A few go in search of a community that, rumor has it, has sprung up somewhere in the vast expanse of the Amargosa.

Those include the book's main characters: a soldier turned drug dealer turned surfer and his girlfriend, a third-tier model who was once the literal poster child for water conservation, back when there was still water to conserve. A sweet, damaged, secret-keeping couple, they squat in the remains of a starlet's mansion, until a neglected and developmentally delayed toddler happens into their life. They rescue her, or kidnap her, then head off to the desert—hoping, like so many parents before them, to make a better life for their child. It is there in the Amargosa that the book comes into its own, as a story about the desert, and about deserters—about those who abdicate responsibility and, conversely, those who lay claim to things to which they have no right, from the child of strangers to the resources of a nation.

In the months before "Gold Fame Citrus" was published, reservoirs built to funnel water from the Colorado River to the Southern California desert had sunk to below half their capacity. The snowpack in the Sierra Nevadas dropped to five per cent of the historical average. The air rippled with record-setting heat above the parched ground, an illusion of water where water used to be. And the U.S. Senate, in a move that Robert FitzRoy would have recognized, voted to reject the scientific consensus that humans are changing the climate.

You could describe "Gold Fame Citrus" as science fiction, but only in the sense that it is fiction borne out by contemporary science. You could also describe it as dystopic, but that would miss the point. As Watkins deftly makes clear while almost never panning away from the desert, the plight of the Mojavs is specific to the region, and functionally ignored by the rest of the world. Ask a Syrian, ask a single mother of six in São Paulo's slums, ask those who are bothering to keep track of the effects of climate change: like the future, dystopia is already here. It's just unevenly distributed.

Our earliest stories about the weather concerned beginnings and endings. What emerged from the cold and darkness of the void will return to it; waters that receded at the origin of the world will rise at its end. It is easy, in grim climatological times, to be drawn to the far pole of these visions. Weather has long been a handmaiden of the apocalypse, and the end of the world is so often presaged or effected by extreme climate shifts—floods, fires, freezing cold—that eschatology sometimes seems like a particularly dark branch of meteorology. Today, it is, if anything, even more difficult to imagine an end of the world that is not driven by a change in the weather. We speak of a "nuclear winter," of the firestorms and the radical temperature drop that would follow an asteroid strike, of global climate change nudging planetary temperatures out of the range of the habitable.

But apocalyptic stories are ultimately escapist fantasies, even if no one escapes. End-times narratives offer the terrible resolution of ultimate destruction. Partial destruction, displacement, hunger, want, weakness, loss, need—these are more difficult stories. That is all the more reason we should be glad writers are beginning to tell them: to help us imagine not dying this way but living this way. To weather something is, after all, to survive.

The End of Ice

Dexter Filkins

April 4, 2016

The journey to the Chhota Shigri Glacier, in the Himalayan peaks of northern India, begins thousands of feet below, in New Delhi—a city of twenty-five million people, where smoke from diesel trucks and cow-dung fires dims the sky and where the temperature on a hot summer day can reach a hundred and fifteen degrees. The route passes through a churning sprawl of lowland cities, home to some fifty million people, until the Himalayas come into view: a steep wall rising above the plains, the product of a tectonic collision that began millions of years ago and is still under way. From there, the road snakes upward, past cows and trucks and three-wheeled taxis and every other kind of moving evidence of India's economic transformation. If you turn around, you can see a great layer of smog, lying over northern India like a dirty shroud. In the mountains, the number of cars drops sharply—limited by government regulation, for fear of what the smog is doing to the ice. The road mostly lacks shoulders; on turns, you look into ravines a thousand feet deep. After the town of Manali, the air cools, and the road cuts through forests of spruce and cedar and fir.

A few months ago, I followed that route with an international group of scientists who were travelling to Chhota Shigri to assess how rapidly it is melting. Six of us were pressed together in a van packed with scientific instruments, cold-weather gear, and enough provisions to last several days. My guides were two Indian scientists, Farooq Azam and Shyam Ranjan. Azam, a thirty-three-year-old former bodybuilding champion, has made more than twenty trips to Chhota Shigri. This time, he would be carrying out measurements for the National Institute of Hydrology, in Roorkee. Ranjan, a large, soft-spoken man who grew up in a village on the plains of North India, had never been on a Himalayan glacier. He was hoping to extract an ice core—a sample from deep inside the glacier, which would provide a detailed picture of the area's past climate. It would be the first such sample to be taken from the Indian Himalayas.

There are a hundred and ninety-eight thousand glaciers in the world, and, while many of them have been studied extensively, the nine thousand in India remain mostly unexamined. On the Chinese side of the Himalayas, researchers have performed thorough surveys, but, according to one American scientist, "the other side of the Himalayas is a black hole." The reasons are largely financial: India is a relatively poor country, and there are scant funds available for research. "To adequately study the Himalayan glaciers, we need thirty to forty times more money than we actually receive," A. L. Ramanathan, a glaciologist at Jawaharlal Nehru University, who oversaw our expedition, told me.

Scientists from other countries have moved in to fill the void. Markus Engelhardt, a German, joined us in Manali, and a second vehicle carried a group of Norwegian glaciologists who were heading to a lake near Chhota Shigri to take samples of sediment dating back as far as twelve thousand years. For the Norwegians, the expedition amounted to a tutorial: they were hoping to teach the Indian scientists how to do similar experiments. "There's a thirty-year lag in India," Jostein Bakke, one of the Norwegians, said. "Without a firm understanding of the long-term dynamics of the climate, making predictions about it is like playing the lottery."

In India, the lack of precise knowledge has caused confusion. Two years ago, an article in *Current Science*, an Indian publication, concluded that "most of the Himalayan glaciers are retreating." Soon afterward, the Indian Space Research Organization found nearly the opposite, that eighty-seven per cent of them were stable. Some scientists expressed doubts about both studies, saying that data gathered only by satellites are not reliable for making such judgments. "You really can't tell anything unless you see the glacier up close," Azam said. "That's why I come up here."

For the people who live on the Indian subcontinent, the future of the high-mountain climate is of more than academic interest. The three great rivers that flow from the Indian Himalayas—the Ganges, the Indus, and the Brahmaputra—provide water for more than seven hundred million people in India, Pakistan, and Bangladesh, and they power numerous hydroelectric plants. Already, villages in India and Pakistan are experiencing more frequent flooding from the melting ice; scientists are predicting even more.

At thirteen thousand feet, our van arrived at a pass known as Rohtang La—"pile of corpses," so called because of the many people who have frozen to death trying to get through. Winter was coming, and in a few days the pass would close for six months. The Norwegians had wanted to come earlier, but they received permission from the Indian government only at the last minute; for researchers hoping to work on India's glaciers, the bureaucracy can be as big an obstacle as the lack of funding. "We do not want to get trapped on the other side for the winter," Bakke said.

When we reached the top of Rohtang La, the horizon appeared: a line of mountains skidding downward half a mile to the valley floor. Zigzagging through switchbacks, we made our way down. A new landscape emerged; instead of forests and grassy hillsides, there were boulders, barren slopes, and expanses of scree. The only signs of human habitation were fallow, neatly marked farm plots that crept up the valley walls at improbable angles.

Near the valley floor, we veered onto a rocky trail that tracked an icy river called the Chandra. Our van halted and a group of men appeared: Nepali porters, who led us to an outcropping on the river's edge. Chhota Shigri—six miles long and shaped like a branching piece of ginger—is considered one of the Himalayas' most accessible glaciers, but our way across was a rickety gondola, an open cage reminiscent of a shopping cart, which runs on a cable over the Chandra. With one of the porters working a pulley, we climbed in and rode across, one by one, while fifty feet below the river rushed through gigantic boulders.

Once we had arrived at the other side, we made our way across a rock-strewn field to get to our base camp, elevation twelve thousand six hundred and thirty-one feet. The sun was setting and whatever warmth was left vanished. In a few minutes, it was dark, and the stars came out, forming a dome of light so bright you could almost read a book.

Annual expeditions to Chhota Shigri began only fourteen years ago, so relatively little is known about its climatic history. Chhota Shigri and the other glaciers of the eastern Himalayas are unusual, in that, unlike the majority of the world's glaciers, which get most of their snow from winter storms, they get much of theirs from the summer monsoons, which tend to insulate them from more rapid melting. (Most of the glaciers of the Karakoram Mountains, in Pakistan, are not receding at all; it's one of the few places in the world where this is the case.)

The data are also limited by the uneven quality of the expeditions. Glaciologists can spend hundreds of thousands of dollars on research trips, but Azam and Ranjan had only a few thousand dollars to buy equipment and to pay porters. Some glacial expeditions extract ice cores using cranes and ferry them home by helicopter. The Indian scientists would transport their cores in dry ice, using a portable cooler, of the kind you might use to chill beer for a picnic, driving them by car back to Ranjan's laboratory, in New Delhi—a sixteen-hour trip. Some of the experiments that they planned to perform on Chhota Shigri seemed comically rudimentary. In one, to measure the volume of meltwater flowing out of the glacier, a graduate assistant would toss a wooden block into the water and time its float downstream.

In the morning, the sun rose over the mountains, but for hours the high-walled valley remained shaded and bitterly cold. Unlike glaciers in other parts of the world—Greenland, say, or the Alps—many of those in the Himalayas lie at the bottom of narrow valleys that get only a few hours of direct sunlight each day. As a result, they are melting more slowly than they would on flatter ground. It was not until 8:20 a.m. that the sun shone on our camp; by midafternoon the valley was in shadow again.

Markus Engelhardt's first task was to check the camp's weather monitor, which had been planted four months earlier, and recorded temperature, solar radiation, and barometric pressure. There was an array of similar instruments installed throughout the camp; one of them, a five-foot-tall aluminum thistle with a crown of flaps, looked like something you might find in a Santa Fe sculpture garden. Engelhardt had two other weather stations on the glacier, and he was eager to download their data, which would allow him to construct a precise record of fluctuations in the local climate. As he watched information scroll across the screen of his laptop, Engelhardt, who had been stoic during our long ascent, could barely contain his enthusiasm. "I want to go back to the office right now and start studying the data," he said.

The team set out into the valley, following a stream that was flowing from the glacier. There were nine of us, including three graduate assistants who'd come with Azam and Ranjan. I had imagined a smooth carpet of ice that led to the top of the glacier. Instead, there was a rough track of boulders, a destructive path that marked Chhota Shigri's retreat. Thousands of years ago, as the glacier moved forward, debris from the valley walls was torn loose by the advancing ice and tumbled onto its face, creating a craggy obstacle course.

Azam had not visited since 2013, when he was completing a doctorate at the University of Grenoble, in France. (His thesis topic: the effect of the climate on Chhota Shigri and the surrounding glaciers.) Like many of the glaciologists I encountered, Azam entered the field not because he was drawn to science but because he loved the outdoors. Born in the plains state of Uttar Pradesh, he grew up seeing the Himalayas on television and dreamt of going there. In college, he took a sensible path, studying chemistry, but he was also athletically inclined; he won several bodybuilding titles, including Mr. Jawaharlal Nehru University. After he finished a master's degree in chemistry, his teachers urged him to go into medical research. But, he said, "I was being pulled by some invisible force."

That same year, he had signed up for a mountaineering course offered by the Indian Army, which took place on the Dokriani Glacier, near the Chinese border. During the course, Azam noticed a series of bamboo rods protruding from the snow: ablation stakes, basic instruments of glaciology. "Until then, I didn't realize you could work on a glacier," he told me. Not long afterward, he went to Grenoble, where he spent the next three years studying ice, making field trips to India every summer. "When I am in the mountains, on the glacier, I feel close to myself—I'm far from everybody, there's no technology, and I can think," Azam said. "Only recently has the science become more important to me."

Ranjan, who is thirty-one, spent years examining glaciers as a graduate student in Switzerland, but he had never been to one in India, where the terrain is much more rugged. On the trail, in his heavy clothes—layers of thermal underwear and fleece and a down jacket—he cut a husky figure. As we started off, he worried that he was not fit enough to complete the expedition. "I am not sure that I can do this," he said. He moved slowly, panting heavily. The porters practically skipped across the rocky ground as they carried several hundred pounds of our equipment, as well as dozens of eggs.

At higher elevations, the valley deepened; the walls rose a thousand feet on either side, in layers of colored sediment, each representing a different mineral and a different epoch. The landscape was desolate, but occasionally there was a surprise: a golden eagle, a butterfly with orange wings. A solitary black crow followed us the length of the glacier.

Rounding a bend in the stream, we arrived at the glacier's snout, a cave of ice with water rushing from the entrance. Behind it, Chhota Shigri spread upward into the peaks, a vast shoehorn of snow and ice covered with sharp-edged boulders, most of them the size of a car. The glaciers of the Himalayas are scattered with geological debris, which, along with the lack of direct sunlight, slows melting. Yet, since Azam's last visit, two years earlier, Chhota Shigri's snout had receded more than sixty feet. At its largest, the glacier sat almost atop the Chandra, slowly filling it with frigid meltwater; now it is barely visible from the banks. "It's going very fast," Azam said, standing on a ridge above it. The shrinking snout had left behind enormous hunks of what glaciologists call "dead ice," which were melting on the glacier's trail. A single glance belied the reports that India's glaciers are stable. After this, all the activity would consist of taking small, precise measurements, to find out exactly what was changing and how much.

The opening of Chhota Shigri's snout was five feet high, large enough for us to enter. Pressing ourselves against the interior walls and shimmying along the narrow banks of the rushing water, we worked our way into a vaulting palace of ice, where ten-foot-long icicles hung from the ceiling like giant fishhooks. Underneath the roar, you could hear the drip of melting ice. In the walls and the ceiling, water and earth streamed behind sheets of clear ice, the sediment tinting the walls orange and pale green. Air bubbled in the water, trapped when the glacier's ice froze around it, more than two hundred and fifty years ago. "It could collapse at any moment," Azam said. "When we come back next year, it will be gone."

On one of Azam's early trips to Chhota Shigri, in 2008, he and a French scientist, accompanied by a porter, trekked to the head of the glacier. When they started back, the next day, Azam fell behind the others. Then the sun went down and the temperature dropped. There was no moon, and the way through the boulders disappeared in the darkness. Alone and disoriented, Azam tripped and fell into the glacial stream. On his knees, he crawled alongside the water—his only clear path—wondering if he would survive. Several hours later, another member of the team found him not far from the base camp, shivering and numb, and helped him make his way back. At the camp, the French scientist apologized for leaving him behind. Azam, worried that his legs were frostbitten, dunked them in a barrel of steaming water. "What I learned was nature is always stronger," he said.

For many glaciologists, the scientific work that they perform on glaciers consumes less time and effort than surviving the journey. There is the cold to consider—temperatures in Antarctica reach seventy degrees below zero—along with steep treks through thin mountain air, and gusts of wind powerful enough to sweep researchers from mountains, not to mention rock slides, marauding polar bears, deep crevasses, and lightning strikes. "Logistics is about ninety per cent of your work," said Aaron Putnam, a glaciologist at the University of Maine who has done field work in Bhutan,

Mongolia, western China, and the Beagle Islands, at the tip of South America. "The science can seem almost incidental."

Glaciology is a diffuse field, encompassing meteorologists, geologists, and physicists. While some researchers spend most of their time in the lab, looking at satellite imagery and readouts from remote sensors, many collect their data in far more challenging environments. Mike Kaplan, a Columbia University geologist who studies glacial and polar ice, has fallen head first between boulders in Patagonia and watched a polar bear destroy his camp in northern Canada. Once, on an expedition to Baffin Island, in the Canadian Arctic, Kaplan drifted out to sea when the engine on his Zodiac boat wouldn't start. "I've never been so miserable in my life," Kaplan told me. "You're just so cold and so uncomfortable. But you've got work to do, so you have to do it."

Lonnie Thompson, a sixty-seven-year-old glaciologist at Ohio State University, has completed sixty-one expeditions to glaciers around the world, conducting research in the Himalayas, the Andes, and the mountains of East Africa, among other places. He's fallen into crevasses in the Andes, and endured seventy-mile-per-hour winds atop a twenty-thousand-foot Peruvian peak, where a pair of Italian climbers were blown to their deaths. A few years ago, he began to have heart trouble, and, rather than retire, he got a transplant. "I may be sixty-seven, but my heart is twenty-five," he said. Last summer, Thompson led a team of sixty to the Guliya Glacier, in Tibet, elevation twenty-two thousand feet; seven tons of equipment had to be hauled in on foot. "There I was, it's minus thirty-five in my tent," Thompson said. "It's not for everybody." But he was able to retrieve samples of ice that was a half million years old. The trip had its pleasures, too. At night, the Tibetan sky was so dark and so clear that Thompson was able to see other galaxies. "I went into geology because I didn't want to sit behind a desk," he said. "I didn't even know what glaciology was. But I'm a tough dude. I can suffer."

Until the last decade or so, glaciology was an obscure field; today it's being flooded with new students. Like many of the recent recruits, Thompson is propelled by the knowledge that the focus of his career is rapidly vanishing. The ice cores that he's collecting make up an archive of the Earth's weather over the past millennia. But the glacial ice is disappearing, and so is the archive itself. "We are trying to document the history of climate," Thompson said. "If it's not done now, it will never be done." Two of the six ice fields he had visited on Mt. Kilimanjaro are gone. By his estimate, the glaciers in New Guinea will disappear in twenty years. "We're on a salvage mission," he said.

Azam had come to Chhota Shigri to measure three things: the mass of the glacier, its thickness, and the speed with which it was moving downhill. Glacial melt is calculated in "mass balance," a measure of how much ice has been gained or lost. According to surveys conducted by Azam and ten other scientists, Chhota Shigri's mass has declined significantly since 2002, losing more than twenty feet across its surface. The glacier has shrunk in fits and starts; its greatest reductions have occurred in years in which the monsoon faltered, depriving the glacier of much of its snowfall. Recently, India's monsoons have become more sporadic, for reasons that many scientists ascribe to the world's changing climate.

Azam usually begins his expeditions by extracting a snow core, which indicates how much fresh snow has accumulated since the last measurement. In 2012, he climbed to seventeen thousand feet to extract a snow core. In a video he took of the operation, he and his assistant stood in a driving snowstorm, rotating the aluminum handle of a tool that looks much like a gigantic corkscrew. The tool pulled loose a foot-long cylinder, which Azam carefully weighed on a digital scale. While he completed the measurement, two porters stood by, unfazed, as snow piled up on their jackets and hats. Finally, Azam said, laughing, "So, for today, it's enough." At the completion of each season's snow core, Azam marks the spot using a G.P.S. device and then places a small beacon—a "reco tablet"—on the snow's surface and marks it with blue powder. When he returns, he locates the beacon with an electronic detector and drills down until he finds the blue powder. "This is the most amazing exercise on the whole glacier," he told me. "I feel like a detective."

One afternoon, we clambered onto the glacier, following a steep path that was covered in snow. The air got thinner, and it was harder to keep going. We were walking in the "ablation zone," the part of the glacier where melting exceeds accumulation; it typically comprises the lower third of a glacier. After several minutes, we came to a bamboo stick poking out of the glacier; this was Ablation Stake 12, one in a network of poles planted across the surface. The stake, buried deep in the ice, had been installed years before, with a steam drill. Azam opened his pack and pulled out his G.P.S. device and a tape measure. "At last I can get to work," he said.

Standing at Stake 12, Azam measured how much of the stick was poking above the snow: about thirty inches. Then he used the G.P.S. device to determine the stick's precise location. He was hoping to learn two things. The first was how much snow had been lost since 2013, when he was last on Chhota Shigri. It's a simple calculation: if there's more snow against the stakes than there was in 2013, then the glacier grew; if not, it shrank. "This seems like a normal amount of melt, but I won't know until I get back to the lab," he said.

The second measurement was the glacier's thickness. On a previous visit, Azam's colleagues, using ground-penetrating radar, had charted the base of the glacier, where the ice meets the earth. Now, by measuring the elevation at various points, he could calculate the glacier's thickness. The data from this trip would take Azam months to sort out. But in previous years the patterns were clear. In 2009, the ice near Stake 12 was four hundred and twenty feet thick. In 2013, it had thinned to three hundred and ninety feet.

Ranjan was far behind us now, moving slowly but waving every so often to signal that he was O.K. At Ablation Stake 11, Azam took another measurement, gauging how much the stake had moved down the glacier. When snow accumulates on the surface of a glacier, its weight pushes the ice forward and down. Using the G.P.S. beacon, Azam calculated the location;

since 2013, Stake 11 had moved about a hundred feet down the glacier. "All these measurements show us that the glacier is shrinking," he said. Indeed, most of the other omens were not good: the Indian monsoon season in 2015 was among the driest in decades, and Chhota Shigri appears to have received less snow.

The center line of the glacier, known as the medial moraine, was strewn with boulders that had tumbled and drifted down from the peak. Around many of them, the snow had melted away, leaving them perched like giant mushrooms on stems of snow. Stopping at one boulder, marked with red paint, Azam lay the G.P.S. device on top and calculated its location and elevation to find the speed of the glacier's flow.

As we trudged up the glacier, Azam stopped using his instruments and simply looked around, searching for clues to how Chhota Shigri was changing. His vision was uncanny; he spotted a pile of boulders that appeared to be of a different mineral than the ones around them. "You see those? They are not from here," Azam said. They had originated high up on the glacier and moved all the way down. At one point, we stopped, and Azam gestured to where one of the glacier's main tributaries jutted off. "It seems to be detaching itself from the main glacier," he said. "That's because the glacier is thinning."

Continuing on, we heard a noise that sounded like a whirlpool. It was coming from a deep gash in the surface, more than a hundred feet long, into which ice was falling and disappearing: a moulin, a hole connected to a river system inside the glacier. The moulin seemed to have no bottom, but we could hear the water rushing perhaps a hundred feet below. "Don't stand too close," Azam said. "The ground around it is not stable." The moulin was not the only hole in the ice; we had ventured into an area of crevasses, many of them hidden by snow. We had to weave back and forth across the surface to avoid them. Azam went back to check on Ranjan, who was stopping frequently to catch his breath. "I will be O.K.," Ranjan said, staring down at the snow. "I think."

The sun was setting behind the peaks as we arrived at the high camp, at nearly sixteen thousand feet, and the horizon glowed deep orange. The porters had set up tents, and were donning headlamps to help prepare the equipment for the next day's ice core. The temperature was dropping fast, into the teens. We ducked inside the main tent and found the rest of the team huddled in the dark around a stove, drinking cups of salty broth. Ranjan arrived just after the sun went down. "I am so happy to have made it!" he said. The camp was just a handful of tents on the glacier's slope, connected by a little stairway carved into the snow. The porters had made a dinner of lentils and chapati, but we were too nauseated from the altitude to eat more than a few bites. That night, we slept in a ragged tent with no tarp, its doors flapping open, directly atop the ice, nine hundred and fifty feet thick.

The sun was remarkably strong when it shone on us; even though we were freezing, our faces were burned dry and pink. A pool of melting ice had formed around a boulder, and a porter crouched and filled his bucket for cooking. At breakfast—tea and more chapati—everyone was frigid but in high spirits. "Did you see the stars last night?" Ranjan asked. "You could see the whole Milky Way."

After breakfast, Ranjan set about collecting the ice core. From the start, nothing seemed to work right. His gear consisted of a large drill, with an engine the size of an outboard motor, and the drill bit, a clear, sharpened tube that could be driven into the ice. The plan was to drill down about forty feet, where a trove of molecular evidence was preserved in what they expected to be century-old ice. Glaciers are uniquely sensitive recorders of changes in climate, and their ice contains indications of past temperature, precipitation, and volcanic activity, as well as the effects of greenhouse gases. "If we can connect what has happened on the glacier to what is happening in the climate, then we should be able to predict what is going to happen," Ranjan said. The glaciers may already be melting, but knowing their precise state will, he hopes, allow him to understand what it will take to save them.

With Ranjan looking on, one of the Nepali porters started the motor and another pushed the drill into the ice. Ranjan exclaimed with delight—and then the engine stalled. The porter started it again, but the drill could go no deeper than a few feet before stalling. A couple of Ranjan's assistants extracted snow samples, each the size of a rice cake, from the drill bit.

One of the difficulties of taking cores is that the drill bit can melt the ice, causing samples of different ages to mix. Several of the scientists I talked to said that an ice core should be taken from a higher elevation, where the colder temperatures protect the ice from the friction of the drill. I wondered if Ranjan had chosen the lower altitude because he was afraid that he wouldn't be able to climb higher. He told me that the problem was the drill. "I think we need a bigger engine," he said.

Later in the day, Ranjan tried again, lower on the glacier, where the snow was not as hard. This time, he was able to drill down about twelve feet, to ice that was some twenty-five years old. It wasn't nearly as much as he'd hoped for; scientists in Antarctica have taken ice cores from more than a mile below the surface. But it was better than nothing. The samples went into the beer cooler. (Miraculously, they made the long drive to New Delhi intact.) "I've learned a lot from this," Ranjan said. "And I'm coming back."

On the last day of the expedition, two of the graduate assistants decided to hike up another fifteen hundred feet to take samples of the ice there. Azam, standing a thousand yards away, could see that they had wandered into an area riven by crevasses. "You're going the wrong way!" he shouted, but they couldn't hear. They made it as far as sixteen thousand seven hundred feet when a faint, high-pitched cry rose up. When the group turned, they saw the head of Teg Bahadur, one of the porters, peeking out over the edge of a crevasse. The team's gear, including the G.P.S. device, had sailed down into the crevasse and disappeared. One of the graduate students poked the snow around Bahadur and it collapsed, revealing the crevasse's multicolored walls and its seemingly bottomless depth. Bahadur, perched on a shelf, trembled in silence. "I've never been married," he said, mournfully. Digging their boots into the snow, the rest of the team managed to pull him to safety. But, despite several descents by one of the students, the gear was lost.

The day before, I had stood with Azam as he prepared for another ascent. Tethered to a lone porter, he planned to climb to seventeen thousand feet and examine the ablation stakes planted there. In the coming year, Azam and other scientists plan to publish a number of papers based on research performed in the region, in the hope of filling the gap in knowledge. There is still little money in India for this kind of work, but the government seems to be slowly coming to appreciate its importance. In the weeks before the recent climate talks in Paris, some Indian politicians insisted that they should not have to restrict their country's energy consumption to fix a problem that was mostly not of their making. Ahead of the conference, though, India agreed to significant reforms, including greater efforts in the Himalayas, and afterward the Prime Minister, Narendra Modi, announced that "climate justice has won."

I asked Azam what he thought would happen to Chhota Shigri, whether it could survive global warming. "I am not going to save this glacier," he said. "I am just going to find out what is happening."

He turned and looked up at the peak in front of him. "Once I do that, the next step will be to decide what has to be done. But these things don't depend on science. They depend on politics."

The New Harpoon

Tom Kizzia

September 12, 2016

The spring hunt started promisingly last year for the village of Point Hope, on the Chukchi Sea in northern Alaska: crews harpooned two bowhead whales and pulled them onto the ice for butchering. But then the winds shifted. Out on the pack where the water opened up, the ice at the edge was what is called *sikuliaq*, too young and unreliable to bear a thirty-ton whale carcass. The hunters could do nothing but watch the shining black backs of bowheads, breathing calmly, almost close enough to touch.

On a trip to the ice edge, Tariek Oviuk, a hunter from Point Hope, felt a strange sensation: the lift of ocean waves beneath his feet. The older men, nervous about the rising wind, hurried back toward shore, but the younger hunters remained, stripping blubber from a few small beluga whales. Then the crack of three warning shots came rolling across the ice, and the hunters scrambled for their snowmobiles. "As soon as we heard those shots, my heart started pounding," Oviuk recalled.

As Oviuk told me the story, a few months later, we were sitting in the kitchen of his friend Steve Oomittuk, a former village mayor, eating strips of *maktaaq*—chewy beluga blubber—off a piece of cardboard that quickly grew sodden with whale oil. Oviuk is thirty-five, tall and square-jawed, a former basketball star for the Point Hope high-school team, the Harpooners, and a member of a local troupe that performs traditional storytelling dances. "That was our way of communication," he told me. "That was our people's iPhones since time immemorial."

Oviuk said that when he heard the shots he started running, then jumped into a passing sled filled with slippery blubber. "That's not a beautiful thing, to be in a sled full of *maktaaq*," he said. Another snowmobile driver swung by to rescue him, and Oviuk clambered aboard. Then they stopped: a gap of blue water, a hundred feet across, had opened between them and the shore-fast ice. The driver, in a parka and ski pants, said, "Hold on." Accelerating, their heavily laden snowmobile leaped off the ice and skipped over the surface of the Chukchi Sea. Others followed, engines screaming, until everyone was across. "I didn't believe in global warming—I'll tell you that straight up," Oviuk said. "But I teared up out there. I was thinking, Every year, we don't know if it's the last time we're going to see the ice."

Point Hope sits at the northwesternmost corner of North America, on one of the oldest continuously settled sites on the continent. Eight hundred people live near the eroding tip of a fifteen-mile gravel spit thrust into the Chukchi Sea, a peninsula that the Inupiat call Tikigaq, or "index finger." For two thousand years, the digit, stuck into coastal migration routes, has provided an ideal hunting perch. Tikigaq was a capital of the pre-contact Arctic, whose prosperity depended on a subtle understanding of the restless plains of ice that surrounded the community in winter.

In Paris last December, a hundred and ninety-five nations agreed to limit greenhouse-gas emissions and slow the warming of the planet. President Obama, speaking at the Paris conference, called for the global economy to move toward a low-carbon future, citing his own recent trip to Alaska, where melting glaciers, crumbling villages, and thawing permafrost were "a glimpse of our children's fate if the climate keeps changing faster than our efforts to address it." The goal in Paris was to hold the average global increase in temperature to less than two degrees Celsius. The Arctic, which is warming at twice the rate of lower latitudes, has already shot beyond that: average annual air temperatures have increased by about three degrees. If trends continue, northern Alaska is expected to warm another six degrees by the end of the century.

These days, the ice disappears so fast in spring that villagers struggle to catch bearded seals, whose skins are traditionally used in Point Hope to cover hunting boats. Ice cellars in the permafrost, packed with frozen whale meat, are filling with water. People are worried about these changes. Like most families in the village, Oomittuk's survives on wild game; much of the living space in his small house was taken up by two big chest freezers. Villages in Alaska's Arctic consume nearly four hundred and fifty pounds of wild game and fish per person each year, according to a recent study. "Without the animals, we wouldn't be who we are," Oomittuk said.

With a warm July wind battering the peninsula, Oomittuk took me on a four-wheeler ride for a glimpse of Point Hope's past glory. Years ago, elders on the tribal council had picked Oomittuk as a kind of tradition bearer. An amiable fifty-four-year-old with a long wisp of chin hair, he had grown rounder and softer since his own whale-hunting days, when he once helped repel a polar bear nosing into his tent by brandishing a cast-iron skillet. He explained that the lumps in the tundra, visible in all directions, were the husks of prehistoric earthen homes.

Along the coast where people hunt and camp, Oomittuk said, there are haunted places where no one ever stops. (In 1981, the ethnographer Ernest Burch identified four such zones, avoided because of "non-empirical phenomena.") Explorers and whalers in the nineteenth century described Point Hope as an open graveyard, with skeletal remains arrayed for miles atop funerary racks of bleached whalebones—essential building materials in a land without trees. Episcopal missionaries at Point Hope eventually persuaded villagers to bury the human remains—as many as twelve hundred skulls, according to one account—in a single mass grave, surrounded by a picket fence of repurposed bowhead mandibles. At an abandoned village site nearby, we found a line of weathered-gray bones, staked into the tundra by missionaries a century ago to help converts find their way through the blizzards to church.

We drove to the beach overlooking the Chukchi Sea, where the evidence of erosion was plain. The peninsula used to extend considerably farther out. Prehistoric settlements have eroded away, and artifacts wash up after fall storms. "I love my way of life," Oomittuk said in a soothing baritone. "My grandfather's life. The cycle of life. The connection to the land, the sea, the sky."

Few Americans are as bound to the natural world as the whale hunters of the Arctic, or as keenly affected by the warming atmosphere. Yet few Americans are so immediately dependent on the continued expansion of the fossil-fuel economy that science says is causing the change. The underground igloo where Oomittuk was born, in 1962, had earthen walls braced with wood scraps and whalebone, and a single electric light bulb. Point Hope today is a grid of small but comfortable homes, laid out around a new school and a diesel-fired power plant—everything provided by a regional municipality with eight thousand permanent residents and an annual budget of four hundred million dollars. Oil drilling in the Arctic has paid for nearly all of it, and Oomittuk does not want to go back.

There is a cost, though. Over the horizon from the beach where we stood, Shell Oil had assembled a floating city. The project was opening an entirely new part of the Arctic Ocean to oil drilling. The dangers posed to the Tikigaq hunting culture by a massive spill were never far from Oomittuk's mind. But he worried, too, about how the village would survive if there was no more oil industry. The trade-offs have racked Alaska's Inupiat communities. For nearly a decade, Point Hope pressed a lawsuit against the offshore leases, becoming a last stronghold of indigenous opposition. Finally, in the spring of 2015, the village dropped the suit. On the day the thin ice nearly carried Tariek Oviuk out to sea, his whaling captain had been in Houston, meeting with Shell officials.

The first oil boom in the Alaskan Arctic was devastating for the Inupiat. It began in 1848, when Yankee whalers, having depleted the sperm whales of the Pacific, discovered an unexploited population of bowheads north of the Bering Strait. In two decades, the fleet killed nearly thirteen thousand of the oil-rich whales, and then it turned to decimating the walrus. Eskimo hunting communities, already struggling with alcohol and diseases brought by the whalers, faced another scourge: hunger. In the early eighteen-eighties, a government revenue cutter that landed on St. Lawrence Island, south of the Bering Strait, found that a thousand inhabitants had died of starvation. At Point Hope, dozens of people starved, but only after eating their dogs and making soup from the skins off their boats.

The second boom came after 1968, when oil was discovered at Prudhoe Bay, and this time the Inupiat were better prepared. A pipeline had to be built across Alaska to a tanker port, and the Inupiat, along with other Alaska natives, asserted rights to the federal land along the way; the question of aboriginal land rights had gone unresolved since Alaska was bought from Russia.

In 1971, Congress awarded the natives a huge settlement: forty-four million acres and nearly a billion dollars. In another age, the settlement might have been used to create reservations, to sequester aspects of a traditional life. Instead, the land went to twelve new for-profit corporations, owned by native shareholders. Some natives were ambivalent about entrusting their future to a corporation, and worried about losing hunting and fishing grounds to sale or bankruptcy. But the more urbanized leaders saw a means of forcing their way into Alaska's modern economy: one activist said that native corporations would be "the new harpoon."

When Steve Oomittuk was growing up, Point Hope, the once-great capital of the Arctic, had receded to the margin of the civilized world. People lived in small frame houses and in a few last underground homes, scavenging materials by dog team from abandoned whaling and military sites. Oomittuk recalled that the most exquisite treat available at the village store was a roll of Life Savers.

In the decade after oil was discovered, regional leaders organized a municipal government and set out to reverse a long history of neglect. The North Slope Borough, encompassing the town of Barrow and seven small villages in an area the size of Minnesota, was given authority to collect property taxes from the new production facilities and pipelines. The borough built power plants, schools with swimming pools, sewer systems with heaters to prevent freezing. Today, the government subsidizes a tribal college, child care, bus service, heating oil, and a thirty-five-million-dollar public-safety department. In 1997, the borough's helicopters rescued a hundred and seventy-three whalers drifting into the fog on a breakaway slab of ice.

In Point Hope, Oomittuk's father served on the local tribal council and helped launch the new government. The entire community was moved two miles from the fast-eroding tip of the peninsula, and the population doubled, as wages and transportation lifted the air of deprivation around village life. Oomittuk began working as a carpenter and started a family.

He was uneasy about some of the changes that the new prosperity brought. During a no-bid construction boom in the nineteen-eighties, he watched a corruption scandal bring down a borough mayor. New tools like outboards and snowmobiles improved hunters' productivity but required cash; Oomittuk had one of the village's last dog teams, until a power line blew down and landed in his dog yard. On the other hand, Oomittuk was on the borough's payroll for a decade as village fire chief—one of many positions that set North Slope communities apart from the two hundred or so other villages in Alaska. Point Hope today has a spotless fire station, with a full-time staff of four and a fire engine, a tanker truck, and an ambulance. By contrast, when a fire last spring in Emmonak, a village in the southwestern part of the state, roared through a fish-processing plant, residents could only stand beside their broken-down equipment and watch.

The cultures of the Arctic were known for being quick to adopt new technology, but subsistence-hunting traditions remained at the heart of Inupiat life. Oomittuk joined the tribal council, and worked as a harpooner in his uncle's *umiak* skin boat. In the North Slope villages, whaling captains continued to serve as leaders of the community. These captains tended to be the best village hunters: shrewd judges of ice and men, affluent enough to support a crew and a camp, passing down their equipment and know-how to generations of whalers. In general, the captains embraced the opportunities of the oil age—as long as the oil was drilled on land, away from the marine hunting grounds.

When oil companies made efforts to drill in the Beaufort Sea, the captains' association, along with the North Slope Borough, raised alarms about the intrusion of industrial traffic and noise in the migratory corridors of the whales and seals. Above all, they feared an uncontrolled oil spill in an icebound ocean, far from cleanup reinforcements. In 2008, Shell Oil bid heavily for federal leases in the Chukchi Sea, and a series of clamorous hearings began on the North Slope. Feelings were particularly strong in Point Hope, which had an unusual history of activism: in the early nineteen-sixties, the village stopped a plan by government scientists to use "peaceful" nuclear weapons to blast a harbor out of a nearby valley. "There were some very harsh words said about oil companies at meetings here," Oomittuk recalled. Villagers invoked memories of the starvation that followed the Yankee whalers. Caroline Cannon, an activist who led delegations to Washington, D.C., said at the time, "It feels as if the government and industry want us to forget who we are . . . as if they hope we will either give up or die fighting. We are not giving up." The North Slope Borough and the whaling captains sued to stop the first federal permits, and Shell was forced to retreat. To succeed, the company realized, it would need to find allies among the Eskimos.

A towering wooden fence, fifteen feet high and a half mile long, runs across the north side of Point Hope, built by the borough to protect against winds that descend from the North Pole. Before the fence was built, Oomittuk's work as fire chief included shovelling houses out of drifts, sometimes relying on their stovepipes to find them. When I returned to Point Hope last March, I walked along the fence, freshly buried in snow, on the way to Oomittuk's house for a dinner of raw whale meat and caribou stew. The sky was blue and the air calm, but ominous drifts tapered to the south of every house. Soon the wind came, and the next morning Oomittuk's house was frigid: the stove had run out of oil. The wind chill was thirty-eight below, according to my phone. Dishes on shelves on the north wall rattled with each gust. Oomittuk went out in the storm and slipped a two-by-four under one end of the fuel tank to get the oil flowing again, then sent his son off with buckets in search of borough-subsidized fossil fuel.

Oomittuk served as Point Hope's mayor for ten years, and in those days he opposed offshore oil development. "It was time for us to take care of the animals," he said. "They've taken care of us since time immemorial." But, as a steward of traditional culture, he was conscious of the Inupiaq principle of *paaqlaktautaiññiq*—the avoidance of community conflict—and he saw that power on the North Slope was shifting. The borough had the money. Tribal councils were losing influence; people had stopped coming to public meetings unless door prizes were offered. And prominent whaling captains had become leaders in business—especially in the land-claims native corporation, the Arctic Slope Regional Corporation.

Starting out in support services for the oil fields that spidered across the region, A.S.R.C. had expanded into construction, refining, oil leasing, and government contracting, growing into the largest company based in Alaska, with ten thousand employees and gross revenues of $2.5 billion. Because its shares can't be traded publicly, A.S.R.C. is subject to few disclosure requirements and can seem opaque. The Alaska Supreme Court recently ruled that it had been unreasonably secretive about executive compensation.

But big dividends have tended to quell concerns; in 2013, the Inupiat shareholders—there were twelve thousand, many of whom lived outside the region—received an average of ten thousand dollars each.

The corporation's president, Rex Rock, is a prominent whaler from Point Hope; last year, he was the captain of Tariek Oviuk's crew. Rock told me it is not a coincidence that many top A.S.R.C. officials are whaling captains. "It's the community's whale," he said. "The captain and crew each know their roles. You work as one to go out and provide for the community. We've taken that into the business world." At the headquarters, a three-story building near the ocean in Barrow, a whaling skin boat provides the center support for a glass-topped boardroom table. But the hunt for profit tells executives which way to steer: though the company runs television ads of squinting Eskimo hunters, declaring "I am Inupiaq," it also took the oil industry's side in a controversial state referendum over oil taxes.

As offshore drilling became a real prospect, A.S.R.C. and the industry pressed the argument that whaling and oil could thrive side by side. Shell sponsored borough-wide projects and village feasts, and agreed to seasonal drilling restrictions that pleased hunters. Meanwhile, A.S.R.C. sought to neutralize village opposition. In

March, I stopped by a run-down former schoolhouse in Point Hope, where a man named Sayers Tuzroyluk was waiting for a computer to be installed in a recently remodelled office. Tuzroyluk, seventy and silver-haired, was the president of Voice of the Arctic Inupiat, a new organization funded by A.S.R.C. and the North Slope Borough. The idea, he said, was to line up all the region's tribes and corporations and city governments to speak with one pro-development voice. People were frustrated by hearing anti-oil activists represent the Inupiat in the media, he said: "You have more power when you speak as one voice. We don't speak as individuals. We speak as the whole North Slope."

The biggest change had come in 2010, when the North Slope Borough dropped its legal battle and started working with Shell—first tentatively, then with greater enthusiasm. Last year, when I went to Barrow to ask about the change, I was ushered into the office of Jacob Adams, the borough mayor's top aide. Before taking office, Adams was the former longtime president of A.S.R.C.; he had come out of retirement to help run the borough as Shell's ships were sailing into the Chukchi.

The previous mayor, Edward Itta, had also been willing to sit down with Shell, but he complained that A.S.R.C. was pushing the borough too far. "The

A.S.R.C. are in cahoots with industry, and they're not amateurs at P.R.," he told me last year. "This campaign up here saying 'I am Inupiaq.' Claiming to be the Voice of the Natives. Well, I'm sorry, they're not."

Like Itta, Adams is an eminent whaling captain. An intense, compact man with gray hair and a crisply pressed business shirt, he told me that the local government could win more safety concessions by negotiating with Shell than by fighting in court. Perhaps more to the point, onshore oil production was declining, and a pipeline coming ashore would mean new roads and facilities, new jobs, more property to tax. The Inupiat, he argued, did not want to go back to hauling lake ice for drinking water, cutting up walrus for dog food, waking up in houses at twenty-five below zero: "We've created, in the past forty years, an infrastructure that our children are enjoying now. So will our grandchildren."

Once the whaling captains' association followed the borough and dropped its legal opposition, the tribal council of Point Hope stood almost alone, as the indigenous lead plaintiff in an environmental lawsuit against Chukchi Sea drilling. But the melting of the Arctic was drawing new international attention to the cause. Opponents from around the world called the Chukchi prospect an "unexploded carbon bomb," better left in the ground. "Kayaktivists"—inflamed by Shell's many mishaps, including a 2012 debacle that ended with a runaway drill rig in the Gulf of Alaska—prepared to blockade the Shell vessels as they passed through the Pacific Northwest. Caroline Cannon, who told Congress that a major oil spill would amount to "genocide," had been awarded the Goldman Environmental Prize, which came with a stipend of a hundred and fifty thousand dollars. Robert Redford narrated a short film about her.

In early 2014, a federal appeals court ruled in favor of Point Hope, blocking Shell's drilling plans for the year, but the pressure on the village only grew. A.S.R.C. seemed ready to declare a moratorium on *paaqlaktautaiññiq*. In an open letter, an A.S.R.C. executive named Richard Glenn accused the tribes of working with outside environmental groups "to close the door on the future of our communities." As if to emphasize the point, A.S.R.C. withdrew several hundred thousand dollars' worth of funding for social and environmental programs run by a borough-wide tribal group that had supported the lawsuit. Glenn is a disarming spokesman, a geologist with a canny sense of how the wider world likes to see Eskimos: impoverished and clinging to noble tradition. "This is not a Western," he told me. "The word 'village' has a quaint image that belies the huge dollar cost of these small cities we have built here. This subsistence life style depends on a lot of money." As he sees it, if environmentalists succeed in closing off the Arctic to oil development, the North Slope Inupiat will become climate-change victims, no less than if the ice melts away. "We're selfish about our region," he said. "If we sacrifice ourselves, if we shut down all the Arctic, someone elsewhere will turn the valve open a little more."

In July, 2014, the new harpoon struck. A.S.R.C. called a press conference in Anchorage to announce a joint venture with Shell, which would grant the small village-based native corporations a royalty interest in Chukchi Sea oil. Point Hope's village corporation, Tikigaq, was in on the deal. It brought in a professional facilitator for a "visioning" session with leaders of the tribe and the city, and all agreed that, without money from offshore oil, their community had no clear path forward. Though anti-oil sentiment remained strong—I was approached many times during my visits by people stressing this—a tribal election was called, and the new officers wrote a letter asking A.S.R.C. for help with a budget deficit. Caroline Cannon, the Goldman prize winner, took a job with the borough mayor. In March, 2015, Point Hope withdrew from the Shell lawsuit.

Curiously, worries about the warming Arctic had hardly figured in the region's long debate. When I asked Jacob Adams about the prospects of subsistence in a future without ice, his answer showed a mixture of cultural pride and a hunter's bravado. The Inupiat had always struggled with scarcity and change, he said. They would adapt. The animals would adapt, too, he predicted. "Nobody knows whether the ice melting is going to threaten any species at all," he said.

In the late nineteen-seventies, when Steve Oomittuk was going to high school, in Barrow, he had a job caring for caged animals at the Naval Arctic Research Lab, north of town. Some of the work he found troubling: wolves and marmots were stressed, dehydrated, and needle-jabbed as the government searched for metabolic secrets that could be useful on the battlefield. But several decades of Cold War research were winding down, and soon the quonset-hut labs were turned over to wildlife biologists working for the North Slope Borough. Under local control, research shifted toward protecting animals that are vital to Inupiat subsistence. Now the borough's Department of Wildlife Management, with an annual budget of more than five million dollars, has a mission to marry traditional knowledge and scientific methodology.

A story about the department's origins has become a kind of creation myth for the borough itself: How Oil Saved Subsistence. In 1977, the new borough was confronted by an international effort to halt Eskimo whaling, as regulators claimed that the bowheads had not recovered from the commercial slaughter a century ago. On the advice of elders, the borough's biologists, funded by oil taxes, undertook sophisticated acoustic studies, proving that much of the population had gone uncounted by swimming under the ice. A compromise was reached, eventually allowing Alaska's villages a maximum of sixty-seven whale strikes a year. (Regulators were also concerned that too many novice hunting crews, funded by the new oil wealth, were striking and losing whales, a detail sometimes overlooked in the retelling of the story.)

The warming Arctic became a focus for the borough's biologists. They drew on insights of elders like Arnold Brower, Sr., the Inupiaq son of a nineteenth-century Yankee whaler. Brower, who turned down schooling in San Francisco in order to spend his boyhood in a reindeer camp, landed more than three dozen bowheads in his lifetime, making him one of the most successful captains of his day. In 2001, he described to Charles Wohlforth, in "The Whale and the Supercomputer," how the weather was undermining traditional knowledge. "You could predict to go out there and hunt all day and not think about getting stranded," he said. "But I think we had a crazy type of change." Brower died a few years later, during an unseasonably late freeze-up. As he travelled alone to his fishing camp, at the age of eighty-six, his snowmobile broke through river ice.

In the fall of 2009, two biologists with the U.S. Geological Survey, on a flight south of Barrow, spotted a sandy beach littered with walrus carcasses: a hundred and thirty-one dead, most of them young, evidently trampled in a stampede. Traditionally, female walrus and their young rested on drifting pack ice over a shallow offshore feeding area in the Chukchi Sea. In the past decade, as the summer ice has disappeared early, they have been forced onto beaches, where herds are easy to panic. More worrying, they now face the possibility of a two-day commute to their feeding grounds.

The difficulty, as the borough's biologists point out, is that no one is sure whether this crazy type of change is causing actual harm. Walrus, which range widely and spend much of their time underwater, are notoriously hard to count. A 2006 regional census tallied 129,000, but conceded that the actual number could be between 55,000 and 507,000. With error intervals so wide, the walrus could practically go extinct before a statistical change was detected, Robert Suydam, a biologist for the North Slope Borough, told me. "We're not in a good place to predict the future, so we're in a very poor place to do anything about it," he said.

It's possible that increased sunlight and productivity in the ice-free waters have helped the walrus by providing more food, but biologists don't know. Similarly, bowheads seem to be thriving, and humpback whales, harbor porpoises, and salmon are expanding their range. But, as the climate grows warmer, good fortune can turn bad. Biologists worry especially about the corrosive effect of carbon entering cold Arctic waters, which could eventually hurt the zooplankton that bowheads travel so far to devour.

Since 1975, on a barrier island near Barrow, the ornithologist George Divoky has tended a pioneer nesting colony of black guillemots. Each year, as the snow melted earlier, the guillemots produced more numerous young. It was a global-warming success story—except that the ice was drawing away from the shore faster every summer, pulling with it the Arctic cod that the guillemots fed their offspring. Around 2002, reproduction rates began to decline—even before polar bears, marooned on the island by retreating ice, started to ravage the nests.

Tools for addressing such slow-developing problems are limited. In 2011, the federal government, citing the effects of lost sea ice, listed the Pacific walrus as a candidate for protection under the Endangered Species Act. But, for regulators looking to preserve wildlife populations, Eskimo hunters offer an easier target than major producers of carbon emissions do. In the Arctic, potential limits on subsistence hunting are met with anger and disdain—not least among business leaders, who use them to argue that villagers and climate-change activists are not natural allies. Rex Rock and Jacob Adams both pointedly recalled for me that environmentalists had tried in the nineteen-seventies to stop indigenous whaling, as an illustration of why the Inupiat should mistrust "outside entities." In recent years, the North Slope Borough and A.S.R.C. joined oil-industry groups in lawsuits that opposed protections for bearded seals and polar bears, which could impinge on future oil facilities.

The village hunters adapt where they can, travelling farther across open water to the broken ice where marine mammals can be found. On St. Lawrence Island, they couldn't reach the walrus last spring; a charity shipped in frozen halibut for replacement protein. In Barrow, where ice in the spring is growing thin and unreliable, the majority of bowheads are now taken during an open-water hunt in the fall. Starting last September, whaling crews in high-speed aluminum boats harpooned fifteen bowheads and towed them back to the sandy beach north of town. Meat and blubber were shared among local families and sent to relatives and friends; crew pennants flew from captains' homes, inviting neighbors to come eat. The Inupiat were adapting, and thriving. But even in the celebrations a note of menace lurked. Several young bowheads had been found dead from attacks by killer whales, ice-averse predators that are expanding their range in the Arctic. The first such killing anyone remembered was only two years ago.

By the time Shell's offshore drilling finally got under way, last summer, the operation had taken on the familiar air of an Alaska gold rush. The cost of the operation was enormous—seven billion dollars—and reports of the expense encouraged rumors of a big find. Why else would Shell have gambled so much? State officials were hoping to refill the Alaska pipeline, which was down to one-quarter capacity. In Barrow, Jacob Adams envisioned tax revenues providing for his grandchildren. In the villages, there was talk of royalties, corporate dividends, and jobs.

Then, at the end of September, Shell delivered a shocking announcement: it had failed to find sufficient oil in its Chukchi Sea exploration well and was withdrawing from drilling in Alaskan waters "for the foreseeable future." A.S.R.C.'s president, Rex Rock, predicted "a fiscal crisis beyond measure" for local communities. He blamed excessive federal regulation—rules that had been largely intended to prevent oil spills and protect wildlife. Environmentalists welcomed the retreat as a sign of a new era, but industry analysts suggested that Shell's decision had less to do with a post-Paris future of carbon budgets and carbon taxes than with conventional economics: the low price of oil, and the high cost and uncertainty of drilling in the Arctic.

The argument will surely continue. In March, the Obama Administration proposed a five-year plan for offshore oil leasing that includes future sales in the Beaufort Sea and the Chukchi Sea. But diminished expectations of an offshore bonanza are now drawing attention to a different scenario, in which the Inupiat no longer struggle to choose between oil and subsistence: instead, they could lose them both.

While I was in Barrow last year, I drove north along a beach road, which runs past the old naval research labs and ends at a small landmark in the world of climate science: a yellow clapboard house on the tundra with a three-story scaffolded tower. In April, 2012, the National Oceanic and Atmospheric Administration's

Barrow observatory was one of the first places on Earth to record a monthly average of four hundred parts per million of carbon dioxide in the atmosphere. That average is now typical for the entire planet. At the Paris climate conference, last December, the threshold that raised concerns was four hundred and fifty.

At the NOAA facility, two young technicians led me to a rooftop platform looking across the tundra, where polar-bear sightings sometimes bring a squad car from the borough police, with a second car for backup. They described how prevailing winds off the ocean provide pure readings of the carbon dioxide that drifts over the pole from Europe. The technicians pointed out the Dobson spectrophotometer: a small silver dome standing alone, like a miniature planetarium, and tracking changes in the earth's ozone layer. Forty years ago, when scientists first installed a dome there, the world was awakening to concerns about ozone-erasing chlorofluorocarbons. The chemicals began to be phased out in 1989, and now, decades later, the stratosphere shows signs of stabilizing. The dome on the tundra seemed a small shrine to hope.

When I returned to Point Hope in March, however, none of the optimism and resolve of the Paris agreements had made its way north. The Arctic had just seen two months of record temperatures, in a winter that researchers were calling "absurdly warm." The polar ice pack was thinner, and its maximum extent was the smallest ever measured. Approaching in a small plane, I could see open water stretching for miles south from the Tikigaq peninsula. Hunters in the village had already spotted a bowhead, a month early, and were wondering if they might have to set up their spring ice camps on the beach.

A cold snap was settling in, however, and a frozen skim had formed beyond the rough cuticle of shorefast ice. In town, snowy streets gleamed like polished marble. I stopped by the school complex to see Steve Oomittuk, who had a new job as shop teacher. He told me he had turned it into Inupiat shop, making tools like the *unaaq*, a staff with a hook and a pick that hunters use to probe the ice pack for holes. The school, which has two hundred and thirty-eight students, was undergoing a forty-one-million-dollar renovation, including a new gym that could seat the entire village on bleachers. This was not just a testament to the popularity of the Harpooners (and Harpoonerettes). Point Hope had been moved, at great borough expense, to a beach six feet above sea level; the gym, on high pilings, will be the safest refuge if a storm crashes through town.

Oomittuk was preparing to join the whaling this year. He serves as a kind of referee after a whale is landed, dividing the catch among crews according to arcane rules that reach back to prehistory. It's an exciting time, but it reminds him of the things that are disappearing, things that may not be recoverable when the oil runs out—not just knowledge about hunting or about the ceremonies passed down by his grandparents but the ice itself. "When all that money goes away, what's going to happen to this next generation?" he said. "They say the native people were nomadic, following the animals. That's not true about the Tikigaq people. The animals came to us. We knew they were coming, to give themselves to us. And the animals go with the ice. If the ice goes away, the animals go away."

During my visit, hunters went out regularly on snowmobiles to watch the sea. If the open water lingered, they might need aluminum powerboats, instead of the quiet skin boats they prefer. "Maybe we're going to have to go farther out into the ocean, take chances," a whaler named Hanko told me. "There's going to be a time in our life when we're hunting in T-shirts and tank tops." Once the north wind abated, however, the *sikuliaq* ice started to return. Oomittuk called elders on his cell phone to organize an Eskimo dance that they hoped would bring the hunters favorable winds.

Full-costume Eskimo drumming and dancing remains popular in Point Hope for big cultural celebrations. But, with these informal dances, which are intended to seek favor from the forces of nature, the animist echoes are a little strong for some; Oomittuk told me that Christian whaling captains tend to stay away. Still, six drummers turned out, and dozens of dancers of all ages. They gathered at the city office, a two-story geodesic dome that doubles as a bingo hall. "We believe if you follow these rituals, the animals will always come to us," Oomittuk said, as he pulled a drum made of whale-liver membrane from a carrying case.

The lead drummer that night was a small, animated man named Leo Kinneeveauk, a retired whaling captain with the angular face of a seabird. He started every song with what sounded like a wail. The male drummers, and the women sitting behind them, sang in Inupiaq style: a first verse, plaintive, then a second, furious and loud, as the men lashed handheld drums. In jeans and sweatshirts, the dancers took turns on the floor, joyously simulating the motions of hunters and their prey. After two hours, Steve Oomittuk was tired and happy. He walked home in the late-evening light of springtime in the Arctic. He would wait to see, in the weeks to come, if the dancing would bring the ice back.

PART II
Hell and High Water
Where We Are

The Sixth Extinction?

Elizabeth Kolbert

May 25, 2009

The town of El Valle de Antón, in central Panama, sits in the middle of a volcanic crater formed about a million years ago. The crater is almost four miles across, but when the weather is clear you can see the jagged hills that surround the town, like the walls of a ruined tower. El Valle has one main street, a police station, and an open-air market that offers, in addition to the usual hats and embroidery, what must be the world's largest selection of golden-frog figurines. There are golden frogs sitting on leaves and—more difficult to understand—golden frogs holding cell phones. There are golden frogs wearing frilly skirts, and golden frogs striking dance poses, and ashtrays featuring golden frogs smoking cigarettes through a holder, after the fashion of F.D.R. The golden frog, which is bright yellow with dark-brown splotches, is endemic to the area around El Valle. It is considered a lucky symbol in Panama—its image is often printed on lottery tickets—though it could just as easily serve as an emblem of disaster.

In the early nineteen-nineties, an American graduate student named Karen Lips established a research site about two hundred miles west of El Valle, in the Talamanca Mountains, just over the border in Costa Rica. Lips was planning to study the local frogs, some of which, she later discovered, had never been identified. In order to get to the site, she had to drive two hours from the nearest town—the last part of the trip required tire chains—and then hike for an hour through the rain forest.

Lips spent two years living in the mountains. "It was a wonderland," she recalled recently. Once she had collected enough data, she left to work on her dissertation. She returned a few months later, and though nothing seemed to have changed, she could hardly find any frogs. Lips couldn't figure out what was happening. She collected all the dead frogs that she came across—there were only a half dozen or so—and sent their bodies to a veterinary pathologist in the United States. The pathologist was also baffled:

the specimens, she told Lips, showed no signs of any known disease.

A few years went by. Lips finished her dissertation and got a teaching job. Since the frogs at her old site had pretty much disappeared, she decided that she needed to find a new location to do research. She picked another isolated spot in the rain forest, this time in western Panama. Initially, the frogs there seemed healthy. But, before long, Lips began to find corpses lying in the streams and moribund animals sitting on the banks. Sometimes she would pick up a frog and it would die in her hands. She sent some specimens to a second pathologist in the U.S., and, once again, the pathologist had no idea what was wrong.

Whatever was killing Lips's frogs continued to move, like a wave, east across Panama. By 2002, most frogs in the streams around Santa Fé, a town in the province of Veraguas, had been wiped out. By 2004, the frogs in the national park of El Copé, in the province of Coclé, had all but disappeared. At that point, golden frogs were still relatively common around El Valle; a creek not far from the town was nicknamed Thousand Frog Stream. Then, in 2006, the wave hit.

Of the many species that have existed on earth—estimates run as high as fifty billion—more than ninety-nine per cent have disappeared. In the light of this, it is sometimes joked that all of life today amounts to little more than a rounding error.

Records of the missing can be found everywhere in the world, often in forms that are difficult to overlook. And yet extinction has been a much contested concept. Throughout the eighteenth century, even as extraordinary fossils were being unearthed and put on exhibit, the prevailing view was that species were fixed, created by God for all eternity. If the bones of a strange creature were found, it must mean that that creature was out there somewhere.

"Such is the economy of nature," Thomas Jefferson wrote, "that no instance can be produced, of her having permitted any one race of her animals to become extinct; of her having formed any link in her great work so weak as to be broken." When, as President, he dispatched Meriwether Lewis and William Clark to the Northwest, Jefferson hoped that they would come upon live mastodons roaming the region.

The French naturalist Georges Cuvier was more skeptical. In 1812, he published an essay on the "Revolutions on the Surface of the Globe," in which he asked, "How can we believe that the immense mastodons, the gigantic megatheriums, whose bones have been found in the earth in the two Americas, still live on this continent?" Cuvier had conducted studies of the fossils found in gypsum mines in Paris, and was convinced that many organisms once common to the area no longer existed. These he referred to as *espèces perdues*, or lost species. Cuvier had no way of knowing how much time had elapsed in forming the fossil record. But, as the record indicated that Paris had, at various points, been under water, he concluded that the *espèces perdues* had been swept away by sudden cataclysms.

"Life on this earth has often been disturbed by dreadful events," he wrote. "Innumerable living creatures have been victims of these catastrophes." Cuvier's essay was translated into English in 1813 and published with an introduction by the Scottish naturalist Robert Jameson, who interpreted it as proof of Noah's flood. It went through five editions in English and six in French before Cuvier's death, in 1832.

Charles Darwin was well acquainted with Cuvier's ideas and the theological spin they had been given. (He had studied natural history with Jameson at the University of Edinburgh.) In his theory of natural selection, Darwin embraced extinction; it was, he realized, essential that some species should die out as new ones were created. But he believed that this happened only slowly. Indeed, he claimed that it took place more gradually even than speciation: "The complete extinction of the species of a group is generally a slower process than their production." In "On the Origin of Species," published in the fall of 1859, Darwin heaped scorn on the catastrophist approach:

> So profound is our ignorance, and so high our presumption, that we marvel when we hear of the extinction of an organic being; and as we do not see the cause, we invoke cataclysms to desolate the world.

By the start of the twentieth century, this view had become dominant, and to be a scientist meant to see extinction as Darwin did. But Darwin, it turns out, was wrong.

Over the past half-billion years, there have been at least twenty mass extinctions, when the diversity of life on earth has suddenly and dramatically contracted. Five of these—the so-called Big Five—were so devastating that they are usually put in their own category. The first took place during the late Ordovician period, nearly four hundred and fifty million years ago, when life was still confined mainly to water. Geological records indicate that more than eighty per cent of marine species died out. The fifth occurred at the end of the Cretaceous period, sixty-five million years ago. The end-Cretaceous event exterminated not just the dinosaurs but seventy-five per cent of all species on earth.

The significance of mass extinctions goes beyond the sheer number of organisms involved. In contrast to ordinary, or so-called background, extinctions, which claim species that, for one reason or another, have become unfit, mass extinctions strike down the fit and the unfit at once. For example, brachiopods, which look like clams but have an entirely different anatomy, dominated the ocean floor for hundreds of millions of years. In the third of the Big Five extinctions—the end-Permian—the hugely successful brachiopods were nearly wiped out, along with trilobites, blastoids, and eurypterids. (In the end-Permian event, more than ninety per cent of marine species and seventy per cent of terrestrial species vanished; the event is sometimes referred to as "the mother of mass extinctions" or "the great dying.")

Once a mass extinction occurs, it takes millions of years for life to recover, and when it does it generally has a new cast of characters; following the end-Cretaceous event, mammals rose up (or crept out) to replace the departed dinosaurs. In this way, mass extinctions, though missing from the original theory of evolution, have played a determining role in evolution's course; as Richard Leakey has put it, such events "restructure the biosphere" and so "create the pattern of life." It is now generally agreed among biologists that another mass extinction is under way. Though it's difficult to put a precise figure on the losses, it is estimated that, if current trends continue, by the end of this century as many as half of earth's species will be gone.

The El Valle Amphibian Conservation Center, known by the acronym EVACC (pronounced "e-vac"), is a short walk from the market where the golden-frog figurines are sold. It consists of a single building about the size of an average suburban house. The place is filled, floor to ceiling, with tanks. There are tall tanks for species that, like the Rabb's fringe-limbed tree frog, live in the forest canopy, and short tanks for species that, like the big-headed robber frog, live on the forest floor. Tanks of horned marsupial frogs, which carry their eggs in a pouch, sit next to tanks of casque-headed frogs, which carry their eggs on their backs.

The director of EVACC is a herpetologist named Edgardo Griffith. Griffith is tall and broad-shouldered, with a round face and a wide smile. He wears a silver ring in each ear and has a large tattoo of a toad's skeleton on his left shin. Griffith grew up in Panama City, and fell in love with amphibians one day in college when a friend invited him to go frog hunting. He collected most of the frogs at EVACC—there are nearly six hundred—in a rush, just as corpses were beginning to show up around El Valle. At that point, the center was little more than a hole in the ground, and so the frogs had to spend several months in temporary tanks at a local hotel. "We got a very good rate," Griffith assured me. While the amphibians were living in rented rooms, Griffith and his wife, a former Peace Corps volunteer, would go out into a nearby field to catch crickets for their dinner. Now EVACC raises bugs for the frogs in what looks like an oversized rabbit hutch.

EVACC is financed largely by the Houston Zoo, which initially pledged twenty thousand dollars to the project and has ended up spending ten times that amount. The tiny center, though, is not an outpost of the zoo. It might be thought of as a preserve, except that, instead of protecting the amphibians in their natural habitat, the center's aim is to isolate them from it. In this way, EVACC represents an ark built for a modern-day deluge. Its goal is to maintain twenty-five males and twenty-five females of each species—just enough for a breeding population.

The first time I visited, Griffith pointed out various tanks containing frogs that have essentially disappeared from the wild. These include the Panamanian golden frog, which, in addition to its extraordinary coloring, is known for its unusual method of communication; the frogs signal to one another using a kind of semaphore. Griffith said that he expected between a third and a half of all Panama's amphibians to be gone within the next five years. Some species, he said, will probably vanish without anyone's realizing it: "Unfortunately, we are losing all these amphibians before we even know that they exist."

Griffith still goes out collecting for EVACC. Since there are hardly any frogs to be found around El Valle, he has to travel farther afield, across the Panama Canal, to the eastern half of the country.

One day this winter, I set out with him on one of his expeditions, along with two American zookeepers who were also visiting EVACC. The four of us spent a night in a town called Cerro Azul and, at dawn the next morning, drove in a truck to the ranger station at the entrance to Chagres National Park. Griffith was hoping to find females of two species that EVACC is short of. He pulled out his collecting permit and presented it to the sleepy officials manning the station. Some underfed dogs came out to sniff around.

Beyond the ranger station, the road turned into a series of craters connected by ruts. Griffith put Jimi

Hendrix on the truck's CD player, and we bounced along to the throbbing beat. (When the driving got particularly gruesome, he would turn down the volume.) Frog collecting requires a lot of supplies, so Griffith had hired two men to help with the carrying. At the very last cluster of houses, in the village of Los Ángeles, they materialized out of the mist. We bounced on until the truck couldn't go any farther; then we all got out and started walking.

The trail wound its way through the rain forest in a slather of red mud. Every few hundred yards, the main path was crossed by a narrower one; these paths had been made by leaf-cutter ants, making millions—perhaps billions—of trips to bring bits of greenery back to their colonies. (The colonies, which look like mounds of sawdust, can cover an area the size of a suburban back yard.) One of the Americans, Chris Bednarski, from the Houston Zoo, warned me to avoid the soldier ants, which will leave their jaws in your shin even after they're dead. "Those'll really mess you up," he observed. The other American, John Chastain, from the Toledo Zoo, was carrying a long hook, for use against venomous snakes. "Fortunately, the ones that can really mess you up are pretty rare," Bednarski said. Howler monkeys screamed in the distance. Someone pointed out jaguar prints in the soft ground.

After about five hours, we emerged into a small clearing. While we were setting up camp, a blue morpho butterfly flitted by, its wings the color of the sky.

That evening, after the sun set, we strapped on headlamps and clambered down to a nearby stream. Many amphibians are nocturnal, and the only way to see them is to go looking in the dark, an exercise that's as tricky as it sounds. I kept slipping, and violating Rule No. 1 of rain-forest safety: never grab onto something if you don't know what it is. After one of my falls, Bednarski showed me a tarantula the size of my fist that he had found on a nearby tree.

One technique for finding amphibians at night is to shine a light into the forest and look for the reflecting glow of their eyes. The first amphibian sighted this way was a San José Cochran frog, perched on top of a leaf. San José Cochran frogs are part of a larger family known as "glass frogs," so named because their translucent skin reveals the outline of their internal organs. This particular glass frog was green, with tiny yellow dots. Griffith pulled a pair of surgical gloves out of his pack. He stood entirely still and then, with a heronlike gesture, darted to scoop up the frog. With his free hand, he took what looked like the end of a Q-tip and swabbed the frog's belly. Finally, he put the Q-tip in a little plastic vial, placed the frog back on the leaf, and pulled out his camera. The frog stared into the lens impassively.

We continued to grope through the blackness. Someone spotted a La Loma robber frog, which is an orangey-red, like the forest floor; someone else spotted a Warzewitsch frog, which is bright green and shaped like a leaf. With every frog, Griffith went through the same routine—snatching it up, swabbing its belly, photographing it. Finally, we came upon a pair of Panamanian robber frogs locked in amplexus—the amphibian version of sex. Griffith left these two alone.

One of the frogs that Griffith was hoping to catch, the horned marsupial frog, has a distinctive call that's been likened to the sound of a champagne bottle being uncorked. As we sloshed along, the call seemed to be emanating from several directions at once. Sometimes it sounded as if it were right nearby, but then, as we approached, it would fall silent. Griffith began imitating the call, making a cork-popping sound with his lips. Eventually, he decided that the rest of us were scaring the frogs with our splashing. He waded ahead, while we stood in the middle of the stream, trying not to move. When Griffith gestured us over, we found him standing in front of a large yellow frog with long toes and an owlish face. It was sitting on a tree limb, just above eye level. Griffith grabbed the frog and turned it over.

Where a female marsupial frog would have a pouch, this one had none. Griffith swabbed it, photographed it, and put it back in the tree.

"You are a beautiful boy," he told the frog.

Amphibians are among the planet's great survivors. The ancestors of today's frogs and toads crawled out of the water some four hundred million years ago, and by two hundred and fifty million years ago the earliest representatives of what became the modern amphibian clades—one includes frogs and toads, a second newts and salamanders—had evolved. This means that amphibians have been around not just longer than mammals, say, or birds; they have been around since before there were dinosaurs. Most amphibians—the word comes from the Greek meaning "double life"—are still closely tied to the aquatic realm from which they emerged. (The ancient Egyptians thought that frogs were produced by the coupling of land and water during the annual flooding of the Nile.) Their eggs, which have no shells, must be kept moist in order to develop. There are frogs that lay their eggs in streams, frogs that lay them in temporary pools, frogs that lay them underground, and frogs that lay them in nests that they construct out of foam. In addition to frogs that carry their eggs on their backs and in pouches, there are frogs that carry them in their vocal sacs, and, until recently, at least, there were frogs that carried their eggs in their stomachs and gave birth through their mouths. Amphibians emerged at a time when all the land on earth was part of one large mass; they have since adapted to conditions on every continent except Antarctica. Worldwide, more than six thousand species have been identified, and while the greatest number are found in the tropical rain forests, there are amphibians that, like the sandhill frog of Australia, can live in the desert, and also amphibians that, like the wood frog, can live above the Arctic Circle. Several common North American frogs, including spring peepers, are able to survive the winter frozen solid.

When, about two decades ago, researchers first noticed that something odd was happening to amphibians, the evidence didn't seem to make sense. David Wake is a biologist at the University of California at Berkeley. In the early nineteen-eighties, his students began returning from frog-collecting trips in the Sierra Nevadas empty-handed. Wake remembered from his own student days that frogs in the Sierras had been difficult to avoid. "You'd be walking through meadows, and you'd inadvertently step on them," he told me. "They were just everywhere." Wake assumed that his students were going to the wrong spots, or that they just didn't know how to look. Then a postdoc with several years of experience collecting amphibians told him that he couldn't find any, either. "I said, 'O.K., I'll go up with you and we'll go out to some proven places,'" Wake recalled. "And I took him out to this proven place and we found, like, two toads."

Around the same time, other researchers, in other parts of the world, reported similar difficulties. In the late nineteen-eighties, a herpetologist named Marty Crump went to Costa Rica to study golden toads; she was forced to change her project because, from one year to the next, the toad essentially vanished. (The golden toad, now regarded as extinct, was actually orange; it is not to be confused with the Panamanian golden frog, which is technically also a toad.) Probably simultaneously, in central Costa Rica the populations of twenty species of frogs and toads suddenly crashed. In Ecuador, the jambato toad, a familiar visitor to backyard gardens, disappeared in a matter of years. And in northeastern Australia biologists noticed that more than a dozen amphibian species, including the southern day frog, one of the more common in the region, were experiencing drastic declines.

But, as the number of examples increased, the evidence only seemed to grow more confounding. Though amphibians in some remote and—relatively speaking—pristine spots seemed to be collapsing, those in other, more obviously disturbed habitats seemed to be doing fine. Meanwhile, in many parts of the world there weren't good data on amphibian populations to begin with, so it was hard to determine what represented terminal descent and what might be just a temporary dip.

"It was very controversial to say that amphibians were disappearing," Andrew Blaustein, a zoology professor at Oregon State University, recalls. Blaustein, who was studying the mating behavior of frogs and toads in the Cascade Mountains, had observed that some long-standing populations simply weren't there anymore. "The debate was whether or not there really was an amphibian population problem, because some people were saying it was just natural variation." At the point that Karen Lips went to look for her first research site, she purposefully tried to steer clear of the controversy.

"I didn't want to work on amphibian decline," she told me. "There were endless debates about whether this was a function of randomness or a true pattern. And the last thing you want to do is get involved when you don't know what's going on."

But the debate was not to be avoided. Even amphibians that had never seen a pond or a forest started dying.

Blue poison-dart frogs, which are native to Suriname, had been raised at the National Zoo, in Washington, D.C., for several generations. Then, suddenly, the zoo's tank-bred frogs were nearly wiped out.

It is difficult to say when, exactly, the current extinction event—sometimes called the sixth extinction—began. What might be thought of as its opening phase appears to have started about fifty thousand years ago. At that time, Australia was home to a fantastic assortment of enormous animals; these included a wombatlike creature the size of a hippo, a land tortoise nearly as big as a VW Beetle, and the giant short-faced kangaroo, which grew to be ten feet tall. Then all of the continent's largest animals disappeared. Every species of marsupial weighing more than two hundred pounds—there were nineteen of them—vanished, as did three species of giant reptiles and a flightless bird with stumpy legs known as *Genyornis newtoni.*

This die-off roughly coincided with the arrival of the first people on the continent, probably from Southeast Asia. Australia is a big place, and there couldn't have been very many early settlers. For a long time, the coincidence was discounted. Yet, thanks to recent work by geologists and paleontologists, a clear global pattern has emerged. About eleven thousand years ago, three-quarters of North America's largest animals—among them mastodons, mammoths, giant beavers, short-faced bears, and sabre-toothed tigers—began to go extinct. This is right around the time the first humans are believed to have wandered onto the continent across the Bering land bridge. In relatively short order, the first humans settled South America as well. Subsequently, more than thirty species of South American "megamammals," including elephant-size ground sloths and rhino-like creatures known as toxodons, died out.

And what goes for Australia and the Americas also goes for many other parts of the world. Humans settled Madagascar around two thousand years ago; the island subsequently lost all mammals weighing more than twenty pounds, including pygmy hippos and giant lemurs. "Substantial losses have occurred throughout near time," Ross MacPhee, a curator at the American Museum of Natural History, in New York, and an expert on extinctions of the recent geological past, has written. "In the majority of cases, these losses occurred when, and only when, people began to expand across areas that had never before experienced their presence." The Maori arrived in New Zealand around eight hundred years ago. They encountered eleven species of moas—huge ostrichlike creatures without wings. Within a few centuries—and possibly within a single century—all eleven moa species were gone. While these "first contact" extinctions were most pronounced among large animals, they were not confined to them. Humans discovered the Hawaiian Islands around fifteen hundred years ago; soon afterward, ninety per cent of Hawaii's native bird species disappeared.

"We expect extinction after people arrive on an island," David Steadman, the curator of ornithology at the Florida Museum of Natural History, has written. "Survival is the exception."

Why was first contact with humans so catastrophic? Some of the animals may have been hunted to death; thousands of moa bones have been found at Maori archeological sites, and man-made artifacts have been uncovered near mammoth and mastodon remains at more than a dozen sites in North America. Hunting, however, seems insufficient to account for so many losses across so many different taxa in so many parts of the globe. A few years ago, researchers analyzed hundreds of bits of emu and *Genyornis newtoni* eggshell, some dating from long before the first people arrived in Australia and some from after. They found that around forty-five thousand years ago, rather abruptly, emus went from eating all sorts of plants to relying mainly on shrubs. The researchers hypothesized that Australia's early settlers periodically set the countryside on fire—perhaps to flush out prey—a practice that would have reduced the variety of plant life. Those animals which, like emus, could cope with a changed landscape survived, while those which, like *Genyornis*, could not died out.

When Australia was first settled, there were maybe half a million people on earth. There are now more than six and a half billion, and it is expected that within the next three years the number will reach seven billion.

Human impacts on the planet have increased proportionately. Farming, logging, and building have transformed between a third and a half of the world's land surface, and even these figures probably understate the effect, since land not being actively exploited may still be fragmented. Most of the world's major waterways have been diverted or dammed or otherwise manipulated—in the United States, only two per cent of rivers run unimpeded—and people now use half the world's readily accessible freshwater runoff. Chemical plants fix more atmospheric nitrogen than all natural terrestrial processes combined, and fisheries remove more than a third of the primary production of the temperate coastal waters of the oceans. Through global trade and international travel, humans have transported countless species into ecosystems that are not prepared for them. We have pumped enough carbon dioxide into the air to alter the climate and to change the chemistry of the oceans.

Amphibians are affected by many—perhaps most—of these disruptions. Habitat destruction is a major factor in their decline, and agricultural chemicals seem to be causing a rash of frog deformities. But the main culprit in the wavelike series of crashes, it's now believed, is a fungus. Ironically, this fungus, which belongs to a group known as chytrids (pronounced "kit-rids"), appears to have been spread by doctors.

Chytrid fungi are older even than amphibians—the first species evolved more than six hundred million years ago—and even more widespread. In a manner of speaking, they can be found—they are microscopic—just about everywhere, from the tops of trees to deep underground. Generally, chytrid fungi feed off dead plants; there are also species that live on algae, species that live on roots, and species that live in the guts of cows, where they help break down cellulose. Until two pathologists, Don Nichols and Allan Pessier, identified a weird microorganism growing on dead frogs from the National Zoo, chytrids had never been known to attack vertebrates. Indeed, the new chytrid was so unusual that an entire genus had to be created to accommodate it. It was named *Batrachochytrium dendrobatidis*—*batrachos* is Greek for "frog"—or Bd for short.

Nichols and Pessier sent samples from the infected frogs to a mycologist at the University of Maine, Joyce Longcore, who managed to culture the Bd fungus. They then exposed healthy blue poison-dart frogs to it. Within three weeks, the animals sickened and died.

The discovery of Bd explained many of the data that had previously seemed so puzzling. Chytrid fungi generate microscopic spores that disperse in water; these could have been carried along by streams, or in the runoff after a rainstorm, producing what in Central America showed up as an eastward-moving scourge. In the case of zoos, the spores could have been brought in on other frogs or on tracked-in soil. Bd seemed to be able to live on just about any frog or toad, but not all amphibians are as susceptible to it, which would account for why some populations succumbed while others appeared to be unaffected.

Rick Speare is an Australian pathologist who identified Bd right around the same time that the National Zoo team did. From the pattern of decline, Speare suspected that Bd had been spread by an amphibian that had been moved around the globe. One of the few species that met this condition was *Xenopus laevis*, commonly known as the African clawed frog. In the early nineteen-thirties, a British zoologist named Lancelot Hogben discovered that female *Xenopus laevis*, when injected with certain types of human hormones, laid eggs. His discovery became the basis for a new kind of pregnancy test and, starting in the late nineteen-thirties, thousands of African clawed frogs were exported out of Cape Town. In the nineteen-forties and fifties, it was not uncommon for obstetricians to keep tanks full of the frogs in their offices.

To test his hypothesis, Speare began collecting samples from live African clawed frogs and also from specimens preserved in museums. He found that specimens dating back to the nineteen-thirties were indeed already carrying the fungus. He also found that live African clawed frogs were widely infected with Bd, but seemed to suffer no ill effects from it. In 2004, he co-authored an influential paper that argued that the transmission route for the fungus began in southern Africa and ran through clinics and hospitals around the world.

"Let's say people were raising African clawed frogs in aquariums, and they just popped the water out," Speare told me. "In most cases when they did that, no frogs got infected, but then, on that hundredth time, one local frog might have been infected. Or people might have said, 'I'm sick of this frog, I'm going to let it go.' And certainly there are populations of African clawed frogs established in a number of countries around the world, to illustrate that that actually did occur."

At this point, Bd appears to be, for all intents and purposes, unstoppable. It can be killed by bleach—Clorox is among the donors to EVACC—but it is impossible to disinfect an entire rain forest. Sometime in the last year or so, the fungus jumped the Panama Canal. (When Edgardo Griffith swabbed the frogs on our trip, he was collecting samples that would eventually be analyzed for it.) It also seems to be heading into Panama from the opposite direction, out of Colombia. It has spread through the highlands of South America, down the eastern coast of Australia, and into New Zealand, and has been detected in Italy, Spain, and France. In the U.S., it appears to have radiated from several points, not so much in a wavelike pattern as in a series of ripples.

In the fossil record, mass extinctions stand out, so sharply that the very language scientists use to describe the earth's history derives from them. In 1840, the British geologist John Phillips divided life into three chapters: the Paleozoic (from the Greek for "ancient life"), the Mesozoic ("middle life"), and the Cenozoic ("new life"). Phillips fixed as the dividing point between the first and second eras what would now be called the end-Permian extinction, and between the second and the third the end-Cretaceous event. The fossils from these eras were so different that Phillips thought they represented three distinct episodes of creation.

Darwin's resistance to catastrophism meant that he couldn't accept what the fossils seemed to be saying. Drawing on the work of the eminent geologist Charles Lyell, a good friend of his, Darwin maintained that the apparent discontinuities in the history of life were really just gaps in the archive. In "On the Origin of Species," he argued:

> With respect to the apparently sudden extermination of whole families or orders, as of Trilobites at the close of the palaeozoic period and of Ammonites at the close of the secondary period, we must remember what has been already said on the probable wide intervals of time between our consecutive formations; and in these intervals there may have been much slow extermination.

All the way into the nineteen-sixties, paleontologists continued to give talks with titles like "The Incompleteness of the Fossil Record." And this view might have persisted even longer had it not been for a remarkable, largely inadvertent discovery made in the following decade.

In the mid-nineteen-seventies, Walter Alvarez, a geologist at the Lamont Doherty Earth Observatory, in New York, was studying the earth's polarity. It had recently been learned that the orientation of the planet's magnetic field reverses, so that every so often, in effect, south becomes north and then vice versa. Alvarez and some colleagues had found that a certain formation of pinkish limestone in Italy, known as the *scaglia rossa*, recorded these occasional reversals. The limestone also contained the fossilized remains of millions of tiny sea creatures called foraminifera. In the course of several trips to Italy, Alvarez became interested in a thin layer of clay in the limestone that seemed to have been laid down around the end of the Cretaceous. Below the layer, certain species of foraminifera—or forams, for short—were preserved. In the clay layer there were no forams. Above the layer, the earlier species disappeared and new forams appeared. Having been taught the uniformitarian view, Alvarez wasn't sure what to make of what he was seeing, because the change, he later recalled, certainly "looked very abrupt."

Alvarez decided to try to find out how long it had taken for the clay layer to be deposited. In 1977, he took a post at the University of California at Berkeley, where his father, the Nobel Prize-winning physicist Luis Alvarez, was also teaching. The older Alvarez suggested using the element iridium to answer the question.

Iridium is extremely rare on the surface of the earth, but more plentiful in meteorites, which, in the form of microscopic grains of cosmic dust, are constantly raining down on the planet. The Alvarezes reasoned that, if the clay layer had taken a significant amount of time to deposit, it would contain detectable levels of iridium, and if it had been deposited in a short time it wouldn't. They enlisted two other scientists, Frank Asaro and Helen Michel, to run the tests, and gave them samples of the clay. Nine months later, they got a phone call. There was something seriously wrong. Much too much iridium was showing up in the samples. Walter Alvarez flew to Denmark to take samples of another layer of exposed clay from the end of the Cretaceous. When they were tested, these samples, too, were way out of line.

The Alvarez hypothesis, as it became known, was that everything—the clay layer from the *scaglia rossa*, the clay from Denmark, the spike in iridium, the shift in the fossils—could be explained by a single event. In 1980, the Alvarezes and their colleagues proposed that a six-mile-wide asteroid had slammed into the earth, killing off not only the forams but the dinosaurs and all the other organisms that went extinct at the end of the Cretaceous. "I can remember working very hard to make that 1980 paper just as solid as it could possibly be," Walter Alvarez recalled recently. Nevertheless, the idea was greeted with incredulity.

"The arrogance of those people is simply unbelievable," one paleontologist told the *Times*.

"Unseen bolides dropping into an unseen sea are not for me," another declared.

Over the next decade, evidence in favor of an enormous impact kept accumulating. Geologists looking at rocks from the end of the Cretaceous in Montana found tiny mineral grains that seemed to have suffered a violent shock. (Such "shocked quartz" is typically found in the immediate vicinity of meteorite craters.) Other geologists, looking in other parts of the world, found small, glasslike spheres of the sort believed to form when molten-rock droplets splash up into the atmosphere. In 1990, a crater large enough to have been formed by the enormous asteroid that the Alvarezes were proposing was found, buried underneath the Yucatán. In 1991, that crater was dated, and discovered to have been formed at precisely the time the dinosaurs died off.

"Those eleven years seemed long at the time, but looking back they seem very brief," Walter Alvarez told me. "Just think about it for a moment. Here you have a challenge to a uniformitarian viewpoint that basically every geologist and paleontologist had been trained in, as had their professors and their professors' professors, all the way back to Lyell. And what you saw was people looking at the evidence. And they gradually did come to change their minds."

Today, it's generally accepted that the asteroid that plowed into the Yucatán led, in very short order, to a mass extinction, but scientists are still uncertain exactly how the process unfolded. One theory holds that the impact raised a cloud of dust that blocked the sun, preventing photosynthesis and causing widespread starvation. According to another theory, the impact kicked up a plume of vaporized rock travelling with so much force that it broke through the atmosphere. The particles in the plume then recondensed, generating, as they fell back to earth, enough thermal energy to, in effect, broil the surface of the planet.

Whatever the mechanism, the Alvarezes' discovery wreaked havoc with the uniformitarian idea of extinction. The fossil record, it turned out, was marked by discontinuities because the history of life was marked by discontinuities.

In the nineteenth century, and then again during the Second World War, the Adirondacks were a major source of iron ore. As a result, the mountains are now riddled with abandoned mines. On a gray day this winter, I went to visit one of the mines (I was asked not to say which) with a wildlife biologist named Al Hicks. Hicks, who is fifty-four, is tall and outgoing, with a barrel chest and ruddy cheeks. He works at the headquarters of the New York State Department of Environmental Conservation, in Albany, and we met in a parking lot not far from his office. From there, we drove almost due north.

Along the way, Hicks explained how, in early 2007, he started to get a lot of strange calls about bats. Sometimes the call would be about a dead bat that had been brought inside by somebody's dog. Sometimes it was about a live—or half-alive—bat flapping around on the driveway. This was in the middle of winter, when any bat in the Northeast should have been hanging by its feet in a state of torpor. Hicks found the calls bizarre, but, beyond that, he didn't know what to make of them. Then, in March, 2007, some colleagues went to do a routine census of hibernating bats in a cave west of Albany. After the survey, they, too, phoned in.

"They said, 'Holy shit, there's dead bats everywhere,'" Hicks recalled. He instructed them to bring some carcasses back to the office, which they did. They also shot photographs of live bats hanging from the cave's ceiling. When Hicks examined the photographs, he saw that the animals looked as if they had been dunked, nose first, in talcum powder. This was something he had never run across before, and he began sending the photographs to all the bat specialists he could think of. None of them could explain it, either.

"We were thinking, Oh, boy, we hope this just goes away," he told me. "It was like the Bush Administration. And, like the Bush Administration, it just wouldn't go away." In the winter of 2008, bats with the white powdery substance were found in thirty-three hibernating spots. Meanwhile, bats kept dying. In some hibernacula, populations plunged by as much as ninety-seven per cent.

That winter, officials at the National Wildlife Health Center, in Madison, Wisconsin, began to look into the situation. They were able to culture the white substance, which was found to be a never before identified fungus that grows only at cold temperatures. The condition became known as white-nose syndrome, or

W.N.S. White nose seemed to be spreading fast; by March, 2008, it had been found on bats in three more states—Vermont, Massachusetts, and Connecticut—and the mortality rate was running above seventy-five per cent. This past winter, white nose was found to have spread to bats in five more states: New Jersey, New Hampshire, Virginia, West Virginia, and Pennsylvania.

In a paper published recently in *Science*, Hicks and several co-authors observed that "parallels can be drawn between the threat posed by W.N.S. and that from chytridiomycosis, a lethal fungal skin infection that has recently caused precipitous global amphibian population declines."

When we arrived at the base of a mountain not far from Lake Champlain, more than a dozen people were standing around in the cold, waiting for us. Most, like Hicks, were from the D.E.C., and had come to help conduct a bat census. In addition, there was a pair of biologists from the U.S. Fish and Wildlife Service and a local novelist who was thinking of incorporating a subplot about white nose into his next book. Everyone put on snowshoes, except for the novelist, who hadn't brought any, and began tromping up the slope toward the mine entrance.

The snow was icy and the going slow, so it took almost half an hour to reach an outlook over the Champlain Valley. While we were waiting for the novelist to catch up—apparently, he was having trouble hiking through the three-foot-deep drifts—the conversation turned to the potential dangers of entering an abandoned mine. These, I was told, included getting crushed by falling rocks, being poisoned by a gas leak, and plunging over a sheer drop of a hundred feet or more.

After another fifteen minutes or so, we reached the mine entrance—essentially, a large hole cut into the hillside. The stones in front of the entrance were white with bird droppings, and the snow was covered with paw prints. Evidently, ravens and coyotes had discovered that the spot was an easy place to pick up dinner.

"Well, shit," Hicks said. Bats were fluttering in and out of the mine, and in some cases crawling on the ground. Hicks went to catch one; it was so lethargic that he grabbed it on the first try. He held it between his thumb and forefinger, snapped its neck, and placed it in a ziplock bag.

"Short survey today," he announced.

At this point, it's not known exactly how the syndrome kills bats. What is known is that bats with the syndrome often wake up from their torpor and fly around, which leads them to die either of starvation or of the cold or to get picked off by predators.

We unstrapped our snowshoes and put on helmets. Hicks handed out headlamps—we were supposed to carry at least one extra—and packages of batteries; then we filed into the mine, down a long, sloping tunnel. Shattered beams littered the ground, and bats flew up at us through the gloom. Hicks cautioned everyone to stay alert. "There's places that if you take a step you won't be stepping back," he warned. The tunnel twisted along, sometimes opening up into concert-hall-size chambers with side tunnels leading out of them.

Over the years, the various sections of the mine had acquired names; when we reached something called the Don Thomas section, we split up into groups to start the survey. The process consisted of photographing as many bats as possible. (Later on, back in Albany, someone would have to count all the bats in the pictures.) I went with Hicks, who was carrying an enormous camera, and one of the biologists from the Fish and Wildlife Service, who had a laser pointer. The biologist would aim the pointer at a cluster of bats hanging from the ceiling. Hicks would then snap a photograph. Most of the bats were little brown bats; these are the most common bats in the U.S. and the ones you are most likely to see flying around on a summer night. There were also Indiana bats, which are on the federal endangered-species list, and small-footed bats, which, at the rate things are going, are likely to end up there. As we moved along, we kept disturbing the bats, which squeaked and started to rustle around, like half-asleep children.

Since white nose grows only in the cold, it's odd to find it living on mammals, which, except when they're hibernating (or dead), maintain a high body temperature. It has been hypothesized that the fungus normally subsists by breaking down organic matter in a chilly place, and that it was transported into bat hibernacula, where it began to break down bats. When news of white nose began to get around, a spelunker sent Hicks photographs that he had shot in Howe's Cave, in central New York. The photographs, which had been taken in 2006, showed bats with clear signs of white nose and are the earliest known record of the syndrome. Howe's Cave is connected to Howe's Caverns, a popular tourist destination.

"It's kind of interesting that the first record we have of this fungus is photographs from a commercial cave in New York that gets about two hundred thousand visits a year," Hicks told me.

Despite the name, white nose is not confined to bats' noses; as we worked our way along, people kept finding bats with freckles of fungus on their wings and ears. Several of these were dispatched, for study purposes, with a thumb and forefinger. Each dead bat was sexed—males can be identified by their tiny penises—and placed in a ziplock bag.

At about 7 p.m., we came to a huge, rusty winch, which, when the mine was operational, had been used to haul ore to the surface. By this point, we were almost down at the bottom of the mountain, except that we were on the inside of it. Below, the path disappeared into a pool of water, like the River Styx. It was impossible to go any further, and we began working our way back up.

Bats, like virtually all other creatures alive today, are masters of adaptation descended from lucky survivors. The earliest bat fossil that has been found dates from fifty-three million years ago, which is to say twelve million years after the impact that ended the Cretaceous. It belongs to an animal that had wings and could fly but had not yet developed the specialized inner ear that, in modern bats, allows for echolocation. Worldwide, there are now more than a thousand bat species, which together make up nearly a fifth of all species of mammals. Most feed on insects; there are also bats that live off fruit, bats that eat fish—they use echolocation to detect minute ripples in the water—and a small but highly celebrated group that consumes blood. Bats are great colonizers—Darwin noted that even New Zealand, which has no other native land mammals, has its own bats—and they can be found as far north as Alaska and as far south as Tierra del Fuego.

In the time that bats have evolved and spread, the world has changed a great deal. Fifty-three million years ago, at the start of the Eocene, the planet was very warm, and tropical palms grew at the latitude of London. The climate cooled, the Antarctic ice sheet began to form, and, eventually, about two million years ago, a period of recurring glaciations began. As recently as fifteen thousand years ago, the Adirondacks were buried under ice.

One of the puzzles of mass extinction is why, at certain junctures, the resourcefulness of life seems to falter. Powerful as the Alvarez hypothesis proved to be, it explains only a single mass extinction.

"I think that, after the evidence became pretty strong for the impact at the end of the Cretaceous, those of us who were working on this naïvely expected that we would go out and find evidence of impacts coinciding with the other events," Walter Alvarez told me. "And, of course, it's turned out to be much more complicated.

We're seeing right now that a mass extinction can be caused by human beings. So it's clear that we do not have a general theory of mass extinction."

Andrew Knoll, a paleontologist at Harvard, has spent most of his career studying the evolution of early life. (Among the many samples he keeps in his office are fossils of microorganisms that lived 2.8 billion years ago.) He has also written about more recent events, like the end-Permian extinction, which took place two hundred and fifty million years ago, and the current extinction event.

Knoll noted that the world can change a lot without producing huge losses; ice ages, for instance, come and go. "What the geological record tells us is that it's time to worry when the rate of change is fast," he told me. In the case of the end-Permian extinction, Knoll and many other researchers believe that the trigger was a sudden burst of volcanic activity; a plume of hot mantle rock from deep in the earth sent nearly a million cubic miles' worth of flood basalts streaming over what is now Siberia. The eruption released enormous quantities of carbon dioxide, which presumably led—then as now—to global warming, and to significant changes in ocean chemistry.

"CO_2 is a paleontologist's dream," Knoll told me. "It can kill things directly, by physiological effects, of which ocean acidification is the best known, and it can kill things by changing the climate. If it gets warmer faster than you can migrate, you're in trouble."

In the end, the most deadly aspect of human activity may simply be the pace of it. Just in the past century, CO_2 levels in the atmosphere have changed by as much—a hundred parts per million—as they normally do in a hundred-thousand-year glacial cycle. Meanwhile, the drop in ocean pH levels that has occurred over the past fifty years may well exceed anything that happened in the seas during the previous fifty million. In a single afternoon, a pathogen like Bd can move, via United or American Airlines, halfway around the world. Before man entered the picture, such a migration would have required hundreds, if not thousands, of years—if, indeed, it could have been completed at all.

Currently, a third of all amphibian species, nearly a third of reef-building corals, a quarter of all mammals, and an eighth of all birds are classified as "threatened with extinction." These estimates do not include the species that humans have already wiped out or the species for which there are insufficient data. Nor do the figures take into account the projected effects of global warming or ocean acidification. Nor, of course, can they anticipate the kinds of sudden, terrible collapses that are becoming almost routine.

I asked Knoll to compare the current situation with past extinction events. He told me that he didn't want to exaggerate recent losses, or to suggest that an extinction on the order of the end-Cretaceous or end-Permian was imminent. At the same time, he noted, when the asteroid hit the Yucatán "it was one terrible afternoon." He went on, "But it was a short-term event, and then things started getting better. Today, it's not like you have a stress and the stress is relieved and recovery starts. It gets bad and then it keeps being bad, because the stress doesn't go away. Because the stress is us."

Aeolus Cave, in Dorset, Vermont, is believed to be the largest bat hibernaculum in New England; it is estimated that, before white nose hit, more than two hundred thousand bats—some from as far away as Ontario and Rhode Island—came to spend the winter there.

In late February, I went with Hicks to visit Aeolus. In the parking lot of the local general store, we met up with officials from the Vermont Fish and Wildlife Department, who had organized the trip. The entrance to Aeolus is about a mile and a half from the nearest road, up a steep, wooded hillside. This time, we approached by snowmobile. The temperature outside was about twenty-five degrees—far too low for bats to be active—but when we got near the entrance we could, once again, see bats fluttering around. The most senior of the Vermont officials, Scott Darling, announced that we'd all have to put on latex gloves and Tyvek suits before proceeding. At first, this seemed to me to be paranoid; soon, however, I came to see the sense of it.

Aeolus is a marble cave that was created by water flow over the course of thousands of years. The entrance is a large, nearly horizontal tunnel at the bottom of a small hollow. To keep people out, the Nature Conservancy, which owns the cave, has blocked off the opening with huge iron slats, so that it looks like the gate of a medieval fortress. With a key, one of the slats can be removed; this creates a narrow gap that can be crawled (or slithered) through. Despite the cold, there was an awful smell emanating from the cave—half game farm, half garbage dump. When it was my turn, I squeezed through the gap and immediately slid on the ice, into a pile of dead bats. The scene, in the dimness, was horrific. There were giant icicles hanging from the ceiling, and from the floor large knobs of ice rose up, like polyps. The ground was covered with dead bats; some of the ice knobs, I noticed, had bats frozen into them. There were torpid bats roosting on the ceiling, and also wide-awake ones, which would take off and fly by or, sometimes, right into us.

Why bat corpses pile up in some places, while in others they get eaten or in some other way disappear, is unclear. Hicks speculated that the weather conditions at Aeolus were so harsh that the bats didn't even make it out of the cave before dropping dead. He and Darling had planned to do a count of the bats in the first chamber of the cave, known as Guano Hall, but this plan was soon abandoned, and it was decided just to collect specimens. Darling explained that the specimens would be going to the American Museum of Natural History, so that there would at least be a record of the bats that had once lived in Aeolus. "This may be one of the last opportunities," he said. In contrast to a mine, which has been around at most for a few centuries, Aeolus, he pointed out, has existed for millennia. It's likely that bats have been hibernating there, generation after generation, since the end of the last ice age.

"That's what makes this so dramatic—it's breaking the evolutionary chain," Darling said.

He and Hicks began picking dead bats off the ground. Those which were too badly decomposed were tossed back; those which were more or less intact were sexed and placed in two-quart plastic bags. I helped out by holding open the bag for females. Soon, it was full and another one was started. It struck me, as I stood there holding a bag filled with several dozen stiff, almost weightless bats, that I was watching mass extinction in action.

Several more bags were collected. When the specimen count hit somewhere around five hundred, Darling decided that it was time to go. Hicks hung back, saying that he wanted to take some pictures. In the hours we had been slipping around the cave, the carnage had grown even more grotesque; many of the dead bats had been crushed and now there was blood oozing out of them. As I made my way up toward the entrance, Hicks called after me: "Don't step on any dead bats." It took me a moment to realize that he was joking.

The Ice Retreat

Fen Montaigne

December 21, 2009

Litchfield Island is an outcropping of granite and diorite rising out of the Southern Ocean off the coast of the northwestern Antarctic Peninsula, which juts toward the tip of South America. With dark, snow-streaked ridges flowing into broad, pebbled beaches, Litchfield is picturesque, but amid the imposing Antarctic landscape the island—not quite three-quarters of a mile long and half a mile wide—hardly merits a second glance. To the north, on Anvers Island, the dome of the Marr Ice Piedmont—a glacier roughly forty miles long and twenty miles across at its widest point—dominates the horizon. To the east, forming the spine of the nine-hundred-mile-long Antarctic Peninsula, lies a towering mountain range of sheer black rockfaces and ice fields streaming to the sea. And to the south and west, stretching to the horizon, are innumerable icebergs, their shapes running from flat-topped slabs to whimsical, castle-like structures.

On an overcast January morning, with the temperature just below freezing, Bill Fraser, an ecologist and penguin expert, eased a rubber Zodiac boat onto Litchfield's shore. Grasping the bowline, I leaped onto the rocks and was quickly joined by Fraser, who lashed the rope around a boulder, flung an iridescent-orange flotation coat onto the stony ground, and began walking across the eastern third of the island. It had been two years since Fraser set foot on Litchfield, and he had come to see what remained of a once thriving rookery of Adélies—the classic tuxedoed penguin, and one of only two penguin species that breed exclusively in Antarctica. (The other is the emperor penguin.)

Fraser paused amid an expanse of pebbles, nearly the size of a football field, that were used by the few remaining Adélie penguins to construct their cup-shaped nests. "Everywhere you see these rocks, there used to be penguins," Fraser said. "All these areas used to be colonies. There was a colony here, there, over there."

Studies of Adélie bones buried under layers of guano have shown that penguins have nested on Litchfield Island since at least the sixteenth century. When Fraser first arrived in the region, in 1974, as a graduate student, the island's penguin rookery had nine hundred breeding pairs. Over the years, the number of Adélies had fallen to a few dozen breeding pairs, and a census conducted earlier that season by a birding team that Fraser led indicated that the rookery was on the verge of disappearing.

Still, he was not prepared for the scene that greeted him on Litchfield's southern shore: only five Adélie nests remained, containing seven fluffy chicks—no taller than a man's hand—and eleven adult Adélies, which reached a person's knee. They were huddled on an oval-shaped patch of stones, twenty-five feet across at its widest point. Some of the chicks had burrowed under their parents, seeking the warmth of the brood patch—a four-inch slit that covers the eggs with bare, blood-engorged skin. (The heat of the brood patch enables Adélies to incubate their eggs at 86 degrees Fahrenheit.) Other chicks, with light-gray down and black heads, stood and stretched their necks skyward, greedily accepting from a parent a meal of regurgitated krill, a shrimplike creature. This last Adélie redoubt on Litchfield Island was known as Colony 8.

Fraser shook his head and said, "This is unbelievable. Colony 8 used to have so many birds you could barely count them."

As we returned to the boat, Fraser said that a few dozen adults, some possibly as old as fifteen to twenty years, kept returning to their ancestral breeding grounds. But Adélies have evolved as an ice-dependent species, and, in the rapidly warming environment of the northwestern Antarctic Peninsula, many of the adults—and nearly all their chicks—were not surviving the winter.

"Litchfield is looking miserable," Fraser said. "When this colony goes, it will be the first time in at least five hundred years that there will be no Adélies on the island."

In the austral summer of 2005–06, I spent nearly five months as a member of Bill Fraser's birding team at Palmer Station, a scrap of civilization grafted onto a rocky spit of land. It is named for an early-nineteenth-century American seal hunter, Nathaniel B. Palmer, who at age twenty left Stonington, Connecticut, in command of a small sloop, the Hero, sailed the length of the planet, crossed the wild stretch of ocean known as the Drake Passage, and entered the foggy, iceberg-covered Antarctic waters in search of fur seals.

The U.S. government research station named in Palmer's honor is a cluster of half a dozen corrugated-metal buildings that, at peak times, house forty scientists and support staff. The tiny base—described by one early visitor as an "insolent blot" on the Antarctic landscape—is situated in the midst of numerous small islands that are home to Adélie-penguin rookeries.

One evening, soon after I arrived at Palmer, I hiked up the Marr Ice Piedmont, just behind the station. Though it was only three hundred feet above sea level, I gazed through the dust-free Antarctic atmosphere at mountain summits a hundred and twenty miles away. Close to shore, the sea ice was loosely packed, and the placid pools of water at the base of the Marr glacier reflected the pale-blue walls of the ice piedmont. Beyond the nearby islands, the sea ice was a cool, blue-white table that ran to the horizon. A closer look revealed that its surface was composed of small slabs of sea ice pushed together by persistent winds. A penguin had recently tobogganed across the glacier, its flippers slicing knifelike slits in the snow.

Aerial photographs taken in 1963 show that the Marr Ice Piedmont once engulfed nearly the entire point on which Palmer Station now stands. Since that time, however, the glacier has retreated roughly fifteen hundred feet. In the nineteen-seventies and eighties, Fraser and others used to climb the Marr Ice Piedmont and ski to places such as Biscoe Point, eight miles away. That is virtually impossible now, as the thinning glacier is riven by crevasses and is falling back from the coastline in large, scalloped chunks.

Fraser has spent most of the past thirty-five years working in Antarctica, more time than almost any other U.S. scientist. He and his fellow-researchers at Palmer are mainly engaged in studying the effects of the rapid warming on the formation of sea ice, on the phytoplankton and Antarctic krill that depend on the sea ice, and on the Adélie penguins that rely on the sea ice and the krill. Fraser's work is part of the Long Term Ecological Research Network, a program, launched in 1980 by the National Science Foundation, that chronicles environmental changes in twenty-six ecosystems around the world.

Much attention has been paid to the warming of the Arctic, where for three decades scientists have been tracking the rapid retreat of summer sea ice and the melting of glaciers. But the northwestern Antarctic Peninsula has heated up faster than almost any other place on earth, as temperate air has streamed in from the north and warming ocean currents have swept along the peninsula. From 1951 to the present, the average annual air temperature in the region has increased by nearly five degrees. Winter temperatures have soared about eleven degrees over the past six decades—five times the global average.

For the moment, rising temperatures are only nibbling at the edges of Antarctica's repository of ice, a vast dome of frozen precipitation—three miles thick in places—covering an area one and a half times the size of the United States. But increasing warmth will likely penetrate more deeply into Antarctica; a recent study forecasts that the entire continent will heat up by more than five degrees this century.

"The poles are very sensitive barometers of warming, and what we're looking at here on the Antarctic Peninsula is an entire ecosystem that is changing," Fraser said. "And it's not changing in hundreds of years—it's changing in thirty to fifty years. To me, this is foretelling the future across major parts of the planet."

In the fall of 1975, Bill Fraser—then a twenty-five-year-old graduate student at the University of Minnesota—sailed from Ushuaia, Argentina, to Antarctica aboard a hundred-and-twenty-five-foot research vessel, the R.V. Hero, named in honor of Nathaniel Palmer's sloop. It was his second trip to the continent. After weathering a gale in the Drake Passage, the Hero approached the relatively tranquil waters of the Bransfield Strait, which is sheltered from the Southern Ocean by the South Shetland Islands, at the northern tip of the Antarctic Peninsula. Exhausted after several rough days of travel, Fraser went to sleep. The next morning, he awoke to unaccustomed stillness. The water was flat, and the only sound he heard was the drone of the Hero's diesel engine.

"I stepped out into this gray world," Fraser recalled. "It had been snowing heavily and the ship was moving through this three-to-four-inch layer of snow and slush on the water. There was snow all over the deck. It was incredibly foggy. And the ship was completely surrounded by snow petrels, one of the most beautiful birds in the world—absolutely white. When you see them, you know the ice is near."

After stopping at King George Island, in the South Shetlands, the Hero steamed toward Anvers Island, where Palmer Station is situated, at 64° S. The ship's progress was soon blocked by thick sea ice. Built for the National Science Foundation to supply Palmer Station and work along the Antarctic Peninsula, the Hero had an oak hull, overlain with a rock-hard layer of greenheart. The forward hull was sheathed in metal. But the Hero could not penetrate the sea ice surrounding the small American base. So Fraser disembarked and began heading toward Palmer Station on skis, carrying provisions and equipment behind him on a sled.

As Fraser glided swiftly over the surface, leopard seals—sleek, spotted predators that devour everything from krill to penguins—cruised under the ice, tracking his movements. Fraser could hear the leopard seals trilling low, droning calls beneath him, and occasionally a seal poked its head through an opening in the ice to observe the skiers. Leopard seals love ice, and they were a common feature around Palmer Station, lounging on floes, attacking Adélie penguins, trailing the rubber Zodiacs used by scientists, and occasionally biting a hole in one.

Fraser said of the peninsula, "It was completely remote and absolutely wild. The rawness and beauty of this place just cannot be described. It was a place where you could still feel inconsequential. You were part of a working system that paid you no mind."

Fraser spent thirteen months in Antarctica. Palmer Station had been in operation for only seven years, and the scientists and support staff lived in isolation, communicating with the outside world by ham radio. During the winter, Fraser and five other men manned the station, and no ships came or went. Fraser worked in the field, observing the behavior and foraging habits of one of the rare midwinter resident seabirds, the kelp gull. For recreation, he skied by himself into the frozen Southern Ocean. When the moon was full, bathing the ice in a ghostly blue light, Fraser skied for miles along the shore. He stopped from time to time and listened to the stillness, broken occasionally by the report of cracking ice.

"You start doing what you know you need to do to get through the winter," Fraser recalled. "You're taking care of your sanity. You knew you were there for the long haul. And the winters were a lot colder then."

During the summer months, Fraser and his fellow graduate students travelled from island to island, taking a census of Adélie-penguin rookeries. One of the largest was on Torgersen Island, half a mile from the station. On a January evening, when many of the adults were on the island feeding their chicks, more than twenty thousand Adélies of all ages mobbed Torgersen, their raspy calls splitting the air with a din that could be heard miles away.

"You'd see thousands of Adélies walking to and from the sea," Fraser said. "It was like ants in the forest—there was a constant stream of birds. Torgersen was an absolute mass of life. It just manifested the incredible productivity of this ocean and its ability to support life."

One of the defining features of Antarctica is the great skirt of sea ice that spreads out from the continent each year. At the end of the Southern Hemisphere's summer, in February, the sea ice girding the continent is roughly a million square miles. By September, the circle of sea ice around the continent has expanded six times—more than doubling the size of Antarctica itself. (Although sea ice has declined markedly along the western Antarctic Peninsula, it is increasing in some regions of Antarctica, most notably in the Ross Sea, because of changing atmospheric patterns.)

Once, on a midwinter cruise in the Weddell Sea, Fraser went to bed as a severe cold snap set in and woke up the next morning to find that the ice edge had spread outward by several miles. On the infrequent occasions when the sun was out—for about two hours around noon—Fraser could stand on the bridge and, as temperatures dropped, watch the sea ice form before his eyes.

"Suddenly, you'd see what looks like grease on the water, and then that would stop moving," Fraser said. "And, before you knew it, the sea was turning gray, gray, gray, and finally you would see an inch or so of ice just appear."

The Intergovernmental Panel on Climate Change has concluded that human-induced global warming began to clearly impact the planet at about the time that Fraser first came to Palmer Station. But in 1974 few scholars were linking human activity with rising global temperatures. One of Fraser's professors at the University of Minnesota did discuss climate change, but told his students that it was unlikely they would witness its effects in their lifetimes.

Fraser, who is fifty-eight, is a lean and handsome man of six feet, with light-brown hair, a ruddy complexion, blue eyes, thin lips, and a cleft in his chin. After three decades of working in Antarctica, he has perfected his field outfit, wearing clothes that protect him from the elements yet do not leave him overheated during swift hikes across the terrain. During my season at Palmer Station, he wore a fleece sweater covered by a royal-blue windbreaker that was faded by the sun and ripped in places by bird beaks and rocks; rugged black rain pants heavily stained by guano and crusty with sea salt; and ankle-high leather hiking boots, deeply scarred by years of scrambling over Antarctic rocks.

At the rookeries on Litchfield and other islands, the Adélies paid us little attention as long as we stayed ten feet away. If you moved closer, you would be subjected to an array of elaborate threat displays, all exhaustively studied by scientists—a walleyed stare, or the raising of the feathers on the head into something resembling a flat-top haircut. If you moved closer still, some Adélies would charge and jam a sharp beak into your leg. (One day, a belligerent penguin bounded up to me and plunged its beak into the flesh below my knee. Like an awl, it penetrated my rain pants, thick fleece pants, long underwear, and skin, drawing blood.)

While Fraser occasionally uses high-tech tools in his research—such as tracking penguins by affixing satellite transmitters to their backs—most of his work involves spending long days in the field performing decidedly low-tech, repetitive tasks. He and his birding team measure snow depth, count penguins and seabirds, and sample guano to check what's in their diets.

He is assisted by a team of five young field workers. Though nearly all of them have a background in biology, Fraser is looking primarily for people who are amiable—working five months in the field, often ten to twelve hours a day, tests a team's cohesiveness—and in excellent physical condition. The birders were an outdoorsy and collegial group who often came to a decision when one member looked at another and said, "Your call."

Fraser is fiercely loyal to his squad of birders; unlike many field biologists, he provides health insurance to his seasonal field-team workers and contributes to their pension plans. At Palmer, he was keenly attuned to the dangers of exhaustion. Once, lost in thought near season's end, I walked too close to some giant petrels and spooked them. A day or two later, I wandered near a group of elephant seals wallowing at the edge of an Adélie colony, prompting two of them to slither rapidly away in panic and nearly crush to death several penguin chicks. As we walked back to our boat, Fraser pulled me aside and said, "I think you're getting burned out. I've noticed it for a while. You came out of the boat yesterday with a rope wrapped around your legs."

I apologized and told him I felt fine.

"You don't even know it's happening to you," he continued. "I saw one person, after four months here, literally step off the boat right into the water. I had to talk to you because when someone is like this he can easily get hurt himself or hurt someone else on the team. And it's no good for morale."

Although Fraser and his team study many seabird species around Palmer Station—including giant petrels, skuas, and kelp gulls—his work is heavily focussed on Adélies. In the thirty-five years since he first arrived at Palmer Station, the population of Adélie penguins on the seven islands that compose his main study area—Torgersen, Humble, Litchfield, Cormorant, Christine, Dream, and Biscoe Point—has declined from between thirty thousand and forty thousand breeding pairs to fifty-six hundred breeding pairs, a drop of more than eighty per cent. Meanwhile, the Adélie's cousin from warmer climes, the gentoo penguin, is moving in, following the heat as it travels down the Antarctic Peninsula. In 1993, there were no gentoo penguins in Fraser's study area. Today, there are twenty-four hundred pairs. As Fraser is fond of saying, "There goes the neighborhood."

Adélies are an easy bird to love, with their upright carriage and industrious demeanor. Their bonding displays are particularly charming: males daintily give gifts of nest pebbles to females, and couples greet each other by issuing full-throated honks and weaving their heads back and forth, in a kind of Adélie air kiss. After they've paired off, Adélie couples sometimes stand side by side, looking like partners at the start of a square dance. Occasionally, one penguin bows to its mate.

What most appeals to Fraser, though, is the Adélies' feistiness. Once, on Humble Island, I watched as an Adélie marched up to a fifteen-hundred-pound elephant seal that had moved too close to the penguin's chicks and jabbed the seal in the mouth. The seal retreated, shimmying back several feet even as it opened its bubblegum-pink mouth in a display of aggression.

Fraser once came across a female Adélie that had been grievously injured in a leopard-seal attack. "The seal had grabbed her by the head and ripped her breastbone off right at the neck and you could look down into her stomach cavity," he said. "You could actually see her lungs working in there. But she spent a few days recouping, hunched over, and I'll be damned if in less than a week she wasn't back in the water, feeding her chicks. She and her mate even pulled a brood off."

From time to time, the team affixed satellite transmitters to the backs of Adélies to determine where they were foraging. Fraser would approach a colony and slowly insinuate his arm among the penguins, some of which would hurl themselves at him and jab his legs or hands. Snatching an Adélie, he passed the bird to me. I grabbed it by the flippers and delivered it to the team members attaching the tags. The penguin's might was electric, its football-shaped body—a solid mass of muscle—rocking my arms as it struggled to free itself.

One evening, we travelled to Torgersen Island to do Adélie diet sampling. This entailed grabbing several penguins, pumping water into their stomachs through plastic tubing until they regurgitated, and then collecting their stomach contents in a white plastic bucket.

A team member, Brett Pickering, noticed drops of blood flowing into the bucket as one of the Adélies regurgitated krill. Pickering released the penguin, which stood for a few seconds, then lay down on the pebbles.

It shook its head, spraying a combination of water, krill, and blood. The bird rose and tried to walk but couldn't. Blood-stained bubbles poured from its mouth. Fraser silently picked up the penguin, walked twenty-five feet away, knelt down, placed the Adélie between his legs, grabbed a rock, and ended its life with a blow to the head.

Several members of the team were on the verge of crying. We walked in silence back to the boat landing.

"Something has changed," Fraser said, as we neared the Zodiac. "I don't know what it is, but something has changed, and I want to find out what it is. We have done hundreds of these diet samples for years with no problems, and now in the last few years some birds have started dying.

"I know that it's very likely that the bird we're diet sampling is probably an old animal, because chick recruitment is so poor now. And it really bothers me that here's a bird that's survived thick and thin and gone through hell and it's an old bird and I come along and it dies because of me."

Like many earlier explorers, Fraser seems happiest in the Antarctic. ("Once you have been to the white unknown, you can never escape the call of the little voices," Frank Wild, Ernest Shackleton's second-in-command on the Endurance expedition, said.) I once asked Fraser if he could live in a city, and he replied, "When I have to spend four or five days at a conference in a big city, I wither away. I become lethargic. It drives me insane. I hate it when I can't see the horizon."

Fraser's many years in the field have been instrumental in helping him understand how a warming climate has decimated the region's Adélies.

"It always seemed intuitive to me that the only way to really understand something is to live in it, to spend a tremendous amount of time in the field, collecting the same data year after year," Fraser told me. "You develop a sense for what the rhythms should be, the flow of things. And that's what has allowed me to pick up things that don't make sense, the anomalies. The anomalous years really cue you in as to how this system is operating."

For eighteen years, beginning in 1987, Fraser spent from three to five months every year in Antarctica—and occasionally seven or eight—as his wife and two daughters stayed at home in the U.S. The marriage ended in divorce, and relations with one of his daughters have been strained. (Once, as he was preparing to depart for another season at Palmer, one of the daughters told him, "Mommy says you love penguins more than you love me.")

Though he worked for several years as an associate professor at Montana State University, Fraser's decision to run a field operation for nearly half the year made it virtually impossible for him to hold a traditional academic job. In 2002, he married one of his team members, Donna Patterson, who shares his commitment to the work at Palmer Station. The couple now run the Polar Oceans Research Group, a non-profit based in their home, a thirty-four-acre farm in southwestern Montana's Ruby Valley. The modest, century-old house owned by Fraser and Patterson sits at five thousand feet and has mountain views in all directions. Madison County, where they live, is half the size of New Jersey but has no stoplights and not quite eight thousand inhabitants.

"He's part of a generation of scientists we won't see again," Polly Penhale, a former program director of Antarctic biology and medicine at the National Science Foundation, told me. "How many people would give up their personal lives for four to five months a year for all those years?" Since having a son in 2004, Fraser has spent less time at Palmer and more time in Montana with his family, increasingly leaving the field work to his team.

Over the past two decades, Fraser and his colleagues have gradually identified two main culprits in the decline of the region's Adélie penguins. One is the steady loss of sea ice along the northwestern Antarctic Peninsula. The second is a related increase in snowfall, which has primarily interfered with the ability of Adélies to successfully breed and incubate their eggs.

"There's this gigantic puzzle and you have all these pieces and it has taken twenty years to figure it out," Fraser said. "The most dangerous thing you can do is assume it is one factor that is causing these changes, when in fact many factors are playing a part."

The decline in sea-ice duration has been dramatic. Satellite imagery shows that sea ice now blankets waters off the peninsula three months less than it did in 1979. On average, sea ice forms fifty-four days later in the autumn and retreats thirty-one days sooner in the spring.

"The whole system revolves around the freezing point, so a slight change in the positive direction has major implications for the entire food web," Fraser said. "The freezing point is truly a threshold."

The importance of sea ice to Adélies first became clear to Fraser during a government-funded cruise in 1988. Fraser was aboard a Coast Guard icebreaker in the Weddell Sea, east of Palmer Station, on the other side of the Antarctic Peninsula. As the cruise unfolded in bitterly cold conditions, with temperatures sometimes plunging to minus 40 degrees Fahrenheit, Fraser noticed tens of thousands of Adélies on the pack ice and the absence of chinstrap penguins. A second research vessel, to the north, in the largely ice-free Scotia Sea, found almost exclusively chinstrap penguins, whose numbers were growing in the region.

The widely accepted explanation for the increase in chinstrap penguins was the decimation of baleen whales in the Southern Ocean by large-scale commercial whaling operations, which continued into the nineteen-eighties. Scientists believed that, with these voracious krill consumers largely removed from the sea, krill populations had rebounded, sparking a resurgence by creatures that ate krill, including chinstrap penguins.

If that hypothesis was true, Fraser wondered, then why were Adélies, which also ate krill, declining in number in some places? Not long after the cruise, he attended a conference in Hobart, Tasmania, during which several papers discussed growing evidence of global warming. "And that is when I sort of put these stories together," Fraser said. "It reeked of habitat change due to a warming climate."

He began to believe that changing penguin populations had less to do with krill abundance than with a warming winter habitat. Adélies were denizens of sea ice, while chinstraps preferred open water. The temperature record showed a steady decrease in the frequency of the cold years that led to extensive sea-ice formation along the Antarctic Peninsula, which occurs when the mean annual air temperature falls below 24 degrees Fahrenheit. At mid-century, roughly four out of five years saw temperatures below that mean. But by the nineteen-eighties the frequency of such winters had fallen to one or two out of every five years.

Some of Fraser's colleagues were skeptical of his thesis about the central role played by declining sea ice. Fraser spoke with scientists at NASA, which had begun using satellites to monitor sea ice along the peninsula. NASA researchers told him that there were no discernible trends in sea ice. Within a few years, however, satellite data clearly showed that sea ice was swiftly disappearing in the region. In 1991, Fraser and three colleagues wrote a paper on the subject, but the prestigious journals *Science* and *Nature* rejected it. A year later, he and his collaborators published the paper in *Polar Biology*, arguing that shrinking sea ice was leading to the growth of chinstrap populations and the decline of Adélies. That year, John Croxall, a prominent scientist with the British Antarctic Survey, published a paper questioning Fraser's thesis, saying that it was "certainly premature" to conclude that changing environmental factors, such as declining sea ice, were the reason behind shifting penguin populations. Croxall, who has retired, now says that, given the accumulated evidence of the last two decades, there is "no doubt" that declining sea ice has been a crucial factor affecting penguin populations.

As his research on the peninsula continued, Fraser became convinced that shrinking sea ice was depriving Adélies of a crucial platform that allowed them to reach rich feeding grounds. In winter, Adélies throughout Antarctica must have access to open water, which on the peninsula is typically at the heads of undersea canyons, where upwelling brings nutrients—and krill—to the surface. Those foraging areas must also be above the Antarctic Circle in midwinter, so that the penguins have sufficient light to see their prey. If a lack of ice prevents the Adélies from easily reaching these feeding grounds, their winter survival rates plummet.

"Without enough sea ice, these birds can't reach what must be very productive areas of the ocean," Fraser said. He had placed satellite tags on Adélies along the northwestern Antarctic Peninsula, which showed that they were foraging in three main locations, at the heads of the large undersea canyons.

One way to look at the Adélies' plight, Fraser said, is through the prism of an old ecological concept: "match-mismatch dynamics." More recently, some scientists and writers have taken to calling it "global weirding." On the most basic level, it means that a species has fallen out of sync with its environment.

In the field one afternoon, Fraser said, "What we're looking at here is an entire ecosystem that is changing, and it's not changing in hundreds of years, which is what we used to be taught. It's changed so quickly that it has encompassed the research lives of a few people who have spent a lifetime here."

The loss of sea ice has also taken a toll on ice-dependent krill—juvenile krill eat phytoplankton embedded in the underside of the ice—and on phytoplankton that bloom when sea ice melts in summer. Recent studies have shown drops in populations of both species along the northwestern Antarctic Peninsula. In addition, populations of ice-dependent Antarctic silverfish—which thirty years ago made up nearly half the diet of Adélie penguins around Palmer Station—have also fallen in the region, meaning that Adélies are overwhelmingly dependent on krill.

Hugh Ducklow, the director of the Ecosystems Center at Woods Hole and the head of the Palmer Long Term Ecological Research Network, said, "As sea ice begins to decline and then fails to form, as is now happening very rapidly, all these populations that depend on the timing and the existence and the extent of sea ice for their successful feeding and breeding will be high and dry."

Fraser has noticed from radio-transmitter data that Adélie penguins have been spending more time foraging at sea in recent years. Average foraging times appear to have increased from roughly eight to thirteen hours per trip, an indication that the Adélies are having a harder time finding the krill they need to feed their chicks—and themselves.

In the early nineties, Fraser began focussing on the impact of increasing snowfall on Adélie penguins. Working one November on Torgersen Island, Fraser walked to the top of the low, rocky ridge that bisects the island. He recalls, "I looked to my right, and that area was almost snow-free and there were thousands of Adélies there. Then I looked to my left—the south side—and it was under a metre of snow, with fewer than a thousand Adélies sitting on the snow, having a hell of a time trying to breed." Adélies have scant success nesting on snow or in snowmelt, as their eggs often become addled—and newly hatched chicks sometimes drown—in meltwater ponds.

Fraser realized that the abundant snow on the south side had to be related to the state of Adélie populations there, which were declining far more rapidly than on the north side of the island. As he stood on the ridge, it also became clear to him why the south side was snowier than the north: this relatively modest crest of crumbling, lichen-covered rocks was still high enough to create a wind barrier that prevented the prevailing northerly winds from blowing the snow off the south side.

"As counter-intuitive as it may sound, Adélie penguins are a snow-intolerant species," Fraser told me. "They evolved in a polar desert. And these birds' life histories are so finely tuned to the environment that tiny differences can affect their survival. If they don't have snow-free habitat by the end of November, their breeding success is catastrophic. You would think the Adélies would delay their breeding a bit because of the snow, but they can't. The birds are just hardwired and they don't adjust. They are hardwired into oblivion."

Snowfall had been increasing along the western Antarctic Peninsula, according to ice-core records and meteorological data from a nearby British base. More snow was consistent with a marine Antarctic environment that was experiencing an influx of warmer air and seeing sea ice shrink—thus exposing more open water, leading to more evaporation, which then formed as precipitation.

The main impact of increased snowfall on Adélies, according to Fraser, is that more eggs are being lost as colonies have to cope with melting snow. I witnessed that on Torgersen Island, where half a dozen eggs in Colony 19-A rested at the bottom of two murky ponds of snowmelt, about six inches deep. The water was a putrid brown color, and under the surface the eggs looked like large golf balls resting in a shallow water hazard. Adélies were standing belly-deep in the water, dunking their heads under the surface to try to move their eggs or rebuild their collapsed nests. Several penguins had managed to rebuild their nests high enough so that they just poked above the ponds, looking like castles in the middle of a lake. One or two even had succeeded in placing an egg atop the pile, after which they attempted to brood the eggs. That colony has since disappeared.

By monitoring the penguins from hatching to fledging, Fraser determined that pairs of Adélie penguins in his study area successfully raised, on average, 1.3 chicks to fledging. This rate is roughly twice as high as the success rate at some other colonies on the continent.

Fraser soon realized that the reason Palmer's Adélies were so good at rearing chicks was that the breeders were almost all older, experienced adults, which have more success raising chicks than younger Adélies. Fraser said, "This high breeding success is actually an indicator of just how bad recruitment of new penguins into the population is, because new breeders always screw up."

Survival rates for fledged Adélie chicks that head to sea are low—by one estimate, between ten and fifteen per cent—so environmental factors that cut down on those odds can push a population of Adélies over the edge.

In February, I spent many days on Torgersen Island watching Adélie chicks plunge tentatively into the Southern Ocean. Looking half-panicked and half-playful, these fledging penguins dog-paddled in the shallows, splashed wildly with their flippers, and honked to their fellow-swimmers. Within a minute, the stronger chicks began heading out to sea. Most travelled in packs, but some headed out alone. I saw a young Adélie struggle through three-foot waves and then, as it cleared the surf, scramble onto a passing ice floe in a desperate attempt to find some semblance of terra firma. The chick then floated away, its ice platform rocking in the swells.

Biscoe Point, on the southeastern corner of Anvers Island, sits in the shadow of Mt. William, a fifty-two-hundred-foot peak whose ridges are cloaked in ice and whose gray face is etched with snow-filled draws. A dozen miles away, on the peninsula, looms a line of tombstone-shaped mountains that plunge into the Southern Ocean. These days, however, Biscoe Point is in need of rechristening. As recently as twenty years ago, the Marr Ice Piedmont covered part of Biscoe. But the retreat of the glacier has revealed that what once looked like a peninsula is actually an island, now separated from the glacier by more than fifty feet of open water. The warmer weather has brought other changes to Biscoe: As Adélie-penguin populations have declined, gentoo penguins—with their fondness for higher temperatures—have arrived in force.

Fraser and I went to Biscoe in mid-January. He had not been there in two years, and as we puttered to our landing site he noted that gentoo-penguin colonies had spread across a series of terraces, all the way to the highest ridge. "Is that a gentoo colony way the hell up there on the ridge?" he asked. "This is just amazing." There were seven hundred and sixty-one nesting pairs of Adélies—down from twenty-eight hundred pairs in 1984—and they were relegated to a few ridges close to the sea. For the first time, more gentoos—nine hundred and two pairs—were nesting on Biscoe than Adélies. As we ranged across the terraces where the gentoos had built their round, high-sided nests, they filled the air with donkey-like braying.

We walked to the end of the point, next to the terminus of the Marr glacier. For two hundred yards, we traversed gradually sloping terrain of smoke-colored stone and shattered rocks, some frost-fractured into thin slices. In the seventies, the area we walked across was buried under a hundred feet of ice. Now the ice patch, the size of a football field just two years before, had been reduced by half, and seemed to be only a few dozen feet thick. On the surface of the ice, fist-size rocks had been heated by the sun, boring holes in the remnant of the glacier. We dipped our hands into these little wells, drinking the sweet water. I could hear, beneath the plate of ice, the trickling of melt-water flowing to the sea. Patches of Antarctica's two vascular plants—a short, tufty, pale-green hair grass, *Deschampsia antarctica*, and a mossy-looking pearlwort, *Colobanthus quitensis*—were colonizing new territory amid the glacial till.

We sat on rocks and gazed at the glacier's face, which was several stories high. Its luminous front was honeycombed, and much of it slumped forward, looking like melted wax. The noise coming from the glacier was constant—sharp cracks, deep rumbles like the sound of furniture being moved across a floor, and from time to time the showering of ice inside a hidden crevasse. The snapping sounds intensified and then, just in front of us, a twenty-five-foot chunk of the glacier calved into the narrow channel separating the Marr and Biscoe. The cascading pieces of the glacier sent waves lapping against nearby rocks. Thousands of ice shards from this and previous calvings hissed as they were buffeted by the waves. The newly exposed face of the glacier was a heavenly shade of blue, attained over thousands of years as the ice was compressed and air pockets were eliminated.

Eighty-seven per cent of the glaciers along the Antarctic Peninsula are in retreat, according to a study by the British Antarctic Survey and the U.S. Geological Survey. Along the peninsula, eight ice shelves have fully or partially disintegrated in the past three decades. The largest collapse was the Larsen B Ice Shelf—once the size of Connecticut—which in March, 2002, shattered into millions of pieces after several summers of unusually high air temperatures in the customarily frigid Weddell Sea. Many glaciers in west Antarctica are now sliding more rapidly to the sea, most notably the Pine Island Glacier, a river of ice—a hundred and ninety miles long and thirty miles wide—moving toward the Amundsen Sea at the rate of a foot an hour. If that glacier and the neighboring Thwaites Glacier unload their mass into the Southern Ocean, sea levels could rise as much as five feet.

"If someone had taken you here ten or twenty years ago and said, 'This will all disappear in ten years,' you would have said, 'You're fucking crazy,'" Fraser said. "The ice cap on Biscoe has been here for thousands of years, and now it's almost gone. What you're seeing here is what you would have seen if you had been standing in Wisconsin fifteen thousand years ago as the glaciers retreated. This is what these landscapes must have looked like as the ice melted. It's the sheer power of the earth—ice and rock. Lesson No. 1 for me has been the realization that ecology and ecosystems can change like that," he said, snapping his fingers. "In geological time, it's a nanosecond."

In all likelihood, the fate that has befallen Adélie penguins around Palmer Station awaits ice-dependent penguins breeding farther south in Antarctica. Two and a half million pairs of Adélie penguins still breed throughout the continent, and some Adélie populations in colder Antarctic regions are actually growing. Such is the case with a rookery of Adélies that Fraser's team monitors on Avian Island, in Marguerite Bay, two hundred and fifty miles south of Palmer Station. Fifty years ago, much of the ocean around Avian Island was covered year-round in ice, making it difficult for large numbers of Adélies to nest there and gain access to the open water needed for foraging. But the warming of the bay region has melted sea ice and provided an optimal mixture of ice and open water. As a result, Fraser estimates that the Adélie-penguin population on Avian Island has doubled in the past thirty-five years, to between fifty thousand and seventy-five thousand pairs. But he believes that the Avian Island population is probably nearing its peak, and that, as sea ice continues to disappear, the rookery will decline.

David Ainley, a penguin expert, who studies Adélie populations on the Ross Sea, recently led a study funded by the National Science Foundation and the conservation group W.W.F. Ainley and his colleagues predict a dire impact on polar penguins if global temperatures rise 2 degrees Celsius (3.6 degrees Fahrenheit) above pre-industrial levels, which now seems inevitable. The researchers conclude that Adélie- and emperor-penguin colonies north of 70° S—comprising half of Antarctica's emperor-penguin colonies and three-quarters of the continent's Adélie colonies—"are in jeopardy of marked decline or disappearance."

Within several decades, Fraser believes, Adélie penguins will disappear from the northwestern Antarctic Peninsula. As he put it, "They're on a decline that has no recovery."

In late January, two members of the birding team and I returned to Litchfield Island to see if any Adélie chicks remained. Walking across the many abandoned colonies, I strained to catch a glimpse of chicks, but as we neared the other side of the island it was clear that the young Adélies were gone. Lacking the protection of a larger colony, they had presumably been eaten by brown skuas.

Brett Pickering, the field-team member, said, "Litchfield is officially over."

"It's the total lack of any life whatsoever on these quite large rookeries that just keeps going through my mind," Fraser said after I returned from Litchfield.

"I have real affection for Adélies, but everything seems to be working against them," Fraser had told me earlier. "Here you have this unbelievably tough little animal, able to deal with anything, succumbing to the large-scale effects of our activities. And that's the one thing they can't deal with, and they're dying because of it. And that's the sad side of that story for the Adélies. It's such a long-distance effect. The industrial nations to the north are having an impact that Adélies are being subjected to down here. That's what sort of pisses me off about the whole picture, that these incredible animals have to take it in the neck because a bunch of humans can't get together to decide what to do about the planet."

Several days later, Fraser returned to Litchfield, wanting to see for himself the end of the island's Adélie rookery. A lone adult stood staidly in Colony 8, flippers at its sides, scarcely noticing us. The penguin seemed to be sleeping, the white rings around its eyes closing to a slit and gently pulsating. The late-summer sun turned the Bismarck Strait a shade of deep indigo. Except for the gentle slosh of the surf and the occasional cry of a brown skua, the island was silent.

The Inferno

Christine Kenneally

October 26, 2009

Earlier this year, on the day now known as Black Saturday, when the worst wildfires in modern Australian history incinerated more than a million acres of the state of Victoria and killed a hundred and seventy-three people, Bruce Ackerman left his house in Marysville to meet up with his regular Saturday lunch group. Marysville, a small town some sixty miles north-east of Melbourne, sprang up in the eighteen-sixties as a stopover for miners on their way to the gold-rush towns farther north. Situated in a cool valley of the Great Dividing Range, the mountainous spine that dominates eastern Australia, the town is a popular tourist destination. Ackerman, a bluff fifty-year-old, is proud of being a fourth-generation inhabitant, and told me that his work as a plumber had taken him inside every house there except one. Over the years, he had served on the water board, the cemetery trust, the school council, the ambulance service, and the volunteer fire brigade.

The lunch group broke up earlier than usual; three of Ackerman's friends, a fire captain and two members of his team, were about to be placed on standby because of the risk of bushfire. Two days earlier, Victoria's state premier, John Brumby, had held a press conference at which he stated that the coming Saturday, near the end of the antipodean summer, would be the "worst day in the history of the state." "The state is just tinder-dry," he said, and told Victorians that it was time to put their fire plan into action.

After lunch, Ackerman drove some seven miles southwest to the small town of Narbethong and had a drink with some friends. While he was there, he became aware of an unusual amount of activity on the road running through the town. Drivers sped through, asking if Narbethong was safe from fire. Amid the gathering panic, Ackerman smelled smoke. He rushed back north to Marysville against a stream of traffic coming the other way. By the time he was a few miles from Marysville, he could see a colossal firewall coming toward him from the southwest. It was three hundred feet high. He raced it all the way back to the town, driving on the wrong side of the road to get through blocked intersections and dodging cars that sped toward him in their effort to flee. The fire behind Ackerman emitted a roar like a jet engine and threw embers and fireballs out ahead of him. Huge patches of trees and grass ignited around the car as he drove. In his mind Ackerman started running through the steps of his fire plan.

Fire plans are a distinctive feature of Australian life. Whereas other regions prone to wildfire, most notably California, typically employ mandatory evacuations to remove citizens from an affected area, Australians are instructed to pick one of two options: leave early or stay and fight the fire. Though risky, the choice to stay is a popular one, both because unprotected houses are often lost to the flames and because of a streak of self-reliance in the Australian character. Australians have had great success saving their homes: a fire front passes in a matter of minutes, during which time the house shields those inside from the heat, and research has shown that houses usually catch fire not as the front passes but, rather, before or after, when wind-borne embers settle on it. Fire plans include removing flammable items from around the house, storing as much water as possible, and being ready to put out flames as a fire front draws near. Many Australians become experts on matters like smoke inhalation and the effects of radiant heat on the body, and public-information ads drive home the importance of wearing long sleeves and trousers and avoiding all synthetic materials, which ignite much faster in radiant heat. The best-prepared people build their houses out of fireproof materials, install sprinkler systems on their roofs, and maintain water tanks attached to professional-grade firefighting pumps and hoses.

At the start of this year, a California study, pointing to a dramatic increase in the number of buildings lost to wildfire in recent decades, recommended the adoption of "more effective" Australian tactics representing "a more sustainable coexistence with fire." But in the months since Black Saturday the Australian approach to fighting fire has come under question. During an ongoing royal commission, experts have pondered how wildfires should be handled, on an increasingly fire-prone continent, and also the extent to which the deaths of Black Saturday were the result not just of the Australian landscape but of an entire mentality.

Australia's worst fires have all occurred in its southeastern states, with Victoria often hit the hardest. In the Black Friday fires of 1939, almost a quarter of the state was burned, and seventy-one people died. The Ash Wednesday fires of 1983 killed seventy-five people and destroyed more than two thousand homes. In the past ten years, the risk of catastrophic fire has increased, because of a pervasive drought, the severest since records began, which is widely thought to be a result of global warming. Victorians are asked to use less than a hundred and fifty-five litres of water per person per day. Gardens can be watered for only a few hours on two days a week, and it is illegal to use a hose when washing your car.

This year, starting on January 28th, Victoria experienced the most severe heat wave in its history. Many elderly people died; steel train tracks buckled; in one Melbourne park a thousand fruit bats fell dead from the trees. By Monday, February 2nd, Claire Yeo, one of Victoria's two fire meteorologists, noted that all the factors that create extreme fire weather were evident: high temperatures; strong, gusty winds; and very little moisture. On Tuesday, the Bureau of Meteorology predicted that the temperature for Saturday the seventh would be thirty-seven degrees Celsius—ninety-eight degrees Fahrenheit—but Yeo told a fire chief that a temperature upward of a hundred and thirteen degrees Fahrenheit was more likely. "I think we are going to break records," she said. By Wednesday, the numbers she was calculating on a system that rates the danger of fire were worse than anything she had ever seen. So terrible was the forecast that, when she had to brief assembled fire chiefs, meteorologists, and other specialists on the situation, she stood at her lectern for some time, hanging her head and unable to speak.

Yeo's predictions were accurate. Saturday the seventh was the hottest day in Melbourne since records began to be kept, in 1855. The temperature reached 46.4 degrees Celsius (115.5 degrees Fahrenheit) in Melbourne and a hundred and twenty degrees Fahrenheit elsewhere in the state. Humidity was at just six per cent, and a strong wind was blowing from the northwest. The six hundred fires that started that day were not just the deadliest that Australia had ever known but among the worst the world had seen for decades.

At around 3 p.m., a watcher in a fire tower at Mount Despair reported a smoke plume in the Murrindindi state forest, a little more than twenty miles northwest of Marysville. From there, burning embers were carried by the wind to Little Wonder, a point at the top of a ridge in the nearby Black Ranges, and soon there was fire on the other side of the ridge, too. The fire was by now travelling at around seven miles an hour. With the northwesterly wind behind it, it spread southeast toward a popular camping spot. A fire crew raced to get to the campsite before the fire did. The firefighters drove two trucks into a nearby river and gathered nineteen campers into the water. Minutes later, the fire caught up with them. Nine children and an adult were crammed into the cabs, and the other campers sheltered beside the trucks, under wet blankets, while the firemen fought the flames. The campsite was incinerated, but everyone survived.

A little later, a front of cooler weather moved into Victoria. Temperatures dropped eighteen degrees Fahrenheit, and all over Melbourne people spilled from their houses in relief. But, while a drop-in temperature can make a fire less likely, it doesn't have much impact on one that is already raging. Worse, the cool front brought about a ninety-degree change in the wind, from northwesterly to southwesterly. Now the long, narrow line of fire that had been spreading southeast became a wide fire front that charged northeast to a string of towns along the Maroondah Highway: first Narbethong, then Marysville, Buxton, Taggerty, and Alexandra.

The fire hit Marysville so fast that no one had good information about what was happening. Not long before, it had seemed that the fire would pass the town by. Ian Bates, who was on duty at the State Emergency Services building, recalls standing in front of the building talking to the local fire captain. He glanced over the captain's shoulder and saw the fire coming at the town. He pointed and said, "Look at that!" At the same moment, the captain pointed over Bates's shoulder and shouted exactly the same thing. Bates looked behind him. The fire was coming at the town from both sides.

Residents, seeing huge clouds of dirty orange-yellow smoke looming over the trees, got in their cars and drove, either to shelter in the local football oval or out of town entirely. The local police, firefighters, and emergency workers scrambled to alert people to the danger. Bates and his deputy drove through residential streets with their lights and sirens on, knocking on doors and yelling through the P.A. system that everyone needed to leave. They put together a convoy that was able to exit the town in less than thirty minutes.

Once Bruce Ackerman got home, he called his daughter and his son's girlfriend, both in their twenties and living in Marysville, and told them to leave town. (His son works as a fireman, and his wife was in Melbourne with friends for the night.) Then he phoned an elderly couple next door and told them to go to the football oval, but they were reluctant to leave. The wife walked over to Ackerman's house. "She was shaking like a dog," he told me. "She said,

'I'm not scared. I've stood up to bigger men than that bloody bit of smoke.' And I said, 'Elsa! Get down to the oval.'" A little later, her husband came over. "We went through the same routine," Ackerman said. "He said, 'I'm tough and I'm not going for no one.' And I took him outside and I made him listen to the roar, and I said, 'Dudley! Piss off!' And he did." Ackerman phoned his next-door neighbors on the other side, Liz Fiske and her thirteen-year-old son. Fiske's husband, a fire captain who was away on duty, had been best man at Ackerman's wedding, and Ackerman had been best man at his. Fiske, too, refused to leave her house.

Between phone calls, Ackerman frantically prepared his home. He took his truck out of the shed at the foot of his front yard and parked it in a carport next to the house. He removed all the doormats from the house, stuffed draft-blockers under the doors, checked that all the windows were shut, and started the sprinklers on his lawn. He grabbed a ladder and checked that the gutters on his house were clear. The light outside became dimmer and dimmer as the sun was blocked out by an enormous pall of smoke. Ackerman filled eight buckets, two watering cans, the dog's bath, the spa bath, the wheelbarrow, and the trash can with water, the house's water pressure falling all the while. He moved out everything that was stored under his veranda, took down the picnic umbrellas in the garden, and removed the gas tank from the barbecue. He turned off the gas to the house. Then the house's electricity failed, everything went dark, and the water supply petered out. Ackerman called Liz Fiske again, hoping that there might still be time for her and her son to leave. This time, her son answered the phone. He told Ackerman that his mother was in the yard, fighting the fire. Ackerman said to him, "Be brave for your mum. You'll be all right, mate. You're a strong man."

At around half past six, the fire front hit Ackerman's house. The fire shone as brightly as daylight, illuminating everything—"Red, just red, pure bright light, red raw," Ackerman said. As the fire burned around his house, Ackerman was in constant motion. "I was upstairs, downstairs, in and out of every room, back and forth, to and fro, looking where the fire was spotting, popping in and out with buckets, dousing it down. The fire came right up to the house, and it was catching," he told me.

The front passed in minutes, but to Ackerman it seemed like days. "I was a volunteer fireman in Ash Wednesday. I've seen a lot, but I've never, ever seen anything like this," he told me. All night, Marysville was lit by a dull, smoky glow as its buildings burned down. Explosions and flares went off constantly as gas tanks, paint cans, and other flammable objects ignited. Just after the front went through, Ackerman was outside again to put out spot fires that had started around his house. Liz Fiske's house was on fire, and he couldn't find her or her son. "I hoped to Christ they had got out," he said. He drove down to the football oval, where his elderly neighbors had taken refuge, along with fifty or so other residents who had missed the convoy out of town. Everyone there was alive, and the grass on the oval had stayed green, even though all the trees that circled it had burned. Then he drove back through the town to check on friends whose house had begun to burn. By the time he got there, the piping to their water pump had melted, their tanks were leaking, and the pump itself had vaporized. They were trying their hardest, Ackerman said, but they were absolutely exhausted. Ackerman put on gloves and pulled a flaming trellis off a wall, saving their house. He went back to Liz Fiske's house, which had burned down completely, and then drove three miles to check on an uncle and aunt: their house was destroyed, too, but, as there was no sign of their car, he felt confident that they had managed to leave. Around half past three in the morning, Ackerman ran into a firefighter who had a satellite phone, and he called his wife to tell her he was alive.

Six days after Black Saturday, I drove to the town of Alexandra, twenty miles north of Marysville, to meet Geoff McClure, of the Department of Sustainability and Environment, and to see the fire, which had now joined up with another large fire and was heading toward Alexandra. McClure was affable and solicitous, but he looked tired. Over the week, the fire had burned through almost three hundred thousand acres, and he had been working fourteen hours a day. He lent me some flame-retardant overalls and a helmet made from plastic with a melting point above three hundred and twenty degrees. Then we drove out toward the fire line.

As we left Alexandra, we saw the Great Dividing Range framed by a mass of smoke that looked like a giant cumulus cloud. The fire was heading downhill in a fire line about twenty-five miles across. Because the wind that day was blowing back against the fire, the smoke spiralled up from within the burning vegetation. The darker the plume and the fuzzier its boundaries, McClure explained, the more intense the fire.

Some eucalyptus, of which Australia has more than seven hundred species, produces significantly more radiant heat when burning than the wood of other trees, and the growth of eucalyptus forests is naturally punctuated by fires. Fire promotes rejuvenation, by removing the buildup of ground litter, opening the canopy, and eliminating competitive species. It is also a crucible in which the buds of many eucalyptus species sprout. Currently, Victoria's Department of Sustainability and Environment is responsible for controlled burning to reduce fuel loads. This doesn't prevent fires, but it does decrease the intensity of the fires that occur.

We passed a police roadblock, and drove up a tree-lined dirt track in the direction of the fire. The goal, McClure explained, was to stop the fire from coming out of the forest; there was little anyone could do to stop its descent down the hill. Hundreds of firefighters were being deployed; in recent days, they had used bulldozers to establish containment lines of bare earth that could interrupt the progress of the fire. In the bush, the bulldozer is as essential to firefighting as the fire truck. The containment lines, only ten feet wide, looked precarious, but although embers carried on the wind can sometimes enable fire to jump a line, the absence of fuel at the line prevents the fire front itself from burning through. We drove to where two young firefighters were strengthening a containment line by back burning—starting a controlled fire that would burn back toward an approaching fire front, depriving it of fuel. Using fuel cannisters with long nozzles, they had established a line of flames some two feet high and intensely hot. My eyes stung as we stood watching, fifteen feet away.

Strange cataclysmic phenomena occur in a huge wildfire. Kevin Tollhurst, a fire ecologist in Melbourne, told me that fires as hot as the one at Marysville—which is thought to have reached a temperature of twenty-two hundred degrees—can produce their own weather. Fires generate convection columns of gas, which may rise as much as forty thousand feet and form pyrocumulus clouds. The clouds can create lightning, which may then start more fires downwind of the original fire. The sound of the gas—like a twig popping in a fireplace, but exponentially louder—creates a wildfire's distinctive roar. The Marysville fire was so hot that gas flared out laterally, acting as a wick, along which the fire caught quickly, crossing the ground in sudden, unpredictable pulses. In the face of such a fire, it is possible to be looking at a front more than a thousand feet away and then, in an instant, to be surrounded by flames. Firefighters described the Black Saturday firestorm as "alive," and said that its behavior was completely unprecedented. In some areas it was apparently cyclonic, coming at them from all sides, burning up a road in one direction and then, minutes later, burning in the opposite direction. Tollhurst told the royal commission that the energy from all the fires that day was the equivalent of fifteen hundred Hiroshimas.

Later, McClure took me to see some of the damage to a cluster of properties just off the road south of Alexandra. The buildings looked as if they had imploded. Glass that had melted like toffee was draped around their edges. In the yard next to the remains of one building, the blackened branches of a tree stretched out and up, and from them dangled pitch-black round baubles. I picked one up from the ground and pushed my thumb through the charcoal outer layer to the yellow core, which was leathery and desiccated. It still smelled faintly of apple.

The devastation of Black Saturday became clear only over many days, unfolding for survivors as a disorienting nightmare. Devastation extended across Victoria. In places where the fire had passed, the countryside was completely silent, with no sign of animal life. In Marysville, Bruce Ackerman's house was one of only fourteen left standing. The town had been annihilated in an afternoon. Police set up roadblocks around the worst affected places, and three towns, including Marysville, were declared crime scenes, after arson, a surprisingly common cause of wildfires, was suspected. No one was allowed back into Marysville for weeks.

The death toll rose from fifteen to a hundred and seventy-three. It was clear that some bodies would never be recovered, having been effectively cremated. Others were found unburned in positions that suggested that they had simply dropped dead while running from the flame; the radiant heat of a bushfire, which can ignite mattresses and curtains through closed windows, can kill at a considerable distance more or less instantaneously. Survivors emerged with ever more excruciating stories. As people escaped, they had seen their neighbors' bodies lying charred in the streets. A few days after the fire, police found a house with the remains of nine bodies: eight adults had formed a protective huddle around a baby.

Nearly two weeks after Black Saturday, I drove from Alexandra to the Marysville golf course, just outside the town, to meet with Bruce Ackerman, who, because his house was intact, had been allowed to stay in Marysville, alongside the Army, the police, and emergency-services crews. As I waited in the parking lot, I walked over to a semi-trailer that looked as if it had been detonated from within. The exhaust stacks drooped outward, and thick rivulets of melted aluminum lay congealed in the gutter beneath. When I tapped one with my foot, it clanged loudly. Ackerman arrived, and we sat on the porch of the clubhouse. In front of us, many trees were burned, but the links were still green. A tiny bird with a long tail and a glowing blue face hopped about on the grass.

For eight days after the fire, Ackerman told me, he had helped the police as they searched for bodies in collapsed houses. During a fire, he explained, people usually take shelter in the bathroom: "And who would know where the bathroom is? The plumber." Thirty-nine people in Marysville lost their lives; Ackerman's neighbors Liz Fiske and her son were among the dead.

Not until six weeks after Black Saturday was Marysville reopened for residents, and a few days later Ackerman took me on a tour of the town, which was still closed to the public. Wire fencing had been erected along most streets, cordoning off the destroyed buildings. We drove along the river to a waterfall, which, like many things in Marysville, had once been invisible through thick foliage but could now be seen from hundreds of yards away. We stopped at Ackerman's house, a large, rustic two-story home, incongruously intact and attractive amid the devastation. Trees that must have been fully alight stood no more than twenty feet from the house. Ackerman's front yard was green and inviting, but a big tin shed at the foot of the lawn was splayed everywhere, as if another tin shed had been dropped on it from a great height. Farther down the hill, the Army had created a car yard, neatly lining up nearly a hundred skeletal cars and buses.

While we drove around the town, Ackerman's phone rang constantly. It sometimes seemed that there was little in the town that he wasn't doing. He had helped reinstate water service, refilling the depleted reservoir. He was head of a committee to establish a temporary village nearby to house residents who were rebuilding their homes. That morning, he had dug a grave.

Nine days after Black Saturday, Victoria set up a year-long commission to investigate the cause of the fires and "all aspects of the government's bushfire strategy." The commission delivered an interim report in August, in time to influence preparations for the coming fire season, which begins next week. The public hearings, streamed live over the Internet, have been avidly followed, and the sense that many Victorians had that they can manage the risks of fire has started to seem dangerously close to complacency.

The commission asked people how they had heard about and tracked the fire. Had they used mobile phones or land lines? Were they listening to the radio or watching the TV? Did they text their neighbors, or check official Web sites, or dial 000 (the Australian equivalent of 911)? Despite a bristling network of communications technology across the state, when the fire came many people found themselves completely alone. Witnesses described emergency calls that were never answered, and one that was answered by an operator a thousand miles away in Queensland, who had no idea that there even were bushfires in Victoria. Communication failures also occurred between agencies and up and down lines of command. Fire crews repeatedly called for backup that never came. A radio station of the Australian Broadcasting Corporation became a kind of emergency information service, but a number of its broadcasts were based on official reports that were inaccurate or obsolete by crucial minutes, or even hours. People checked their televisions and radios for news of a fire that they—and in many cases *only* they—could see coming toward them.

Victoria's administration is now grappling with new ways of alerting its citizens to large-scale danger. One expert has suggested making fire information available in real time via Twitter. Australia has one of the highest cell-phone-user populations in the world, and, in early March, I (along with millions of other cell-phone users in Victoria) received a text message from the Victorian police, warning of extreme fire weather over the next two days. But other government-level inefficiencies remain. For instance, the Department of

Sustainability and Environment is supposed to conduct controlled fires to lower fuel loads in dense forests, but some witnesses described the incineration of streets that had seen no controlled burning for more than forty years. Residents complained that local councils required them to get permits before uprooting native plants, even trees overhanging their houses, despite fire guidelines which state that all houses in fire-prone areas need large cleared areas around them. (New laws allow Victorians to clear any plants within ten metres of their houses.)

Throughout the commission's earliest hearings, one issue dominated: the "stay or go" policy. Culturally, this is a sensitive issue; protecting a home against fire has been an aspect of Australian life in the bush for a hundred and fifty years. But over weeks of testimony it became apparent that many residents treat "stay or go" more like "wait and see," either because they are complacent or because they panic once they hear the fire's roar and see the sky go dark. Leaving at the last minute is the worst possible strategy. By the time you can see the firewall, your exits may be blocked, and by fleeing you forfeit the heat shield of your home. Often, bodies are found between house and car, keys waiting in the ignition. Before Black Saturday, a study of deaths in a hundred years of Australian fires showed that the majority occurred during late evacuation. The experts behind this study told the commission that they still believed in "stay or go"—that, as one slogan has it, "houses protect people, and people protect houses." But police testimony given at the royal commission indicated a worrying reversal of the usual trend: of the hundred and seventy-three deaths, a hundred and thirteen occurred inside houses. No evidence has yet been released to indicate whether these people were actively defending their homes or passively sheltering from the fire. If the latter, the option of staying and fighting remains at least theoretically viable. But, if investigation reveals that many people died while still fighting the fire, Black Saturday's most significant legacy may be to have exposed a fundamental limitation of the "stay or go" policy. In California, officials currently preparing for the height of fire season have backed away from their consideration of "stay and go," and unanimously advocated evacuation. Even in Australia, "stay or go" now finds fewer supporters: recently, a new scale for rating fire danger was announced, and, at the highest rating—"code red (catastrophic)"—people are told that the safest option is simply to leave.

According to some experts, even if all systems had worked perfectly on Black Saturday, many houses might have been undefendable. The fire ecologist Kevin Tollhurst, giving evidence before the commission, said that Australian fire science tends to be based on observation of smaller fires, but that it seems increasingly likely that fire behavior is partly a function of scale. Many experts predict that global warming—an issue that the commission will formally examine—will make fires on the scale of Black Saturday's more and more common. If that turns out to be the case, it may be that models for understanding a fire as large as the one that destroyed Marysville don't yet exist.

On my last day with Ackerman, as he left me in the clubhouse of the golf course, I asked him whether his own experience defending his home had changed his opinion about the feasibility of staying to fight a fire. He turned to me. "Should I stay or should I go?" he said, and held up a finger. "Go! Believe me, next time I'll be the first one out."

The End of the End of the World

Jonathan Franzen

May 23, 2016

Two years ago, a lawyer in Indiana sent me a check for seventy-eight thousand dollars. The money was from my uncle Walt, who had died six months earlier. I hadn't been expecting any money from Walt, still less counting on it. So I thought I should earmark my inheritance for something special, to honor Walt's memory.

It happened that my longtime girlfriend, a native Californian, had promised to join me on a big vacation. She'd been feeling grateful to me for understanding why she had to return full time to Santa Cruz and look after her mother, who was ninety-four and losing her short-term memory. She'd said to me, impulsively, "I will take a trip with you anywhere in the world you've always wanted to go." To this I'd replied, for reasons I'm at a loss to reconstruct, "Antarctica?" Her eyes widened in a way that I should have paid closer attention to. But a promise was a promise.

Hoping to make Antarctica more palatable to my temperate Californian, I decided to spend Walt's money on the most deluxe of bookings—a three-week Lindblad National Geographic expedition to Antarctica, South Georgia island, and the Falklands. I paid a deposit, and the Californian and I proceeded to joke, uneasily, when the topic arose, about the nasty cold weather and the heaving South Polar seas to which she'd consented to subject herself. I kept reassuring her that as soon as she saw a penguin she'd be happy she'd made the trip. But when it came time to pay the balance, she asked if we might postpone by a year. Her mother's situation was unstable, and she was loath to put herself so irretrievably far from home.

By this point, I, too, had developed a vague aversion to the trip, an inability to recall why I'd proposed Antarctica in the first place. The idea of "seeing it before it melts" was dismal and self-cancelling: why not just wait for it to melt and cross itself off the list of travel destinations? I was also put off by the seventh continent's status as a trophy, too remote and expensive for the common tourist to set foot on. It was true that there were extraordinary birds to be seen, not just penguins but oddities like the snowy sheathbill and the world's southernmost-breeding songbird, the South Georgia pipit. But the number of Antarctic species is fairly small, and I'd already reconciled myself to never seeing every bird species in the world. The best reason I could think of for going to Antarctica was that it was absolutely not the kind of thing the Californian and I did; we'd learned that our ideal getaway lasts three days. I thought that if she and I were at sea for three weeks, with no possibility of escape, we might discover new capacities in ourselves. We would do a thing together that we would then, for the rest of our lives, have done together.

And so I agreed to a year's postponement. I relocated to Santa Cruz myself. Then the Californian's mother had a worrisome fall, and the Californian became even more afraid of leaving her alone. Recognizing, finally, that it wasn't my job to make her life more difficult, I excused her from the trip. Luckily, my brother Tom, the only other person with whom I could imagine sharing a small cabin for three weeks, had just retired and was available to take her place. I changed the booking from a queen-size bed to twin beds, and I ordered insulated rubber boots and a richly illustrated guide to Antarctic wildlife.

Even then, though, as the departure date approached, I couldn't bring myself to say that I was going to Antarctica. I kept saying, "It appears that I'm going to Antarctica." Tom reported being excited, but my own sense of unreality, of failure to pleasurably anticipate, grew only stronger. Maybe it was that Antarctica reminded me of death—the ecological death with which global warming is threatening it, or the deadline for seeing it that my own death represented. But I became acutely appreciative of the ordinary rhythm of life with the Californian, the sight of her face in the morning, the sound of the garage door when she returned from her evening visit to her mother. When I packed my suitcase, it was as if I were doing the bidding of the money I'd paid.

In St. Louis, in August, 1976, on an evening cool enough that my parents and I were eating dinner on the porch, my mother got up to answer the phone in the kitchen and immediately summoned my father. "It's Irma," she said. Irma was my father's sister, who lived with Walt in Dover, Delaware. It must have been clear that something terrible had happened, because I remember being in the kitchen, standing near my mother, when my father interrupted whatever Irma was saying to him and shouted into the telephone, as if in anger, "*Irma, my God, is she dead?*"

Irma and Walt were my godparents, but I didn't know them well. My mother couldn't stand Irma—she maintained that Irma had been terminally spoiled by her parents, at my father's expense—and although Walt was felt to be much the more likable of the two, a retired Air Force colonel who'd become a high-school guidance counsellor, I knew him mainly from a self-published volume of golf wisdom that he'd sent us, "Eclectic Golf," which, because I read everything, I'd read. The person I'd seen more of was Walt and Irma's only child, Gail. She was a tall and pretty and adventurous young woman who'd gone to college in Missouri and often stopped to see us. She'd graduated the previous year and had taken a job as a silversmith's apprentice at Colonial Williamsburg, in Virginia. What Irma was calling to tell us was that Gail, while driving alone, overnight, in heavy rain, to a rock concert in Ohio, had lost control of her car on one of West Virginia's narrow, winding highways. Although Irma apparently couldn't bring herself to say the words, Gail was dead.

I was sixteen and understood what death was. And yet, perhaps because my parents didn't bring me along to the funeral, I didn't cry or grieve for Gail. What I had, instead, was a feeling that her death was somehow inside my head—as if my network of memories of her had been cauterized by some hideous needle and now constituted a zone of nullity, a zone of essential, bad truth. The zone was too forbidding to enter consciously, but I could sense it there, behind a mental cordon, the irreversibility of my lovely cousin's death.

A year and a half after the accident, when I was a college freshman in Pennsylvania, my mother conveyed to me an invitation from Irma and Walt to come to Dover for a weekend, along with her own strict instruction that I say yes. In my imagination, the house in Dover was an embodiment of the zone of bad truth in my head. I went there with a dread which the house proceeded to justify. It had the uncluttered, oppressively clean formality of an official residence. The floor-length curtains, their stiffness, the precision of their folds, seemed to say that no breath or movement of Gail's would ever stir them. My aunt's hair was pure white and looked as stiff as the curtains. The whiteness of her face was intensified by crimson lipstick and heavy eyeliner.

I learned that only my parents called Irma Irma; to everyone else, she was Fran, a shortening of her maiden name. I'd dreaded a scene of open grief, but Fran filled the minutes and the hours by talking to me incessantly, in a strained and overloud voice. The talk—about her house's décor, about her acquaintance with Delaware's governor, about the direction the nation had taken—was exquisitely boring in its remoteness from ordinary feeling. By and by, she spoke of Gail in the same way: the essential nature of Gail's personality, the quality of Gail's artistic talents, the high idealism of Gail's plans for the future. I said very little, as did Walt. My aunt's droning was unbearable, but I may already have understood that the zone she was inhabiting was itself unbearable, and that talking loftily about nothing, non-stop, was how a person might survive in it; how, indeed, she might enable a visitor to survive in it. Basically, I saw that Fran was adaptively out of her mind. My only respite from her that weekend was the auto tour Walt gave me of Dover and its Air Force base. Walt was a lean, tall man, ethnically Slovenian, with a beak of a nose and hair persisting only behind his ears. His nickname was Baldy.

I visited him and Fran twice more while I was in college, and they came to my graduation and to my wedding, and then, for many years, I had little contact with them beyond birthday cards and my mother's reporting (always colored by her dislike of Fran) on the dutiful stops that she and my father made in Boynton Beach, Florida, where Fran and Walt had moved into a golf-centered condominium complex. But then, after my father died, and while my mother was losing her battle with cancer, a funny thing happened: Walt became smitten with my mother.

Fran by now was straightforwardly demented, with Alzheimer's, and had entered a nursing home. Since my father had also had Alzheimer's, Walt had reached out by telephone to my mother for advice and commiseration. According to her, he'd then travelled by himself to St. Louis, where the two of them, finding themselves alone together for the first time, had uncovered so much common ground—each was an optimistic lover of life, long married to a rigid and depressive Franzen—that they'd fallen into a dizzying kind of ease with each other, an incipiently romantic intimacy. Walt had taken her downtown to her favorite restaurant, and afterward, at the wheel of her car, he'd scraped a fender on the wall of a parking garage; the two of them, giggling, a little bit drunk, had agreed to split the repair cost and tell no one. (Walt did eventually tell me.) Soon after his visit, my mother's health worsened, and she went to Seattle to spend her remaining days in my brother Tom's house. But Walt made plans to come and see her and continue what they'd started. Of the feelings they had for each other, his were still forward-looking. Hers were more bittersweet, the sadness of opportunities she knew she'd missed.

It was my mother who opened my eyes to what a gem Walt was, and it was Walt's dismay and sorrow, after she'd died suddenly, before he could see her again, that opened the door to my friendship with him. He needed someone to know that he'd begun to fall in love with her, the joyous surprise of that, and to appreciate how keenly he therefore felt the loss of her. Because I, too, in the last few years of my mother's life, had experienced a surprising upsurge of admiration and affection for her, and because I had a lot of time on my hands—I was childless, divorced, underemployed, and now parentless—I became the person Walt could talk to.

During my first visit to him, a few months after my mother died, we did the essential South Florida things: nine holes of golf at his condominium complex, two rubbers of bridge with two friends in their nineties in Delray Beach, and a stop at the nursing home where my aunt dwelled. We found her lying in bed in a tight fetal position. Walt tenderly fed her a dish of ice cream and a dish of pudding. When a nurse came in to change a Band-Aid on her hip, Fran burst into tears, her face contorting like a baby's, and wailed that it hurt, it hurt, it was horrible, it wasn't fair.

We left her with the nurse and returned to his apartment. Many of Fran's formal furnishings had come along from Dover, but now a bachelor scattering of magazines and cereal boxes had loosened their death grip. Walt spoke to me with plain emotion about the loss of Gail and the question of her old belongings. Would I like to have some of her drawings? Would I take the Pentax SLR he'd once given her? The drawings had the look of school projects, and I didn't need a camera, but I sensed that Walt was looking for a way to disencumber himself of things he couldn't bear to simply donate to Goodwill. I said I'd be very happy to take them.

In Santiago, the night before our charter flight to the southern tip of Argentina, Tom and I attended Lindblad's welcoming reception in a Ritz-Carlton function room. Because berths on our ship, the National Geographic Orion, started at twenty-two thousand dollars and went up to almost double that, I'd pre-stereotyped my fellow-passengers as plutocratic nature lovers—leather-skinned retirees with trophy spouses and tax-haven home addresses, maybe a face or two I recognized from television. But I'd done the math wrong. There turn out to be special yachts for that clientele. The crowd in the function room was less glamorous than I'd expected, and less octogenarian. A plurality of the hundred of us were merely physicians or attorneys, and I could see only one man in pants hiked up around his stomach.

My third-biggest fear about the expedition, after seasickness and disturbing my brother with my snoring, was that insufficient diligence would be devoted to finding the bird species unique to the Antarctic. After a Lindblad staffer, an Australian whose luggage for the trip had been lost by his airline, had greeted us and taken some questions from the crowd, I raised my hand and said I was a birder and asked who else was. I was hoping to establish the existence of a powerful constituency, but I saw only two hands go up. The Australian, who'd praised each of the earlier questions as "excellent," did not praise mine. He said, rather vaguely, that there would be staff members on the ship who knew their birds.

I soon learned that the two raised hands had belonged to the only two passengers who hadn't paid full fare. They were a conservationist couple in their fifties, Chris and Ada, from Mount Shasta, California. Ada has a sister who works for Lindblad, and they'd been offered a slashed-rate stateroom ten days before departure, owing to a cancellation. This added to my feeling of kinship with them. Although I could afford to pay full fare, I wouldn't have chosen a cruise line like Lindblad for my own sake; I'd done it for the Californian, to soften the blow of Antarctica, and was feeling like an accidental luxury tourist myself.

The next day, at the airport in Ushuaia, Argentina, Tom and I found ourselves near the rear of a slow line for passport control. At the urgent instruction of Lindblad, before leaving home, I'd paid the "reciprocity fee" that Argentina charged American tourists, but Tom had been in Argentina three years earlier. The government's Web site hadn't let him pay his fee again, so he'd printed a copy of its refusal and taken it with him, figuring that the printout, plus the Argentine stamps in his passport, would get him over the border. They didn't get him over the border. While the other Lindblad passengers boarded the buses that were taking us to a lunchtime cruise on a catamaran, we stood and pleaded with an immigration officer. Half an hour passed. A further twenty minutes passed. The Lindblad handlers were tearing their hair. Finally, when it looked as if Tom would be allowed to pay his fee a second time, I ran outside and boarded a bus and charged into a sea of dirty looks. The trip hadn't even started, and Tom and I were already the problem passengers.

On board the Orion, our expedition leader, Doug, summoned everyone to the ship's lounge and greeted us energetically. Doug was burly and white-bearded, a former theatrical designer. "I love this trip!" he said into his microphone. "This is the greatest trip, by the greatest company, to the greatest destination in the world. I'm at least as excited as any of you are." The trip, he hastened to add, was not a cruise. It was an expedition, and he wanted us to know that he was the kind of expedition leader who, if he and the captain spied the right opportunity, would *tear up the plan*, throw it out the window, and *go chase great adventure.*

Throughout the trip, Doug continued, two staffers would give photography lessons and work individually with passengers who wanted to improve their images. Two other staffers would go diving wherever possible, to supply us with additional images. The Australian who'd lost his luggage had not lost the late-model drone, with a high-definition video camera, that he'd worked for nine months to get the permits to use on our trip. The drone would be supplying images, too. And then there was the full-time videographer, who would create a DVD that we could all buy at trip's end. I got the impression that other people in the lounge had a clearer grasp than I of the point of coming to Antarctica. Evidently, the point was to bring home images. The National Geographic brand had led me to expect science where I should have been thinking of pictures. My sense of being a problem passenger deepened.

In the days that followed, I was taught what to ask when you meet a person on a Lindblad ship: "Is this your first Lindblad?" Or, alternately, "Have you done a Lindblad before?" I found these locutions unsettling, as if "a Lindblad" were something vaguely but expensively spiritual. Doug typically began his evening recap, in the lounge, by asking, "Was this a great day or was this a great day?" and then waiting for a cheer. He made sure we knew that we'd been specially blessed by a smooth crossing of the Drake Passage, which had saved us enough time to land in our Zodiac dinghies on Barrientos Island, near the Antarctic Peninsula. This was a very special landing, not something every Lindblad expedition got to do.

It was late in the nesting season for the gentoo and chinstrap penguins on Barrientos. Some of the chicks had fledged and followed their parents back into the sea, which is the preferred element of penguins and their only source of food. But thousands of birds remained. Downy gray chicks chased after any adult that was plausibly their parent, begging for a regurgitated meal, or banded together for safety from the gull-like skuas that preyed on the orphaned and the failing-to-thrive. Many of the adults had retreated uphill to molt, a process that involves standing still for several weeks, itchy and hungry, while new feathers push out old feathers. The patience of the molters, their silent endurance, was impossible not to admire in human terms. Although the colony was everywhere smeared with nitric-smelling shit, and the doomed orphan chicks were a piteous sight, I was already glad I'd come.

The scopolamine patches that Tom and I were wearing on our necks had dispelled my two biggest fears. With the help of the patch and calm waters, I wasn't getting seasick, and, with the help of the snore-muffling noise that we blasted on our clock radio, Tom was getting ten hours of deep scopolamine sleep every night. My third fear, however, had been on target. At no point did a Lindblad naturalist join Chris and Ada and me to watch seabirds from the observation deck. There wasn't even a good field guide to Antarctic wildlife in the Orion's library. Instead, there were dozens of books about South Polar explorers, notably Ernest Shackleton—a figure scarcely less fetishized onboard than the Lindblad experience itself. Sewed onto the left sleeve of my company-issued orange parka was a badge with Shackleton's portrait, commemorating the centennial of his epic open-boat voyage from Elephant Island. We were given a book about Shackleton, PowerPoint lectures about Shackleton, special tours to Shackleton-related sites, a screening of a long film about a re-creation of Shackleton's voyage, and a chance to hike three miles of the arduous trail that Shackleton had survived at the end of it. (Late in the trip, under the gaze of our videographer, we would all be herded to the grave of Shackleton, handed shot glasses of Irish whiskey, and invited to join in a toast to him.) The message seemed to be that we, on our Lindblad, were not un-Shackletonian ourselves. Failing to feel heroic on the Orion was a recipe for loneliness. I was grateful that I at least had two compatriots with whom to study the wildlife guides we'd brought, and to puzzle out the field marks of the Antarctic prion (a small seabird), and to try to discern the species-distinguishing hue of the bill of a fast-flying giant petrel.

As we progressed down the peninsula, Doug began dangling the possibility of exciting news. Finally, he gathered us in the lounge and revealed that it was actually happening: because of favorable winds, he and the captain had *thrown out the plan.* We had a very special opportunity to cross below the Antarctic Circle, and would now be steaming hard to the south.

The night before we reached the circle, Doug warned us that he might come on the intercom fairly early in the morning to wake those passengers who wanted to look outside and see the "magenta line" (he was joking) as we crossed it. And wake us he did, at six-thirty, with another joke about the magenta line. As the ship bore down on it, Doug dramatically counted down from five. Then he congratulated "every person onboard," and Tom and I went back to sleep. Only later did we learn that the Orion had approached the Antarctic Circle much earlier than six-thirty—at an hour when a person hesitates to wake up millionaires, an hour too dark for taking a picture. Chris, it turned out, had been awake before dawn and had followed the ship's coördinates on his cabin's TV screen. He'd watched as the ship slowed down, tacked west, and then executed a fishhook turn and steamed due north to buy time.

Although Doug came off as the chief simulacrum manager for a brand with cultish aspects, I had sympathy for him. He was finishing his first season as a Lindblad expedition leader, was clearly exhausted, and was under intense pressure to deliver the trip of a lifetime to customers who, not being plutocrats after all, expected value for their money. Doug was also, as far as I could determine, the only person on the ship besides me who'd been a birder serious enough to keep a list of the species he'd seen. He'd given up listing, but in one of his nightly recaps he told the amusing story of his desperation and failure to find a pipit on his first trip to South Georgia. If he hadn't been frantically catering to a shipful of image seekers, I would have liked to get to know him.

It should also be said that Antarctica lived up to Doug's enthusiasm. I'd never before had the experience of beholding scenic beauty so dazzling that I couldn't process it, couldn't get it to register as something real. A trip that had seemed unreal to me beforehand had taken me to a place that likewise seemed unreal, albeit in a better way. Global warming may be endangering the continent's western ice sheet, but Antarctica is still far from having melted. On either side of the Lemaire Channel were spiky black mountains, extremely tall but still not so tall as to be merely snow-covered; they were *buried* in wind-carved snowdrift, all the way to their peaks, with rock exposed only on the most vertical cliffs. Sheltered from wind, the water was glassy, and under a solidly gray sky it was absolutely black, pristinely black, like outer space. Amid the monochromes, the endless black and white and gray, was the jarring blue of glacial ice. No matter the shade of it—the bluish tinge of the growlers bobbing in our wake, the intensely deep blue of the arched and chambered floating ice castles, the Styrofoamish powder blue of calving glaciers—I couldn't make my eyes believe that they were seeing a color from nature. Again and again, I nearly laughed in disbelief. Immanuel Kant had connected the sublime with terror, but as I experienced it in Antarctica, from the safe vantage of a ship with a glass-and-brass elevator and first-rate espresso, it was more like a mixture of beauty and absurdity.

The Orion sailed on through eerily glassy seas. Nothing man-made could be seen on land or ice or water, no building or other ship, and up on the forward observation deck the Orion's engines were inaudible. Standing there in the silence with Chris and Ada, scanning for petrels, I felt as if we were alone in the world and being pulled forward toward the end of it, like the Dawn Treader in Narnia, by some irresistible invisible current. But when we entered an area of pack ice and became surrounded by it, images were needed. A Zodiac was noisily launched, the Australian's drone unleashed.

Late in the day, in Lallemand Fjord, near the southernmost latitude we reached, Doug announced another "operation." The captain would ram the ship into the huge ice field at the head of the fjord, and we could then choose between paddling around in sea kayaks or taking a walk on the ice. I knew that the fjord was our last hope for seeing an emperor penguin; seven other penguin species were likely on the trip, but the emperors rarely venture north of the Antarctic Circle. While the rest of the passengers hurried to their rooms to put on their life jackets and adventure boots, I set up a telescope on the observation deck. Scanning the ice field, which was dotted with crabeater seals and small Adélie penguins, I immediately caught a glimpse of a bird that looked unfamiliar. It seemed to have a patch of color behind its ear and a blush of yellow on its breast. *Emperor penguin?* The magnified image was dim and unsteady, and most of the bird's body was hidden by a little iceberg, and either the ship or the iceberg was drifting. Before I could get a proper look, the iceberg had obscured the bird altogether.

What to do? Emperor penguins may be the world's greatest bird. Four feet tall, the stars of "March of the Penguins," they incubate their eggs in the Antarctic winter as far as a hundred miles from the sea, the males huddling together for warmth, the females waddling or tobogganing to open water for food, every one of them as heroic as Shackleton. But the bird I'd glimpsed was easily half a mile away, and I was aware of being a problem passenger who'd already been involved in one lengthy delay of the group. I was also aware of my distressing history of incorrect bird identifications. What were the chances of randomly pointing a scope at the ice and instantly spotting the most sought-after species of the trip? I didn't feel as if I'd *made up* the yellow blush and the patch of color. But sometimes the birder's eye sees what it hopes to see.

After an existentialist moment, conscious of deciding my fate, I ran down to the bridge deck and found my favorite staff naturalist hurrying in the direction of Doug's operation. I grabbed his sleeve and said I thought I'd seen an emperor penguin.

"An emperor? You sure?"

"Ninety per cent sure.

"We'll check it out," he said, pulling away from me.

He didn't sound as though he meant it, so I ran down to Chris and Ada's cabin, banged on their door, and gave them my news. God bless them for believing it. They took off their life jackets and followed me back up to the observation deck. By now, unfortunately, I'd lost track of the penguin spot; there were so many little icebergs. I went down to the bridge itself, where a different staffer, a Dutch woman, gave me a more satisfactory response: "Emperor penguin! That's a key species for us, we have to tell the captain right away."

Captain Graser was a skinny, peppy German probably older than he looked. He wanted to know exactly where the bird was. I pointed at my best guess, and he got on the radio with Doug and told him that we had to move the ship. I could hear Doug's exasperation on the radio. He was in the middle of an operation! The captain instructed him to suspend it.

As the ship began to move, and I considered how annoyed Doug would be if I'd been wrong about the bird, I rediscovered the little iceberg. Chris and Ada and I stood at the rail and watched it through our binoculars. But there was nothing behind it now, at least nothing that we could see before the ship stopped and turned around. Radios were squawking impatiently.

After the captain had rammed us into the ice, Chris spotted a promising bird that quickly dived into the water. But then Ada thought she saw it come flopping back onto the ice. Chris put the scope on it, had a long look, and turned to me with a deadpan expression. "I concur," he said.

We high-fived. I fetched Captain Graser, who took one look through the scope and let out a whoop. "*Ja, ja,*" he said, "emperor penguin! Emperor penguin! Just like I was hoping!" He said he'd trusted my report because, on a previous trip, he'd seen a lone emperor in the same area. Emitting further whoops, he danced a jig, an actual jig, and then hurried off to the Zodiacs to have a closer look.

The emperor he'd seen earlier had been exceptionally friendly or inquisitive, and it appeared that I'd refound the same bird, because as soon as the captain approached it we saw it flop down on its belly and toboggan toward him eagerly. Doug, on the intercom, announced that the captain had made an exciting discovery and the plan had changed. Hikers already on the ice bent their steps toward the bird, the rest of us piled into Zodiacs. By the time I arrived on the scene, thirty orange-jacketed photographers were standing or kneeling and training their lenses on a very tall and very handsome penguin, very close to them.

I'd already made a quiet, alienated resolution not to take a single picture on the trip. And here was an image so indelible that no camera was needed to capture it: the emperor penguin appeared to be holding a press conference. While a cluster of Adélies came up from behind it, observing like support staff, the emperor faced the press corps in a posture of calm dignity. After a while, it gave its neck a leisurely stretch. Demonstrating its masterly balance and flexibility, and yet without seeming to show off, it scratched behind its ear with one foot while standing fully erect on the other. And then, as if to underline how comfortable it felt with us, it fell asleep.

At the following evening's recap, Captain Graser warmly thanked the birders. He'd reserved a special table for us in the dining room, with free wine on offer. A card on the table read "KING EMPEROR." Ordinarily, the ship's waiters, who were mostly Filipino, addressed Tom as Sir Tom and me as Sir Jon, which made me feel like John Falstaff. But that evening I really was feeling like King Emperor. All day long, passengers I hadn't even met had stopped me in hallways to thank me or cheer me for finding the penguin. I finally had an inkling of how it must feel to be a high-school athlete and come to school after scoring a season-saving touchdown. For forty years, in large social groups, I'd accustomed myself to feeling like the problem. To be a group's game-winning hero, if only for a day, was a complete, disorienting novelty. I wondered if, all my life, in my refusal to be a joiner, I'd missed out on some essential human thing.

My uncle, the Air Force veteran, now buried in the ranks at Arlington, was a lifelong joiner. Walt never ceased to be passionately loyal to his home town of Chisholm, in Minnesota's Iron Range, where he'd grown up with little money. He'd been a college hockey player and then a bomber pilot in the Second World War, flying thirty-five missions in North Africa and South Asia. He was a self-taught pianist, able to play any standard by ear; the elements of his golf swing were eclectic. He wrote two memoirs devoted to the many great friends he'd made in life. He was also a liberal Democrat who'd married a stringent Republican. He could strike up a lively conversation with almost anyone, and I could imagine the unfettered fun that my mother could imagine having had if she'd been with a regular guy like Walt and not my father.

One night, at the restaurant in the South Florida complex, over several cocktails, Walt told me the story not only of him and my mother but of him and Fran and Gail. After retiring from combat, he said, and after leading an officer's social life with Fran at various overseas bases, he'd realized that he'd made a mistake in marrying her. It wasn't just that her parents had spoiled her; she was an implacable social striver who hated and denied her backwoods Minnesota background as much as he loved and celebrated his own; she was unbearable. "I was weak," he said. "I should have left her, but I was weak."

They had their only child when Fran was in her mid-thirties, and Fran quickly became so obsessed with Gail, and so opposed to sex with Walt, that he felt driven to seek comfort elsewhere. "There were other women," he told me. "I had affairs. But I always made it clear that I was a family man and wasn't leaving Fran. On Sundays, my buddies and I would get loaded up on liquor and drive over to Baltimore to watch Johnny Unitas and the Colts." At home, Fran grew ever more micromanagerial in her attention to Gail's personal appearance, to her schoolwork, to her art projects. Gail seemed to be all Fran could talk about or think about. Her four years at college had brought some relief, but as soon as she returned to the East Coast, and went to work in Williamsburg, Fran redoubled her intrusions into Gail's life. Walt could see that something was terribly wrong; that Gail was being driven crazy by her mother but didn't know how to escape.

By early August, 1976, he'd become so desperate that he did the only thing he could do. He announced to Fran that he was going back to Minnesota, back to his beloved Chisholm, and that he wasn't going to live with her again—couldn't be married to her—unless she curtailed her obsession with their daughter. Then he packed a bag and drove to Minnesota. He was there, in Chisholm, ten days later, when Gail set out to drive through the night in bad weather across West Virginia. Gail was aware, he said, that he'd made a break with her mother. He'd told her himself.

Walt ended his story there, and we spoke of other things—his wish to find a girlfriend among the other residents of the complex, his clearness of conscience regarding this wish, now that my mother was dead and Fran was in a nursing home, and his worry that he was too much of a country boy, too unpolished, for the stylish widows at the complex. I wondered if he'd omitted the coda to his story because it went without saying: how, after an accident in West Virginia that could never be untangled from his flight to Minnesota, and after Fran had lost the one person in the world who mattered to her, becoming locked forever in brittle posthumous monomania, a world of pain, he'd had no choice but to return to her and devote himself henceforth to caring for her.

I saw that Gail's death hadn't merely been "tragic" in the hackneyed sense. It had partaken of the irony and inevitability of dramatic tragedy, compounded by the twenty-plus years that Walt had then devoted to listening to Fran, leavened only by the tenderness of his solicitude toward her. He really was a nice guy. He had a heart full of love and had given it to his broken wife, and I was moved not only by the tragedy but by the ordinary humanity of the man at the center of it. I had a sense of astonishment as well. Concealed in plain sight, my whole life, amid the moral rigidity and Swedish standoffishness of my father's family, had been a regular guy who had affairs and drove to Baltimore with his buddies and manfully accepted his fate. I wondered if my mother had seen in him what I'd now seen, and had loved him for it, as I now did.

The following afternoon, Walt's friend Ed called and asked him to come to his house with jumper cables. Arriving at the house, we found Ed standing in the street beside an enormous American car. Ed looked nearly dead—his skin was a terrible yellow and he was swaying on his feet. He said he'd been sick for a month and was feeling much better. But when Walt connected the jumper cables to Ed's car and asked him to try turning over the engine, Ed reminded him that he was too weak to turn the ignition key. (He had, however, been hoping to drive the car.) I got into Ed's car myself. As soon as I tried the key, I could tell that the car's problem was worse than a dead battery. Ed's car was utterly nonresponsive, and I said so. But Walt wasn't happy with how the jumper cables were connected. He backed his own car away and snagged a cable on the pavement. Before I could stop him, he'd torn off the cable's gripper, and the person he became upset with was me. I worked to reattach the gripper with a screwdriver, but he didn't like how I was doing it. He tried to grab it away from me, and he barked at me, shouted at me. "God damn it, Jonathan! God *damn* it! That's not right! Give it here! God damn it!" Ed, now sitting in the passenger seat, had slumped sideways and was listing downward. Walt and I tussled over the screwdriver, which I wouldn't let go of; I was angry at him, too. When we'd calmed down, and I'd repaired the cable to his satisfaction, I turned the key of Ed's car again. The car was nonresponsive.

After that first visit, I tried to get to Florida every year to see Walt, and to call him every few months. He did eventually find a girlfriend, a sterling one. Even when his hearing worsened and his mind began to cloud, I could sustain a conversation with him. We continued to have moments of intensity, like the time he told me how important it was to him that I someday tell his story, and I promised that I would. But it seems to me that we were never closer than the day he'd shouted at me about the jumper cable. There was something uncanny about that shouting. It was as if he'd forgotten—had been made to forget, perhaps by the overt mortality of Ed and his car, perhaps by the refraction of his love for my mother through the person of me—that he and I didn't have a real history together; had spent, in our lives, no more than a cumulative week with each other. He'd shouted at me the way a father might have shouted at a son.

The Californian had been right to fear the weather, which was colder than I'd led her to believe. But I'd been right about the penguins. From the Antarctic Peninsula, where their numbers were impressive, the Orion's route took us north again and then far east, to South Georgia island, where their numbers were staggering. South Georgia is a principal breeding site for the king penguin, a species nearly as tall as the emperor and even more dramatically plumaged. To see a king penguin in the wild seemed to me, in itself, sufficient reason not only to have made the journey; it seemed reason enough to have been born on this planet. Admittedly, I love birds. But I believe that a visitor from any other planet, observing a king penguin alongside even the most perfect human specimen, with vision unclouded by the possibility of sexual attraction, would declare the penguin the obviously more beautiful species. And it's not just the hypothetical extraterrestrial. Everybody loves penguins. In the erectness of their bearing, and in their readiness to drop down on their bellies, the flinging way they gesture with their armlike flippers, the shortness of the strides with which they walk or boldly scamper on their fleshy feet, they resemble human children more closely than does any other animal, not excepting the great apes.

Having evolved on remote coastlines, Antarctic penguins are also the rare animal with absolutely no fear of us. When I sat on the ground, the king penguins came so close to me that I could have stroked their gleaming, furlike feathers. Their plumage had the hypercrispness of pattern, the hypervividness of color, that you can normally experience only by taking drugs. The colonies of gentoos and chinstraps had not been great for sitting down, because of the excrement. But the king penguins were, as one Lindblad naturalist put it, more tidy. At St. Andrews Bay, on South Georgia, where half a million adult kings and fluffy king chicks were gathered tightly together, all I smelled was sea and alpine air.

Though every penguin species has its charms—the glam-rock head-streamers of the macaroni penguin, the little parallel-footed jumps with which a rock-hopper patiently climbs or descends a steep slope—I loved the kings above all others. They combined untoppable aesthetic splendor with the intently social energies of children at play. After porpoising toward shore, a group of kings would come running headlong up from the breakers, their flippers outstretched and fluttering, as if the water had got too cold for them. Or a lone bird would stand in shallow surf and gaze out to sea for so long that you wondered what thought was in its head. Or a pair of young males, excitedly tottering after an undecided female, would pause to see which of them was the more impressive craner of its neck, or to whap at each other ineffectually with their flippers. They had viciously sharp bills but sparred instead with punchless wings.

At St. Andrews, the activity was mostly on the outskirts of the colony. Because so many of the birds were incubating eggs or molting, the main colony itself seemed strikingly peaceful. The view of it from above reminded me of Los Angeles as seen from Griffith Park very early on a weekend morning. A drowsy megalopolis of upright penguins. Patrolling the thoroughfares were the sheathbills, strange snow-white birds with the body of a pigeon and the habits of a vulture. Even the amazing sound the kings made—a spiralling festive bray that was sort of like bagpipes, sort of like a holiday noisemaker, and sort of like the "woofing dog" sound on certain airplanes, but really like nothing on earth I'd ever heard—had a soothing effect when thousands of distant penguins were making it together.

In the twentieth century, human beings did penguins a favor by all but extirpating many of the whales and seals that they competed with for food. Penguin populations rose, and South Georgia has lately become even more hospitable to them, because the rapid retreat of its glaciers is exposing land suitable for nesting. But humanity's benefit to penguins may be short-lived. If climate change continues to acidify the oceans, the water will reach a pH at which ocean invertebrates can't grow their shells; one of these invertebrates, krill, is a dietary staple of many penguin species. Climate change is also rapidly diminishing the Antarctic Peninsula's encircling ice, which provides a platform for the algae on which krill feed in winter, and which has hitherto protected krill from large-scale commercial exploitation. Supertanker-size factory ships may soon be coming from China, from Norway, from South Korea, to vacuum up the food on which not only the penguins but many whales and seals depend.

Krill are pinkie-size, pinkie-colored crustaceans. Estimating the total amount of them in the Antarctic is difficult, but a frequently cited figure, five hundred million metric tons, could make the species the world's largest repository of animal biomass. Unfortunately for penguins, many countries consider krill good eating, both for humans (the taste is said to be acquirable) and especially for farm fish and livestock. Currently, the total reported annual take of krill is less than half a million tons, with Norway leading the list of harvesters. China, however, has announced its intention to increase its harvest to as much as two million tons a year, and has begun building the ships needed to do it. As the chairman of China's National Agricultural Development Group has explained, "Krill provides very good quality protein that can be processed into food and medicine. The Antarctic is a treasure house for all human beings, and China should go there and share."

The Antarctic marine ecosystem is indeed the richest in the world; it's also the last remaining substantially intact one. Commercial use of it is monitored and regulated, at least nominally, by the Commission for the Conservation of Antarctic Marine Living Resources. But decisions by the commission may be vetoed by any of its twenty-five members, and one of them, China, has a history of resisting the designation of some large marine protected areas. Another, Russia, has lately become openly intransigent, not only vetoing the establishment of new protected areas but questioning the very authority of the convention to establish them. Thus the future of krill, and with it the future of many penguin species, depends on uncertainties multiplied by uncertainties: how much krill there really is, how resiliently it can respond to climate change, whether any of it can now be harvested without starving other wildlife, whether such a harvest can even be regulated, and whether international coöperation on Antarctica can withstand new geopolitical conflicts. What isn't uncertain is that global temperatures, global population, and global demand for animal protein are all rising fast.

Mealtimes on the Orion inevitably put me in mind of the sanatorium in "The Magic Mountain": the thrice-daily rush for the dining room, the hermetic isolation from the world, the unchanging faces at the tables. Instead of Frau Stöhr, dropping the name of Beethoven's "Erotica," there was the Donald Trump supporter and his wife. There was the merry alcoholic couple. There was the Dutch rheumatologist, her rheumatologist second husband, her rheumatologist daughter, and the daughter's rheumatologist boyfriend. There was the pair of couples who, whenever the Zodiacs were being loaded, maneuvered their way to the front of the line. There was the man who, by special permission, had brought along ham-radio equipment and was spending his vacation in the ship's library, trying to contact fellow-hobbyists. There were the Australians who mostly didn't mix.

By way of mealtime small talk, I asked people why they'd come to Antarctica. I learned that many were simply devotees of Lindblad. Some had heard, while on a different Lindblad, that a Lindblad to Antarctica was the best Lindblad, possibly excepting a Sea of Cortez Lindblad. One couple whom I liked very much, a doctor and a nurse, Bob and Gigi, had come to celebrate their twenty-fifth anniversary one year late. Another man, a retired chemist, told me that he'd chosen Antarctica only because he'd run out of other places he hadn't been. I was glad that nobody mentioned seeing Antarctica before it melts. The surprise was that, for nearly the entire trip, not one staff person or passenger even uttered the words "climate change" in my hearing.

Granted, I was skipping many of the onboard lectures. To prove myself a hardest-core birder, I needed to be up on the observation deck. The hardest-core birder stands all day in biting wind and salt spray, staring into fog or glare in the hope of glimpsing something unusual. Even when your intuition is telling you that nothing's out there, the only way to know for sure is to put in the hours and examine every speck of bird life out to the horizon, every Antarctic prion (might be a fairy prion) darting among waves whose color it matches exactly, every wandering albatross (might be a royal albatross) deciding whether the ship's wake is worth following. Seawatching is sometimes nauseating, often freezing, and almost always punishingly dull. After I'd racked up thirty hours of it and tallied exactly one seabird of note, a Kerguelen petrel, I dialled back and devoted myself to the more sociable compulsion of playing bridge.

The other players, Diana and Nancy and Jacq, came from Seattle and belonged to a book club that had several other members on the ship. Along with Chris and Ada, they became my friends. In one of the early hands we played, I made a stupid discard, and Diana, a formidable bankruptcy attorney, laughed at me and said, "That was a *terrible* play." I liked her for this. I liked the foulness of the language at the table. When my partner, Nancy, who owns a forklift dealership, was playing her first slam contract of the trip, and I'd pointed out that the rest of the tricks were hers, she snapped at me, "Let me play the cards, you shit." She told me she'd meant it affectionately. The third player, Jacq, also an attorney, told me that she'd written a stage play about a Thanksgiving dinner she'd attended at Diana's, in the course of which Diana's ailing husband had died in bed in the family room. Jacq had the only tattoo I noticed on any passenger.

As in "The Magic Mountain," the early days of the expedition were long and memorable, the later ones more of an accelerating blur. As soon as I'd had a rewarding encounter with South Georgia pipits (they were gorgeous and confiding), I lost interest in visiting abandoned whaling stations. Even in Doug's voice, on our fifth day at South Georgia, a weariness was audible when he said, "So I think we'll do another sea kayak." He sounded like Vladimir and Estragon when they decide, late in "Godot," after exhausting every other conceivable distraction, to "do the tree."

Toward the end of the trip's final day, which I'd mostly spent at the bridge table while hundreds of potentially interesting seabirds wheeled around outside, I went down to the lounge for a lecture on climate change. The lecture was delivered by the drone-flying Australian, whose name is Adam, and was attended by fewer than half the passengers. I wondered why Lindblad had postponed such an important lecture until the last day. The charitable explanation was that Lindblad, which prides itself on its environmental consciousness, wanted to send us home fired up to do more to protect the natural splendor we'd enjoyed.

Adam's opening plea suggested other explanations. "Passenger-comment cards," he said, "are not the place to voice your beliefs about climate change." He laughed uneasily. "Don't shoot the messenger." He proceeded to ask how many of us believed the earth's climate was changing. Everyone in the lounge raised a hand. And how many of us believed that human activity was causing it? Again, most hands were raised, but not the Donald Trump supporter's, not the ham hobbyist's. From the very back of the lounge came the curmudgeonly voice of Chris: "What about the people who think it isn't a matter of belief?"

"Excellent question," Adam said.

His lecture was a barn-burning reprise of "An Inconvenient Truth," including the famous "hockey stick" graph of spiking temperatures, the famous map of an America castrated of its Florida by the coming rise in sea level. But the picture Adam painted was even darker than Al Gore's, because the planet is heating up so much faster than even the pessimists expected ten years ago. Adam cited the recent snowless start of the Iditarod, the sickeningly hot winter that Alaska was having, the possibility of an ice-free North Pole in the summer of 2020. He noted that whereas, ten years ago, only eighty-seven per cent of the Antarctic Peninsula's glaciers were known to be shrinking, the figure now seems to be a hundred per cent. But his darkest point was that climate scientists, being scientists, must confine themselves to making claims that have a high degree of statistical probability. When they model future climate scenarios and predict the rise in global temperature, they have to pick a lowball temperature, one reached in ninety-plus per cent of all cases, rather than the temperature that's reached in the average scenario. Thus, the scientist who confidently predicts a five-degree (Celsius) warming by the end of the century might tell you in private, over beers, that she really expects it to be nine degrees.

Thinking in Fahrenheit—sixteen degrees—I felt very sad for the penguins. But then, as so often happens in climate-change discussions when the talk turns from diagnosis to remedies, the darkness became the blackness of black comedy. Sitting in the lounge of a ship burning three and a half gallons of fuel per minute, we listened to Adam extoll the benefits of shopping at farmers' markets and changing our incandescent bulbs to L.E.D. bulbs. He also suggested that universal education for women would lower the global birth rate, and that ridding the world of war would free up enough money to convert the global economy to renewable energy. Then he called for questions or comments. The climate-change skeptics weren't interested in arguing, but a believer stood up to say that he managed a lot of residential properties, and that he'd noticed that his federally subsidized tenants always kept their homes too hot in the winter and too cold in the summer, because they didn't pay for their utilities, and that one way to combat climate change would be to make them pay. To this, a woman quietly responded, "I think the ultra-wealthy waste far more than people in subsidized housing." The discussion broke up quickly after that—we all had bags to pack.

At six o'clock, the lounge filled up again, more tightly, for the climax of the expedition: the screening of a slide show to which passengers had been invited to contribute their three or four finest images. The photography instructor who was hosting it apologized in advance to anyone who didn't like the songs he'd chosen for its soundtrack. The music—"Here Comes the Sun," "Build Me Up Buttercup"—certainly didn't help. But the whole show was dispiriting. There was the sense of diminishment I always get from our culture of images: no matter how finely you chop life into a sequence of photographs, no matter how closely in time the photographs are spaced, what the sequence always ends up conveying to me most strongly is what it leaves out.

It was also sadly evident that three weeks of National Geographic instruction hadn't produced National Geographic freshness of vision. And the cumulative effect was painfully wishful. The slide show purported to capture an adventure we'd had as a community, like the community of Shackleton and his men. But there had been no long Antarctic winter, no months of sharing seal meat. The vertical relationship between Lindblad and its customers had been too insistent to encourage the forging of horizontal bonds. And so the slide show came off as an amateur commercial for Lindblad. Its wishful context spoiled even the images that should have mattered to me, the way any amateur photograph matters: by recording the face of what we love. When my brother privately showed me a picture he'd taken of Chris and Ada sitting in a Zodiac (Chris failing to maintain complete disgruntlement, Ada outright smiling), it reminded me of my happiness at having found them on the ship. The picture was full of meaning—to me. Upload it onto Lindblad's Web site, and its meaning collapses into advertising.

So what had been the point of coming to Antarctica? For me, it turned out, the point was to experience penguins, be blown away by the scenery, make some new friends, add thirty-one bird species to my life list, and celebrate my uncle's memory. Was this enough to justify the money and the carbon it had cost? You tell me. But the slide show did perform a kind of backhanded service, by directing my attention to all the unphotographed minutes I'd been alive on the trip—how much better it was to be bored and frozen by seawatching than to be dead. A related service emerged the next morning, after the Orion had docked in Ushuaia and Tom and I were set free to wander the streets by ourselves. I discovered that three weeks on the Orion, looking at the same faces every day, had made me intensely receptive to any face that hadn't been on it, especially to the younger ones. I felt like throwing my arms around every young Argentine I saw.

It's true that the most effective single action that most human beings can take, not only to combat climate change but to preserve a world of biodiversity, is to not have children. It may also be true that nothing can stop the logic of human priority: if people want meat and there are krill for the taking, krill will be taken. It may even be true that penguins, in their resemblance to children, offer the most promising bridge to a better way of thinking about species endangered by the human logic: They, too, are our children. They, too, deserve our care.

And yet to imagine a world without young people is to imagine living on a Lindblad ship forever. My godmother had had a life like that, after her only child was killed. I remember the half-mad smile with which she once confided to me the dollar value of her Wedgwood china. But Fran had been nutty even before Gail died; she'd been obsessed with a biological replica of herself. Life is precarious, and you can crush it by holding on too tightly, or you can love it the way my godfather did. Walt lost his daughter, his war buddies, his wife, and my mother, but he never stopped improvising. I see him at a piano in South Florida, flashing his big smile while he banged out old show tunes and the widows at his complex danced. Even in a world of dying, new loves continue to be born.

The Emergency

Ben Taub

December 4, 2017

Chad was named for a mistake. In the eighteen-hundreds, European explorers arrived at the marshy banks of a vast body of freshwater in Central Africa. Because locals referred to the area as *chad*, the Europeans called the wetland Lake Chad, and drew it on maps. But *chad* simply meant "lake" in a local dialect. To the lake's east, there was a swath of sparsely populated territory—home to several African kingdoms and more than a hundred and fifty ethnic groups. It was mostly desert. In the early nineteen-hundreds, France conquered the area, called it Chad, and declared it part of French Equatorial Africa.

A few years later, a French Army captain described Lake Chad, which was dotted with hundreds of islands, as an ecological wonder and its inhabitants as "dreaded islanders, whose daring flotillas spread terror" along the mainland. "Their audacious robberies gave them the reputation of being terrible warriors," he wrote. After his expeditions, the islanders were largely ignored. "There was never a connection between the people who live in the islands and the rest of Chad," Dimouya Souapebe, a government official in the Lake Region, told me.

Moussa Mainakinay was born in 1949 on Bougourmi, a dusty sliver in the lake's southern basin. Throughout his childhood and teen-age years, he never went hungry. The cows were full of milk. The islands were thick with vegetation. The lake was so deep that he couldn't swim to the bottom, and there were so many fish that he could grab them with his hands. The lake had given Mainakinay and his ancestors everything—they drank from it, bathed in it, fished in it, and wove mats and baskets and huts from its reeds.

In the seventies, Mainakinay noticed that the lake was receding. There had always been dramatic fluctuations in water level between the rainy and the dry seasons, but now it was clear that the mainland was encroaching. Floating masses of reeds and water lilies began to clog the remaining waterways, making it impossible to navigate old trading routes between the islands.

Lake Chad is the principal life source of the Sahel, a semiarid band that spans the width of Africa and separates the Sahara, in the north, from the savanna, in the south. Around a hundred million people live there. For the next two decades, the entire region was stricken with drought and famine. The rivers feeding into Lake Chad dried up, and the islanders noticed a permanent decline in the size and the number of fish.

Then a plague of tsetse flies descended on the islands. They feasted on the cows, transmitting a disease that made them sickly and infertile, and unable to produce milk. For the first time in Mainakinay's life, the islanders didn't have enough to eat. The local medicine man couldn't make butter, which he would heat up and pour into people's nostrils as a remedy for common ailments. Now, when the islanders were sick or malnourished, he wrote Quranic verses in charcoal on wooden boards, rinsed God's words into a cup of lake water, and gave them the cloudy mixture to drink. By the end of the nineties, the lake, once the size of New Jersey, had shrunk by roughly ninety-five per cent, and much of the northern basin was lost to the desert. People started dying of hunger.

In 2003, when Mainakinay was fifty-four years old, he became the chief of Bougourmi. He was proud of his position, but not that proud; his grandfather had presided over more than four hundred islands—until the government stripped the Mainakinays of their authority as Chiefs of the Canton, a position that they had held for more than two hundred years. The center of power was moved to the town of Bol, on the mainland. The islanders were of the Boudouma tribe; the mainlanders were Kanembou. They didn't get along.

Other political developments were more disruptive. Colonial administrators had drawn the boundaries of Chad, Cameroon, Nigeria, and Niger right through tiny circles of huts on the islands. When these nations enforced their borders, the fishermen and cattle herders of Bougourmi, which is in Chad, were cut off from the lake's biggest market, which is in Baga, on the Nigerian shoreline. In the mid-aughts, hungry and desperate, they turned to foraging in the bush for fruit and nuts. Then they began to run out of fruit and nuts.

"These were our problems before," Mainakinay told me, in late July, as he sat on the ground inside a reed hut. He wore a white robe over his bony shoulders, and his dark-brown eyes were turning blue at the edges, fading with age. "It was only recently that our real suffering began."

One night in 2015, Mainakinay saw flames coming from the huts on Médi Kouta, less than a mile away.

For the past several years, Boko Haram had sought to establish a caliphate in northeastern Nigeria. Mainakinay had heard of the group on his shortwave radio. Now, after spreading out along the lake, into southern Niger and northern Cameroon, Boko Haram had come to the Chadian islands and begun kidnapping entire villages, replenishing its military ranks and collecting new wives, children, farmers, and fishermen to sustain its campaigns. At dawn, Mainakinay led the people of Bougourmi to a neighboring island to hide. But Boko Haram continued its attacks, and so, for the first time, Mainakinay's people sought refuge on the mainland, leaving their cattle and belongings behind.

The jihadis encountered little resistance in Lake Chad. Most islands had no more than a couple of hundred inhabitants, and their machetes and fishing tools were no match for Boko Haram's grenades and assault rifles. When the militants arrived on Médi Kouta, they set fire to the mosque and beheaded a few men; after that, the terrified islanders followed the fighters into wooden boats and paddled west, to Nigeria and Niger. As they moved farther away from the Chadian side of the lake, the captives noticed that some islands were already flying the jihadis' black flag.

That spring, a few thousand Boudouma fled to the Chadian mainland, near Bol. The United Nations, anticipating military operations in the islands by Chad against Boko Haram, contacted the government. "We met with the minister of defense and the chief of the Army, and urged them to let us know what they're planning," Florent Méhaule, the head of the U.N.'s Office for the Coordination of Humanitarian Affairs in Chad, told me.

Chad is a weak state with a strong military, known for its brutal treatment of combatants and civilians. In late July, without notifying the U.N., the Chadian Army ordered an evacuation of all islands in the southern basin, warning that anyone who was still there in a week would be considered a member of Boko Haram. Around fifty-five thousand islanders rushed to the mainland. The Boudouma have an extensive history of raiding the Kanembou, and the Chief of the Canton did not allow them into the towns. According to Méhaule, "He just told them, 'Go stay in the empty land between villages. The humanitarians, in their white vans, will come.'"

The evacuation of the southern basin took place just before the harvest, so Boko Haram collected whatever millet, wheat, and maize the islanders had left behind. By the end of November, the Chadian Army had swept through the northern basin, forcibly displacing more than a hundred and ten thousand people in total. They ended up scattered among roughly a hundred and forty spontaneous sites across a vast, inhospitable terrain. "People were everywhere—in places we did not know," Méhaule said. Because Boko Haram had used boats to attack the islands, the Chadian government banned the use of fishing boats, so the Boudouma had virtually nothing to eat. Without sufficient pasture, many of the Boudoumas' cattle died.

Méhaule and his colleagues set off in convoys of white Toyota Highlanders, searching for the displacement sites. There were no roads or signs, no paths to follow. "All of our maps were wrong, because they were from the nineteen-seventies," Méhaule told me. "We were driving through areas that should have been underwater, but we couldn't even see the lake."

In recent years, the Lake Chad region has become the setting of the world's most complex humanitarian disaster, devastated by converging scourges of climate change, violent extremism, food insecurity, population explosion, disease, poverty, weak statehood, and corruption.

The battle against Boko Haram spans the borders of four struggling countries. It is being waged by soldiers who answer to separate chains of command and don't speak the same languages as one another, or as their enemies, or as the civilians, in the least developed and least educated region on earth.

Across the Sahel, millions of people are displaced, and millions more are unable to find work. The desert is expanding; water is becoming more scarce, and so is arable land. According to the U.N., the region's population, which has doubled in the past few decades, is expected to double again in the next twenty years.

The Sahel is rife with weapons and insurgencies, and some states are beginning to collapse. In recent years, cattle herders and farmers have started killing one another over access to shrinking pastures—the number of deaths exceeds fifteen thousand, rivalling that inflicted by Boko Haram.

Western countries and the United Nations have been trying to stabilize local governments. Since the early aughts, the U.S. has spent hundreds of millions of dollars on strengthening Sahelian security forces, in a bid to limit the spread of jihadism in the region's vast, ungoverned spaces. But this strategy fails to take into account the complex cruelties of colonialism and the predatory nature of the regimes that have developed in its place. Across the Sahel, many people experience no benefits from statehood, only neglect and violence. "What we are actually doing is making the predator more capable," a European security official told me. "And that's just stunningly shortsighted."

After France took over Chad, it learned that the territory lacked the riches that colonial powers had discovered elsewhere in West and Central Africa. France sent its least experienced and worst-behaved officers there—often as a kind of punishment—and, in the ensuing decades, French military campaigns disrupted trade routes and local economies, contributing to the deaths of hundreds of thousands of people from famine. The French focussed their attention on the forced production of cotton, in a fertile part of southern Chad that they referred to as "le Tchad Utile"—Useful Chad.

In 1958, French Equatorial Africa split up, and two years later Chad became an independent state. The country's borders had been determined by colonial agreements, and many Chadians couldn't communicate with one another—there were at least a hundred and twenty indigenous languages. Some Chadians in remote areas were unaware that their villages now belonged to a state.

The country spent the next several decades "suspended between creation and destruction," as the South

African historian Sam Nolutshungu writes, in "Limits of Anarchy: Intervention and State Formation in Chad." It was "aberrant, marginal, a fictive state," a country that existed, "even in its peaceful moments, alternately under a cloud of contingent anarchy or tyranny."

Chad was constantly threatened from the north. Libya's dictator, Muammar Qaddafi, aimed to form what he called a "Great Islamic State of the Sahel," and he repeatedly sponsored attempts to topple Chad's leaders. The French usually supported whichever autocrat or warlord was in power. Chad's institutions were propped up by French investors and advisers, and hardly extended beyond the capital, N'Djamena. The illusion of Chadian statehood was useful for France and the United States, who saw a strong Chadian Army as a means with which to cripple Qaddafi's ambitions.

Hissène Habré became President in 1982, in a revolt sponsored by the C.I.A. He had led three violent rebellions and held Europeans hostage, and yet the moment he took the capital he inherited all the international structures of legitimacy afforded to any head of state. Habré ran a vicious security state, with secret detention centers, that tortured and executed tens of thousands of its citizens. But Habré despised Libya's leader; because of this, the U.S., under Ronald Reagan, supplied him with hundreds of millions of dollars' worth of weapons. In 1983, Qaddafi invaded northern Chad, using Soviet tanks. In theory, France and the U.S. were no longer backing a warlord. They were helping a President preserve the territorial integrity of his nation.

Habré's soldiers fought Qaddafi's forces in a fleet of Toyota Hiluxes supplied by the C.I.A. In what became known as the Toyota War, the Chadian Army killed thousands of Libyan fighters.

In 1987, as Qaddafi withdrew his troops, Reagan invited Habré to the White House and praised his commitment to "building a better life for the Chadian people." Then Habré resumed slaughtering ethnic minorities who protested his rule. He also accused three of his highest-ranking officials of plotting a military coup. Two of them were captured and killed. The third, a young colonel named Idriss Déby, fled east to Sudan, and recruited others to join him in a rebellion. Déby also went to Libya, where Qaddafi supplied him with cash and weapons. The next year, Déby's group drove back across the desert. Habré fled into Cameroon, and Déby became the President of Chad.

It was December 2, 1990. To be Chadian was to be born into a territory where you had a fifteen-per-cent chance of dying before your first birthday. In a country of five million, there were five hospitals, and a few dozen qualified doctors. People routinely died of malaria, cholera, and starvation. The average citizen lived to thirty-nine.

Two days later, Déby gave his first public speech. "I have brought you neither gold nor silver but liberty!" he said. "No more military campaigns. No more prisons." He claimed that he was "determined to lead Chad, with the participation of all its citizens, to the system of government longed for by all: a system of government based on democracy." He paused. "I mean, democracy *in its fullest sense.*" It was a telling slip: a new constitution enshrined freedoms of religion, expression, demonstration, and the press, but, in the years that followed, people who tried to exercise those rights often disappeared.

Chad's economy was nearly nonexistent. Many Chadians—including former and current rebels, soldiers, and police officers—resorted to highway banditry to survive. After Déby told gendarmes that bandits "should be shot down like a dog," some officials held public mass executions of suspected criminals, without trials. One day in 1996, gendarmes in N'Djamena arrested an elementary-school student who had stolen food from his neighbors. They put a bag over his head, shot him, and abandoned his body on the banks of the Chari River.

Ten days later, Déby met with representatives from several fledgling Chadian human-rights organizations. He told them that the killings were "in accordance with the wishes of public opinion." Otherwise, Déby has left his citizens in total neglect. The *Irish Times* reported that, outside the capital, Chadians were eating boiled leaves and animal feed, and digging up anthills to search for whatever grains the insects had dragged home. "This country is a bit of a police state, but mostly a pirate ship," the European security official told me. "That's the sense that I get when I'm here—that I'm on a pirate ship, and the captain is always drunk."

Déby's security forces used military planes supplied by the U.S. and France, and maintained by American and French technicians, to transport political prisoners. The French also supplied Déby's regime with money, trucks, fuel, communications systems, and handcuffs—resources that, according to Amnesty International, had been "diverted from their original purpose to be used for execution and torture." Although French military advisers stationed at Chadian outposts had witnessed human-rights abuses, they did not intervene, saying that it was not their responsibility to come between the state and its people.

Since Déby took power, his forces have put down numerous rebellions and coups. In 2006, Déby and the President of Sudan sponsored insurrections against each other. The Chadian rebels made it all the way to N'Djamena. French soldiers helped stabilize the capital, but near the Sudanese border the Chadian Army forcibly conscripted children. "Déby has trouble finding soldiers who are willing to fight for him," a senior Chadian military officer told Human Rights Watch. "Child soldiers are ideal, because they don't complain, they don't expect to be paid, and, if you tell them to kill, they kill."

In 2008, Congress passed a law that banned American military support for governments that used child soldiers. But President Barack Obama secured a waiver for Chad, arguing that it was "in the national interest" of the United States to train and equip Chad's military. Al Qaeda's message was taking root in parts of Africa where nation-states had been sloppily crafted and poorly ruled. The war on terror had reached the Sahel.

Around that time, Mohammed Yusuf, a young Salafi preacher in northeastern Nigeria, was delivering sermons about the ruinous legacy of colonialism and the corruption of Nigeria's élites. After decades of political turbulence and military coups, oil extraction had made Nigeria the richest country in Africa, and yet the percentage of people living in total poverty was growing each year. "The Europeans created the situation in which we find ourselves today," Yusuf said. It was easy to appeal to the existential grievances of northern Nigeria's marginalized, unemployed youth. Yusuf told them that the only way forward was to install a caliphate in Nigeria. His followers, who became known as Boko Haram, revived a tradition of jihadism in northern Nigeria that goes back hundreds of years.

On June 11, 2009, police officers at a checkpoint in Nigeria stopped a group of Boko Haram members on their way to a funeral; in the confrontation that followed, officers opened fire and injured seventeen jihadis. The next month, around sixty Boko Haram members attacked a police station. Gun battles erupted in several towns, and Yusuf was arrested. A few hours later, the police executed him and dumped his body outside the station. A video of the mutilated corpse, still in handcuffs, went viral. Violence exploded all over northern Nigeria: at least seven hundred people were killed in the first week. Yusuf's deputy, Abubakar Shekau, became the leader of Boko Haram.

Shekau dispatched some of his followers to train with Al Qaeda in the Islamic Maghreb. When they returned, the group detonated car bombs in Nigeria's capital, Abuja. Propaganda videos show Shekau double-fisting Kalashnikovs and screaming incoherently as he fires bullets into the sky. In 2012, after leading a rampage in the Nigerian city of Kano, he said, "I enjoy killing anyone whom God commands me to kill, the way I enjoy killing chickens and rams." Shekau's fighters terrorized remote villages, tossing grenades into huts and burning down mosques. They raped women, slaughtered men, and kidnapped children, whom they forced to carry out suicide bombings. Shekau pledged his group's allegiance to the Islamic State, but his battlefield tactics were so depraved that ISIS eventually disowned him. Young men who joined Boko Haram were sent back to their villages to recruit their families. As a teen-age fighter explained to me, "If your family doesn't come, you have to kill them, because they have chosen to be infidels."

The Nigerian security forces responded with a series of massacres that drove villagers into the insurgency. One day in 2013, after Boko Haram killed a Nigerian soldier near Baga, on the muddy western shores of Lake Chad, government troops stormed into the town, lit thatched huts on fire, and shot villagers as they tried to escape. Some villagers tried to swim to the islands and drowned in the lake. Roughly two hundred people are thought to have died, and more than two thousand structures were burned.

On January 3, 2015, Boko Haram returned to Baga and attacked a local military base. The soldiers shed their uniforms and fled into the bush, leaving behind weapons, vehicles, and ammunition. During the next four days, Boko Haram slaughtered civilians in Baga and the surrounding villages. "It was impossible to know how many people they killed," a survivor told me. "I just saw bodies in the streets. Everyone was running." Thousands of people made for the islands of Lake Chad. Boko Haram followed them.

Many islanders were open to Boko Haram. The Boudouma used Nigerian currency, and for decades those who could afford to had been sending their children to study with Quranic tutors in northern Nigeria. A few years ago, some of those children started calling their siblings and friends, urging them to leave the islands and join Boko Haram. "They told me that if I join them I will go to paradise," a sixteen-year-old Boudouma told me. "They also said that, at their camp in Nigeria, there are buckets full of money, and you can just take as much as you want. So I followed them."

Dimouya Souapebe, the government official, said that "it was easy for Boko Haram to come in from Nigeria and poison people's minds," by promising access to basic services and Islamic education. "The islanders never had a school. They've never had sanitation. They drink the same lake water they defecate in. Out in the islands, there is nothing."

As a first line of defense on the mainland, the Chadian governor of the Lake Region set up "vigilance committees," with the help of tribal chiefs. Given Chad's history of rebellion, the governor was wary of allowing vigilantes to carry weapons, but he distributed cell phones to young men so that they could alert the authorities.

On a Tuesday evening in December, 2015, some months after Moussa Mainakinay and his villagers fled Bougourmi, a contact living on a jihadi-controlled island warned him that Boko Haram was planning to attack the market in Bol, the biggest town on the Chadian side of the lake. That night, Mainakinay and a group of vigilantes stood guard, looking for boats. Eventually, they spotted a canoe moving toward them, containing several men and women. When it reached the shore, a few of the passengers detonated suicide vests, killing most of the others. Flesh and cloth rained down. A teen-age girl collapsed screaming; the explosion had mangled her legs.

The vigilantes took her to the hospital in Bol, a small concrete building with rudimentary supplies. Someone roused Sam Koulmini, one of only two doctors in Bol. (The other is his wife.) "Boko Haram told the girl that the explosives wouldn't hurt her—that, if she killed some people in the market, they'd give her money when she got back," Koulmini told me. That night, he amputated her right leg and one of her damaged fingers.

For the next several days, she was kept in an isolated room, guarded by gendarmes. During her time with Boko Haram, she had become addicted to tramadol, an opioid painkiller that is widely abused in the region. "A macro-dose of tramadol makes you feel as if you're in the clouds," Koulmini explained. "You're afraid of nothing. Pretty much all the young people are taking it." The injured girl showed severe symptoms of withdrawal; held in isolation, he says, "she became psychotic." On the second day, she started smearing feces all over her body.

The gendarmes wouldn't let Koulmini visit his patient more than once a day, and they rushed him as he changed the dressings on her wounds. By the time he noticed that there were still pieces of shrapnel inside her left leg, it was too late; the limb was gangrenous. He had to cut that leg off, too.

Before Boko Haram invaded the islands, humanitarian groups in Chad were preoccupied elsewhere, dealing with nationwide health and malnutrition crises, and with refugee crises near Chad's borders with Sudan and the Central African Republic. "We all had to open very quickly," Méhaule, of the U.N.'s Office for the Coordination of Humanitarian Affairs, told me. But the Lake Region was so poor and undeveloped that it was hard to distinguish the needs of the displaced islanders from those of mainland villagers. "Some of the displaced, for example, own a huge quantity of cattle," he said. In the short term, they may be better off than people who have shelter but no food.

In late July, I flew from N'Djamena on an aging propeller plane that, twice a week, takes humanitarians and supplies to Bol. From the sky, I could see black smudges, clustered circles in the sand—remnants of burned-out villages. The movement of people and cows had left faint tracks across the islands and through the reeds and lily pads that filled the waterways between them.

For the next week, I travelled through the Lake Region with two UNICEF employees and the photographer Paolo Pellegrin. The Chadian government was kept informed of our activities and movements; each time we arrived in a different jurisdiction, UNICEF scheduled a "courtesy visit" with the local authorities, and gave them copies of our papers and photographs.

There are no roads in the region, so we followed in the tracks of vehicles belonging to the military and to other N.G.O.s—up sandy hills, past millet patches and goat pens made of gnarled roots and thorny vines. The pens rarely had any goats in them. We rattled downhill into flat, dark-brown depressions that, until recently, had been part of the lake bed. There, people were digging for natron, a mineral ash that's put into camel feed; the going rate for a hundred-pound bag of natron is less than three dollars. According to Méhaule, the soil left by the lake's receding waters is so rich that, properly farmed, it could grow enough food to sustain everyone who's currently starving next to it. But the Chief of the Canton has refused to give displaced Boudouma access to arable plots.

As we drove, we would suddenly come across a reed hut. Then another would come into view, then a hundred more—each big enough for three or four people to lie down in. We would pass a metal billboard announcing the name of the displacement site, as well as the organizations and the countries who were funding some of its needs. Here was the UNICEF-backed well. Over there was the Oxfam one, where children were using their full weight to pull down the lever. Then there was the sign announcing a "joint education project," paid for by Handicap International, Cooperazione Internazionale, and the Swiss

Confederation—although we saw no schools. In certain ways, the displacement sites offer an improvement over life in the islands. Every so often, representatives of international organizations arrive to vaccinate the cows; build a well; talk to the women about health issues; and weigh the children and measure their upper arms, to calculate the level of malnutrition. At one site, I met a woman whose hands had been blown off when Boko Haram threw a grenade into her hut. She had been nine months pregnant, with her seventh child, but after the attack she miscarried. When her husband saw her condition, he left her, sold their last cow, and remarried within two weeks. Having fled the islands, she is now regularly visited by a mental-health professional.

Still, on the Chadian side of the lake, it is better to be any nationality but Chadian. Just outside Baga Sola, the second-largest town in the Lake Region, there is a camp for displaced Nigerians. Around five thousand people live there, most of whom fled the massacre in Baga, across the lake. Because they crossed an international border, they have refugee status, which makes them eligible for funding and legal protections that are not available to the internally displaced. The Nigerian camp has a school with large classrooms, blackboards, chalk, and wooden desks. There's an outdoor basketball court, with floodlights; indoor latrines are a short walk away. Police officers patrol the camp, which is divided into sixteen neatly spaced blocks of roomy, waterproof tents. "The security is good here, and we have enough food," a teen-age girl told me. "I am getting an education for the first time. Why would I want to go back to Nigeria?"

No such amenities exist in the sites for internally displaced Chadians, or even in the Kanembou villages. One morning, a heavy rainstorm flooded Baga Sola. Goats huddled against a mud wall to avoid the wind. When the storm lifted, I went to the Kafia displacement site, which is home to a few thousand Boudouma. Everyone was soaked. Several reed huts had collapsed; their remnants were strewn across a desolate, sandy landscape. A camel gnawed at a prickly tree. "We eat one meal per day, at most," a Boudouma chief told me. "Often we eat nothing."

After Boko Haram attacked Baga, each country bordering the lake supplied a couple of thousand soldiers to an effort called the Multi-National Joint Task Force, which receives intelligence from Western partners. But coöperation among the countries is fragile. One day, at the task-force base in N'Djamena, I met the commander, a wiry, deadly serious Nigerian major general named Leo Irabor. He sat at an imposing desk, with a wall of maps behind him. I told him that I hoped to embed with his troops and to travel north of the lake, west through the desert into Niger, and then south into Nigeria. "It is a fine idea," Irabor said. "But it is not possible," because the M.N.J.T.F. doesn't conduct cross-border operations. The military sectors are divided along national boundaries, and the countries have a history of mistrust—especially between the government of anglophone Nigeria and those of the other countries, which are francophone. Soldiers can pursue militants across borders, if necessary, but only Boko Haram fights as if the borders don't exist. "None of the partner countries want to end up shouldering most of the burden," a Western military adviser to the M.N.J.T.F. told me, with a shrug. "We can't want it to work more than they do."

The M.N.J.T.F. doesn't fight for new ground in the islands during the rainy season—the weather can damage vehicles and leave fighters stranded—but, in the past two dry seasons, it has taken significant territory back from Boko Haram, spurring defections. Last year, on August 25th, seven Boudouma men and one woman showed up at an M.N.J.T.F. checkpoint in Chad, near the border with Nigeria. For more than a year, they had been living with Boko Haram; now they wanted to come home. By the end of 2016, around three hundred men and seven hundred women and children had returned. They were kept in military detention in Baga Sola while the government figured out what to do with them.

Eventually, all the women and children were let go, and it fell to the U.N. to reunite them with other people from their villages among the scattered displacement sites. "There was a big risk that they wouldn't be welcome—that they would be stigmatized or retaliated against," Méhaule, of OCHA, said. "But the reintegration was surprisingly easy."

One afternoon, at the Mélea displacement site, near Bol, I met a twelve-year-old boy whom I'll call Aboudou. He looked about half his age. He wore ragged green pants, a filthy shirt, and, despite the scorching temperatures, a yellow woollen hat, which he pulled down over his eyes whenever he started to cry. His face was marked with the traditional scarification of the Boudouma—a deep cut down the center of the nose, and diagonal marks on each cheek—and his skin was so taut that you could see his jaw muscles move when he spoke.

Aboudou, his parents, and his four younger siblings had been kidnapped by Boko Haram, during the attack on Médi Kouta, near Bougourmi. The family had canoed with the jihadis for two weeks, until they reached the island of Boka, in southern Niger, where his mother and father built a house out of red water lilies. Each day, Aboudou and several hundred other children were given religious lessons by a man named Mal Moussa, who, Aboudou said, taught them that "if you kill an infidel, you will go to paradise."

Life on Boka was hard. Sometimes Aboudou's mother would try to talk to him about the abduction, but if someone else came near she quickly changed the subject. Most nights, his father disappeared, and he didn't know why. People who disobeyed orders were beheaded. Eventually, the island ran out of food, and they moved to another one, to harvest maize.

One day, airplanes came and bombed all the huts. A piece of shrapnel pierced Aboudou's shoulder. He had fifteen friends on the island, but when the attack was over all of them were dead. After that, his family fled to Chad, where they were detained by the military. He was much more afraid of the uniformed soldiers than he was of Boko Haram.

Aboudou's mother confirmed his account. When I asked whether her husband had participated in jihadi raids on the nights he disappeared, she said she didn't know. "He never told me what he did, and I never asked," she said. When I asked to speak with him, she said that it was impossible—he was on his way to the market. But he had brought Aboudou to me a few hours earlier, and now I saw him, about fifty feet away, staring at us from his hut.

The Chadian military didn't know what to do with the returning men. Many of them had received weapons training from Boko Haram, and some had carried out attacks. Méhaule advised the government to screen them, identify and prosecute perpetrators of crimes, and let the others go. "But the government had no capacity to do this," he told me. "It's expensive to feed three hundred people, so, in January, they just released them. All of them."

This year, several hundred more people have returned. "Those who had left to join Boko Haram learned that the humanitarian community is here, giving people food to eat, giving people money," Souapebe, the government official, told me. "That's why people started coming back." To encourage further defections, he said, "I buy phone credit for the local boys. Then they call their friends in Boko Haram and tell them, 'We're O.K. We have food. We have shelter. The humanitarians have given us blankets.'" He continued, "When someone is no longer hungry, he is no longer dangerous."

One of the boys who had voluntarily joined Boko Haram came back to Chad because, he told me, "Boko

Haram lied to us about the money. All I saw was poverty and death."

On the morning of July 22nd, we set off by boat in the direction of Médi Kouta. The chief of the island, a seventy-two-year-old Boudouma named Hassan Mbomi, met us at the shoreline and guided us uphill, through a grove of charred palm trees. He had returned to the island twenty days earlier, to try to grow millet, because he was starving on the mainland. About two hundred people had followed him. "When we got back, everything was burned," he said. "We have to build our village from scratch." A large group of men were waiting for us in a dusty clearing, but Mbomi said I couldn't speak to them. He said that they had been kidnapped by Boko Haram and forcibly conscripted into the jihad before escaping.

To comply with U.N. safety rules, we were accompanied into the islands by a Chadian soldier named Suliman. He seemed ill at ease on Médi Kouta, and the people there eyed him with suspicion. When we left the island, Suliman told me that he didn't accept the chief's explanation. "Sometimes they go away, sometimes they come back," he said. "But they are all complicit." Some jihadis have a branding on their back—a circle with a diagonal line through it—but, in most cases, "we can't distinguish who is Boko Haram and who isn't," Suliman said.

For two years, Suliman had been fighting in the islands. The Army had no boats. Sometimes his group commandeered fishermen's pirogues, and he had come to believe that many fishermen worked as spies, alerting Boko Haram to the military's movements. Like most soldiers, he grew up speaking Chadian Arabic, and cannot communicate with people in the Lake Region. We passed another island lined with burned palm trees. "The jihadis used to come to these islands at night, and we couldn't see them," Suliman said. "So we would light the trees on fire, so they wouldn't come back." He had torched the trees on Médi Kouta.

While I was in Baga Sola, six thousand people showed up near the Dar Nahim displacement site, a few miles from town. They belonged to a nomadic Arab community that has been in Chadian territory for hundreds of years, but they had just come from Niger, where some of them had fled during the Habré regime. Ordinarily, humanitarian workers would classify them as Chadian "returnees." But national governments get the final word on status, and Déby's regime insisted that they were Nigérien refugees, deflecting responsibility and costs to the U.N.

Because governments can decide where the U.N. operates within their national territory, humanitarians are routinely compelled to enter into morally fraught arrangements. In Syria, the U.N. is almost never allowed to deliver food or medicine to besieged civilians who oppose the regime. In Ethiopia, the government has spent the past several decades pressuring N.G.O.s into complying with its coverups of epidemics and a possible famine. In Chad, the International Committee of the Red Cross is the only international organization that has access to prisons. In order to maintain access, the I.C.R.C. keeps any atrocities it sees confidential.

But more common is the scenario in Lake Chad—in a neglected patch of territory, the international community ends up fulfilling the unwanted obligations of statehood. The regime reaps the benefit: the threats that arise from its failure to govern are mitigated, and its leader is left to focus on the task of strengthening the security apparatus that keeps him in power. As Linda Polman writes, in "The Crisis Caravan," from 2010, "If you use enough violence, aid will arrive, and if you use even more violence even more aid will arrive."

Once a humanitarian emergency has been stabilized, it is usually followed by extensive development projects, funded by international donors and institutions. "Baga Sola is three times bigger than it was when we all moved in," Méhaule said. "It's amazing how the money flowing in from humanitarian assistance has changed the city." Jihadi activity in areas susceptible to recruitment tends to attract the major Western donor countries, who see development as an instrument of stabilization. But, if Boko Haram were to suddenly disappear, the humanitarian emergency funds would be directed to other crises, in other desperate parts of the world. The people of the lake would be no less vulnerable to environmental degradation and all its consequences, but, as before, they would largely be on their own. Travelling through the Lake Region, I got the impression that almost everyone there—and especially those in the Presidential palace—has a stake in Boko Haram's continued existence as a distant, manageable threat.

Even so, the assistance has fallen short of the need. This year, the U.N. appealed for a hundred and twenty-one million dollars in aid for the Chadian side of the lake, but only a third of that has appeared. In the northern basin, I met a twenty-seven-year-old man who, three days earlier, had become so hungry that he walked six hours into contested territory in search of fish. Fifteen other men had gone with him; all of them were captured or killed by Boko Haram.

This past spring, Boudouma men started heading back to the islands in the southern basin, to plant millet and maize before the rainy season. Méhaule's team followed them. For two years, the Chadian Army had been telling the U.N. that the islands were empty and off limits, but now, he said, "we realized that there were about forty thousand people living there."

When Moussa Mainakinay, the chief of Bougourmi, went home, he could tell that the lake had receded several hundred feet since he left. There were more reeds and water lilies, and more mosquitoes. The lake water was so shallow and full of sediment that drinking it gave him a stomach ache. Boko Haram had taken the cows and the cooking pots, and either the jihadis or the military had burned all the huts. There was nothing left. But, on the mainland, many from Bougourmi had died of malnutrition, and Mainakinay was determined to bring the remaining people home.

After a few months, a community worker from Bougourmi named Bokoï Saleh arranged for a vaccine delivery to the island. The vaccines would inoculate children against measles, tetanus, polio, and tuberculosis. UNICEF paid for the transport logistics.

Saleh picked up the vaccines at the hospital in Bol and, with the help of a friend, hauled them into a pirogue. The load weighed more than six hundred pounds, split among several large coolers. The two men rowed the pirogue from the mainland to a nearby island called Yga, where they borrowed five donkeys to transport the vaccines to the other side of the island. The walk took forty minutes. At the shoreline, there were two hippopotamuses, making it too dangerous to leave. They spent the night there, sleeping on the ground.

Saleh got up at five o'clock in the morning. It was hot and humid, and he was worried that the coolers might not keep the vaccines fresh for much longer. The hippos were gone, but, during the night, a floating island of reeds and water lilies had blocked the area around his boat. Saleh and his friend started hacking through the reeds with machetes. But they were too thick, so the men abandoned the boat and started dragging the coolers across the top of the floating island. It took half an hour to go a couple of hundred feet, and they had to keep moving to prevent the vaccines from sinking. When they reached the other side, the water was up to their waists. They waved over another pirogue, loaded up the vaccines, and paddled the rest of the way to Bougourmi.

Saleh relayed this story to me later that day, outside a hut on Bougourmi. He was drooping with exhaustion. A few feet away, the vaccination campaign was in progress. At least a hundred women and children were sitting in the shade under the island's biggest tree, waiting their turn.

I asked the translator to tell Saleh that I admired what he had done. Saleh frowned. "If UNICEF had rented me a pirogue with a motor, I could have avoided the hippos and the reeds," he said. "It would have taken twenty minutes to get here from the mainland."

He was right. That's how UNICEF had brought me.

In N'Djamena, I came to know a Chadian spy who is close to the President. He sought me out, wanting to confide state secrets; I was nervous that it was a trap. But, at our third meeting, he described the structure of a group of military-intelligence agents whose job it is to spy not *for* the military but *on* the military—to look out for anyone plotting a coup. "Things are not good here," he said. "The soldiers are unhappy."

Déby's hold on power is reliant on Western support. The European security official told me that "Déby basically blackmails us, saying, 'If I fall, there's a direct line between ISIS in the north, Boko Haram in the south, and Al Qaeda in the west.'" In a bid to prove his worth to international backers, Déby has been renting out the Army to international coalitions, sending Chadians to fight with U.N. forces in Mali and the Central African Republic. But many of those soldiers have not been paid.

The Chadian economy has become so bad that, in the past year, Déby has repeatedly cut civil servants' salaries. This summer, he also cut the salaries of his troops, a move that was quickly followed by a carjacking and a spate of armed robberies and shootings in N'Djamena; one humanitarian worker's stolen handbag turned up on the M.N.J.T.F. base. Chadian infantrymen are now being paid around fifty-eight dollars a month. "The Chadian people are starving for food and for freedom," the spy said. "I work for the President, but my loyalty is to the Chadian people."

August 11th was Chad's fifty-seventh Independence Day, and there was a military parade in the Place de la Nation, a vast public square that rarely has any people in it. Déby didn't show up until three hours after the festivities began. The next time I met up with the Chadian spy, he said that Déby's military-intelligence officers had inspected every weapon on display before the President's arrival, to make sure that none of the guns had secretly been loaded.

Since the country's independence, the French government has routinely sponsored Chadian military officers to train or study in France—including Déby, during the

Habré era. Officially, the purpose of the training has been to help professionalize the Army. But it also seems like a kind of insurance policy, a guarantee that whichever officer leads the next successful rebellion will also have some loyalty to France. In recent months, Chadian rebels linked to Déby's nephew, Timane Erdimi—who led a failed rebellion in 2008, and has been living in exile ever since—have been massing near the Libyan border. "Everyone dreams of being the President," the spy told me. "I am just biding my time."

On most mornings in N'Djamena, French fighter jets roar out of the international airport, to bomb Al Qaeda militants in Mali. The United States has special-operations forces in most Sahelian states, including Chad. The crises of the Sahel have "so many variables that, even in the short term, we don't have a handle on things," an American military officer told me.

In recent months, I have asked many American diplomatic and military officials to define a coherent long-term strategy for the region, but none of them have been able to articulate more than a vague wish: that by improving local governments and institutions, encouraging democratic tendencies, and facilitating development, the international community can defeat terrorism. In Chad, the security-based approach mistakes the strengthening of Déby's regime for the stabilization of the Chadian state. The strategy is a paradox: in pursuing stability, it strengthens the autocrat, but, in strengthening the autocrat, it enables him to further abuse his position, exacerbating the conditions that lead people to take up arms.

As part of international antiterror partnerships, security forces are increasingly coming into contact with communities of people who cross international borders every day. Many who fall into this category are nomadic herders; their way of life is fundamentally at odds with the enforcement of legal boundaries, and they are indifferent to the existence of nation-states. If they are denied the freedom to move with the seasons, their cattle will die. In recent years, as the Sahelian climate has worsened, many herders who had bought weapons to protect their animals have turned to jihad.

It seems likely that, even if Boko Haram is defeated, the rationales for insurgent violence will broaden beyond religion. I asked the European security official whether he thought that, in the future, there will be terrorist groups in the Sahel that carry out attacks in the name of equality instead of jihad. He smiled, and said, "If you examine the lacquer on a wooden table—I think your question is, how thin is that lacquer?"

"Yes."

"I think it's pretty thin."

The Day the Great Plains Burned

Ian Frazier

November 5, 2018

Slapout, Oklahoma, at the intersection of a county road and a much used east-west state highway, has a population of five. The town's name used to be Nye, but in the nineteen-fifties its residents, who were then more numerous, changed it. Locals explain that there was a store in Nye where, if you asked the owner for a particular item, he often went to look for it and came back and said, "We're slap out of it!" How this inspired a name change no one knows. Now the store is gone, and the town consists of a single building that's a combination gas station, truck stop, convenience mart, café, and improvised community center. In the dark of early morning, it's jumping with truckers, oil-field workers, guys who drive the county road graders, and farmers who have been baling hay all night. A hand-lettered sign on the door reads "Please hang on to the door." This is so the howling prairie wind won't keep yanking it open and undoing the feeling of comfort inside.

Just to the south on the county road stands the Slapout firehouse, a metal building with three bay doors and six enormous fire trucks behind them. These vehicles, acquired from the military and the forest service, have been modified for prairie firefighting by the firemen themselves, all of whom are volunteers. Charlie Starbuck is the fire chief. He objects to being called Charles; it's Charlie on his birth certificate. Starbuck has a drooping, Emiliano Zapata mustache and green eyes and he wears overalls, end-of-the-nose spectacles, and a rumpled Army-fatigue hat. His father was a fire chief before him. On his muscled forearms are multiple reddish burns made by sparks from welding, a regular occupation at his ranch, not far from town, as well as at the firehouse.

On the morning of March 6, 2017, Starbuck's pager went off at ten-fifty-three and informed him that a grass fire was burning in the Mocane oil field, by County Road 141 in Beaver County, north of Slapout. Its cause was a downed power line. With three trucks and eight of his crew, he drove to the fire and saw that it had already blown up to a size where, given the conditions, he was going to need help. He called neighboring fire departments. Texas County, just to the west, sent trucks—"I will praise Texas County till the day I die," Starbuck says. He also called Mark Goeller, the director of Oklahoma Forestry Services, who, needing a name for the fire, used Starbuck's. Sometime afterward, Starbuck's sister in Virginia called him to ask about the Starbuck Fire she had heard mentioned on the news. That was the first he learned of his fame.

For weeks, the National Weather Service out of Norman, Oklahoma, Amarillo, Texas, and Dodge City, Kansas, had been sending alerts. The conditions were perfect for wildfires. There had been almost no precipitation for six months; before that, however, a lot of rain had fallen, and now the plentiful prairie grasses stood up tall and tinder-dry. On some days, like this one, the winds blew at fifty-plus miles an hour, while the humidity dipped down into the single digits. An ice storm in January had damaged scores of power lines, making them more vulnerable. Often, the Weather Service alerts are mainly precautionary. But on this day the south-central Great Plains did indeed catch fire. Huge wildfires spread over the Texas and Oklahoma panhandles and in western Kansas, with a smaller burn in Colorado.

This is a part of the world where extreme weather hangs out. Meteorologists refer to the prevailing late-winter "dry line," a phenomenon found almost nowhere else, which in this case is produced by hot, dry air from the Mexican plateau colliding with moist air moving up from the Gulf of Mexico. In March and April, if there's an incoming storm system as a trigger, the combination of wet and dry explodes above the plains. High winds generated by the dry line contributed to some of the dust storms of the nineteen-thirties. Plowed ground on the plains blew away during those years; later, the government encouraged agriculture that returned the land to grass. The big dust storms haven't reappeared since, but now when the winds come the grass holding the soil in place is sometimes thoroughly ready to burn.

A megafire is considered to be one that burns more than a hundred thousand acres. In Oklahoma, the total burned in the March 6th fires was seven hundred and eighty-one thousand acres. Moving northeast with the wind, the Starbuck Fire soon crossed into Kansas. Eventually, the Starbuck and other nearby fires would be given an official, bureaucratic handle, the Northwest Complex Fire, but, to the people most affected, the fires that burned six hundred and eight thousand

Kansas acres are still called by the name of a fire chief in Oklahoma. In the Texas Panhandle, fires burned four hundred and eighty-two thousand acres. Seven people are thought to have died in the March 6th fires. The expanse of burned land on the south-central Great Plains amounted to almost two million acres—roughly three thousand square miles. Rhode Island, that useful state for comparing geographic measure, covers about a thousand square miles of land. The March 6th fires burned an area about the size of three Rhode Islands.

Ashland, Kansas, is almost fifty miles from Slapout, if you follow a straight line across the prairie. As the county seat of Clark County, Ashland grew along the gentle valley of a creek, with ambitious streets as wide as big-city avenues. After going through a familiar Great Plains cycle of boom and not quite bust, the town today has a population of about eight hundred and fifty. Two water towers rise at either end of town; swallows swoop and dive around the tall, white grain elevators at Ashland Feed & Seed, by the train tracks at the end of Main Street. A twenty-foot-high bas-relief map on the front of the courthouse shows the county's historic sites, including the place where one of Coronado's men lost a bridle bit when they were looking for the rumored Cities of Gold in the vicinity in 1541.

The intricate, rusted, ancient object is on display in a glass case next to the district-court clerk's office.

Millie Fudge, Clark County's head of emergency-management operations, is in her late sixties. Mildred Barnes was her name growing up; she has lived her entire life in the western part of the state. Dark-eyed, with short, light-brown hair and black-rimmed glasses, she walks leaning forward a bit, as if successfully towing a great weight. Her twanging, slightly gravelly voice projects calm, and her silences have formidable presence. Everyday attire for her includes a dark-blue T-shirt, bluejeans, running shoes, and a camouflage holster on her belt containing a radio. From time to time the radio squawks, and she answers it. She works with the volunteer fire departments and the police, not only in Ashland but also in the county's two other towns, both of which are smaller. Her office is in the ambulance building, because she is also the E.M.S. director for the county.

Millie's husband, Gary Fudge, is a retired truck driver. The couple married in 1970 and moved to Ashland in 1979. They had three boys and a girl. Their third son, Brannon, born in 1981, suffered from a brain disease called Rasmussen's encephalitis, which caused him to have as many as three hundred seizures a day. When Brannon was seven, his parents took him to the

Johns Hopkins Hospital, in Baltimore, where Dr. Ben Carson (now the Secretary of Housing and Urban Development) performed a brain operation called a hemispherectomy, which cured the seizures and allowed the boy to function. Brannon learned how to talk and take care of himself, and did well until his late teens, when another operation was necessary. He graduated from high school in 2000, but because he required monitoring he stayed in Ashland and opened an ice-cream shop called Fudge Man's. "He and I ran it together," Millie says. "He loved people and they loved him. In fact, he enjoyed visiting with customers so much that he sometimes ignored business and I'd have to remind him about it."

In 2002, because of complications resulting from a hairline fracture in his skull, Brannon went to a hospital in Wichita, and while recovering from surgery he lapsed into a coma. After thirty-three days, his family made the decision to take him off life support. The experience led his mother into despair. She got through this period only by the grace of God, she says. Her family's ordeal left her with love for her Ashland neighbors for patronizing Brannon's shop and supporting her family and being with them as they grieved. In a way, Brannon was why she got into public service in the first place. She had signed up for her first E.M.S.

course back in 1983, partly so she would be able to take care of him and her other children.

On the morning of the fires, Millie was checking the Weather Service updates. At about eleven-thirty, she got a notification about the Starbuck Fire and a request for Clark County to send whoever was available down to Oklahoma as reinforcements. With her daughter, Brandy Fleming, who is the assistant emergency manager, she set out in the department's pickup, but before they'd gone twenty-five miles it became obvious that they would have their own problems closer to home.

They stopped near Englewood, two miles from the state line, the county's southern border. By that time, the fire was approaching a ranch owned by a man named Frosty Ediger, who has raised cattle and wheat there for fifty years. The Englewood volunteer firemen had already given up trying to stop the ranks of flames rolling across the prairie and were focussed on saving people and structures. An Englewood fireman saw Frosty Ediger on his tractor trying to plow a firebreak around his house. The fireman looked at the oncoming inferno and thought, That old man is going to die.

In conversation, people out here bring up God a lot, but that could be because a self-evident mighty power—the sky above—demands attention almost every day. At any minute, this humongous sky might smash you with hail or whirl you away in a tornado or bless you with rainbows and cloud-piercing sunbeams evocative of celestial choirs and the angels ascending and descending. Now, to the southwest, the gray and black smoke was boiling up toward altitudes where airplanes are tiny white X shapes with pipe-cleaner contrails. The smoke mounted in gray cumulus-like eruptions or redacted everything above the horizon line to black, while the underside of the billows glowed orange from the flames. Embers flew through the air, and the fierce heat added its own force to the wind, which blew with such a noise that people standing four feet apart had to shout to talk.

On a quirk of the wind, the fire jumped over Frosty Ediger and his house and outbuildings. In Englewood, directly in its path, some trees already blazed. The Englewood firemen retreated to defend the town. Millie saw the arm of flames sweeping north and realized that she and her team would have to manage the disaster from Ashland. They drove back on Highway 283, along which the telephone poles were soon burning. The Englewood firemen had advised her to go, but as they watched the flashing lights recede they felt completely alone.

Millie Fudge did, too. She had called the sheriff's office in Ashland and asked them to call Mutual Aid, an organization of neighboring counties that help one another in emergencies. But fires caused by sparking or downed power lines had broken out all over. When the sheriff's office called her back, they said that all trucks had gone to other fires and none were available. Millie then asked them to call counties farther out, but the same answer came back. Eleven separate fires burned in Kansas that day, though none the size of what was approaching Ashland. It hit her that Clark County's three volunteer fire departments (in Englewood, Ashland, and Minneola) would be defending their towns and county entirely on their own. For Millie, this was the most frightening moment.

Garth Gardiner, a rancher who raises cattle and quarter horses west of Ashland, watched through the window of his ranch's office as the smoke plumed skyward and figured the fire would miss him. He got in his pickup and drove west on a dirt road for a better look, then pulled over and gauged the smoke's distance—still pretty far off, he thought. Suddenly, a wall of flames came leaping over a ridge about three hundred yards away. "I saw it and hauled butt out of there," he said. Speeding back toward his house, he saw the smoke engulf his family's cattle operation, to the southwest, causing the photo-sensor floodlights above the corrals to turn on. Soon, only those lights, tiny pinpoints, were visible. Within the smoke's blackness, Garth's brothers, Mark and Greg, and Mark's wife, Eva, each escaped without knowing if the others had made it. Mark drove out on a lane so dark with smoke that he had to hold the truck door open so that he could follow the gravel road edge below him.

On folding tables in Ashland's ambulance garage, Millie Fudge set up the Emergency Operations Center for Clark County. She gave her daughter and two volunteer assistants the job of listening to all the police and fire dispatch calls and writing them down. Computer records would do for later; now, in the rush of events, she trusted paper. Other volunteers recorded the comings and goings of first responders on T-cards, which they inserted in a multi-pocket organizer that hung on the wall so the cards were visible at a glance. At three-eleven in the afternoon, Millie ordered the evacuation of the nursing home and the Ashland Health Center. She also ordered the evacuation of the entire town of Englewood and, about twenty minutes later, the evacuation of Ashland itself. Residents grabbed what they could and drove southeast, to Buffalo, Oklahoma, or east, to Protection, Kansas. (The town's name comes from its Republican founders, who espoused protective tariffs; it was later among the first towns in the United States to be inoculated with the Salk polio vaccine.)

Still the fire came on. Burning tumbleweeds flew forty feet above the ground, and the red cedars in the hollows roared as their resinous boughs ignited like kerosene. The wind swept up the dry grass until the air itself was on fire. Ashland's firefighters had never seen a blaze that could not be outflanked and subdued. "But what could you do against this monster?" Millie asked. Like the Englewood firemen, Ashland's tried to save structures and people. In outlying areas, they hosed houses with a flame-retardant foam. Some houses could not be saved. Here on the prairie, fires are fought from trucks, not on foot. Bumping over rough ground, the trucks threw the firemen around, banging them up and bruising them as burning sparks went down their necks. Several times, the fire's front line jumped over the trucks, and the firemen kept from burning by spraying a mist around themselves.

Cattle, for no known reason, sometimes ran into the flames. A man and his eleven-year-old son became separated while trying to move their cows to safety. The father, bouncing on an all-terrain vehicle, lost his cell phone; the son, driving a pickup, couldn't reach him and thought he had died. The son almost went back into the flames to try to save him; in another vehicle, his mother, who still had her phone and kept her head, insisted that the boy not do that. Horrible minutes passed before both father and son made it out of the smoke O.K.

Several ranchers set out to plow firebreaks, as Frosty Ediger had done. Mike Harden, a farmer and rancher, got his tractor and heavy-duty disk plow and began to plow all around Ashland. He disked along the state and county roads on both shoulders, and along fences, and all around the health center, on the town's west side, and around the southwest side of the high school, and around the house of his former math teacher, and around piles of hay bales. When a transmission hose on his tractor broke, he started blading dirt and making firebreaks with a road grader. Once, Harden graded so close to the flames that the dirt he threw put them out. He also bladed a firebreak around his tractor and his disk plow so they wouldn't burn. He drove the grader to his house, quickly ate a supper his wife had fixed for him, and went out again.

At about four in the afternoon, the wind shifted from out of the west or southwest to out of the north. David Redger, the Ashland fire chief, conferred with Millie, decided that Ashland was most vulnerable from the northwest, and sent trucks there. At Garth Gardiner's, the firemen told Gardiner they would try to save his house, and they did. It touched him to know that some of these men, neighbors of his, helped him when at the same moment they were losing property of their own. The fire department now had more reports of houses on fire than it had trucks. Nothing was burning yet in town. The last evacuees made it out on roads with fire so close on either side that it blistered the paint on their cars. By 6:48 p.m., fire surrounded Ashland on all sides, and it was impossible to enter or leave the town. Those evacuated to Buffalo and Protection had to be evacuated farther, to the towns of Woodward, Oklahoma, and Coldwater, Kansas. As the flames closed in, Millie considered moving the Emergency Operations Center to a field of new wheat on the town's north side, just past the golf course. Too green to burn, the field could provide a refuge.

In Englewood, Bernnie Smith, the fire chief, was looking for water. A fire truck had run over a hydrant, draining the town's water tower, and he couldn't pump from wells, because the electricity had gone out. Meanwhile, his daughter was trying to bring his wife, who suffers from severe asthma, out of the smoke and to a hospital. Without treatment, she could die. Smith told his daughter to drive her south, to Woodward; half an hour later, his daughter called back to report that fire was blocking the roads in that direction. He told her to try the town of Beaver, Oklahoma, northwest of there, but when she got to the clinic it didn't have the necessary medications. She turned south, toward Perryton, Texas, but another fire, later called the Perryton Fire—also caused by power-line sparks—ruled that out. Then he told her to go north again, to the town of Liberal, Kansas, even if she had to drive through fire to get there. She finally reached the Liberal hospital, and her mother received treatment.

On the front line in Oklahoma, Charlie Starbuck and his crew fought on. When the wind changed to out of the north, flames suddenly surrounded them on three sides. Starbuck always drives with both windows of the truck cab open so he will feel the same heat as those riding on the back. Now the encroaching flames leaped through the cab, in one window and out the other. Starbuck ducked. By misting around themselves and blasting the fire with their hoses, they reached safer ground upwind. "We almost got overrun," he told me later. "To this day, I'm amazed that we didn't end up going to a lot of firefighters' funerals."

In Texas, the wind shift led to three of the state's deaths. A rancher, a cowboy who worked for him, and the cowboy's girlfriend were trying to rescue cattle when flames from an unexpected direction suddenly caught up with them and brought down their horse and four-wheeler. The three burned to death as they tried to escape on foot.

As night fell in Ashland, the few people remaining in town looked out at flames in every direction. Not all were sure they would ever see their families again.

People give different explanations for what saved the town. Some point to the firebreaks that Mike Harden disked and graded. In places, you could see where the fire had come up to one of those and stopped. The wheat field where Millie had thought of moving the Emergency Operations Center seemed to have been crucial. An aerial photograph from a day or two later shows the soot black of the incinerated prairie meeting the spring green of the four-hundred-acre field in a straight, uncompromised line. The fire kept threatening the town into the next day, and Ashland's firefighters stayed on the job without sleeping, some of them for thirty or forty hours. Millie remained at her command post, getting only a few hours' rest during the same period. On nature's part, the humidity increased, and the wind died down at night.

The only fire fatality in the county (or the state) on March 6th occurred when a truck driver on Highway 34 tried to turn around, jackknifed his truck, got out of the cab, and died of smoke inhalation. His name was Corey P. Holt, and he came from Oklahoma City. In the low visibility, two cars then crashed into the truck; the cars' occupants were injured, but no one died. Englewood lost about a dozen houses, nine in the town itself, and an Englewood man whose house burned down died of a heart attack two weeks later.

Losing buildings and fences and vehicles and stored-up hay was bad, but the suffering of the cattle grieved the ranchers' souls. Thousands of cattle died in the fire, but thousands wretchedly survived—blinded, their ears gone, ear tags melted, udders burned off. Many had little hair left and their feet were burned so badly that they walked out of their hooves. Herds stood swaying slightly, moaning or mute in agony. Shooting cattle occupied the ashy days afterward. Ranchers whose guns and ammunition had burned up had to borrow them or ask neighbors to do the killing. Bulldozers and backhoes dug pits for mass burials.

Ashlanders said God had spared the town and its residents. Miraculous escapes were attributed to God's plan. Even the few older citizens who had been around for the Dust Bowl storms declared they'd never seen anything as awe-inspiring as this. Everybody said they hoped never to experience anything like it again.

But, in a sense, they soon did. Immediately after the fire—even as it still burned—unsolicited and generous aid started arriving from around the country. The outpouring amazed them even more than the fire had.

News outlets did not cover the prairie fires extensively, the way they do California fires. But rural America found out about the March 6th fires on social media and followed their progress in real time. That day was a Monday. By Wednesday, hay to feed the cattle now without pastures started to arrive. For weeks afterward, convoys of flatbeds loaded with large cylindrical hay bales, up to five thousand dollars' worth of hay per truck, all decked out with American flags and hand-lettered messages of support, rolled in, night and day. In Englewood, the fire department couldn't unload all the trucks at late hours and left a skid-loader and a sign by the firehouse asking the truckers to please unload their bales themselves. In the mornings, new piles of bales had appeared.

Replacing a mile of fence costs ten thousand dollars. The Gardiner ranch, for example, lost more than two hundred and seventy miles of fence. Trucks from Iowa and Michigan arrived with donated fenceposts, corner posts, and wire. Volunteer crews slept in the Ashland High School gymnasium and worked ten-hour days on fence lines. Kids from a college in Oregon spent their spring break pitching in. Cajun chefs from Louisiana arrived with food and mobile kitchens and served free meals. Another cook brought his own chuck wagon. Local residents' old friends, retired folks with extra time, came in motor homes and lived in them while helping to rebuild. Donors sent so much bottled water it would have been enough to put out the fire all by itself, people said. A young man from Ohio raised four thousand dollars in cash and drove out and gave it to the Ashland Volunteer Fire Department, according to the *Clark County Gazette.* The young man said that God had told him to; the fireman who accepted the donation said that four thousand was exactly what it was going to cost to repair the transmission of a truck that had failed in the fire, and both he and the young man cried.

Farm and ranch organizations and an association of rodeo cowboys gave tens of thousands of dollars to fire sufferers. Residents of Ashland, who had farsightedly established their own 501(c)3 foundation several years earlier, could accept the donations and distribute them without having to route them through another nonprofit, such as the Red Cross. The president of the Stockgrowers State Bank, Kendal Kay, who is also the town's mayor, offered low-interest loans so ranchers could re-start their operations. But, as one rancher noted, all those who rebuilt acquired extra zeros on their debt line.

Even more contributions arrived: newborn-calf formula, veterinary medications, protein cake for cattle, winter clothes, frozen casseroles with scriptural messages taped to them, a shipment of cheese curds from Wisconsin, and more hay, of all kinds, in bales whose quantities had never been seen in the region before. After the 24/7 task of managing the hundreds of fire trucks and crews that showed up in the days after the fires, Millie Fudge turned to making note of all the donations, so that people could be acknowledged and thanked. Kansas's governor, Sam Brownback, made a visit; nobody chided him for the state's recent to-the-bone budget cuts. The fire had moved so fast that no agency outside Clark County could have done much anyway, Millie believed. In fact, the state's Incident Management Team had showed up on March 7th and provided excellent practical assistance, she pointed out (although the county had been careful to retain local control).

A young woman whose family's ranch houses had burned told a livestock-association meeting in Wichita, "The government didn't help us, but America did." From the point of view of Clark County as a whole, the government did play a part: the National Weather Service sent warnings about the wind shift; the states of Colorado, Nebraska, and South Dakota dispatched firefighting teams and aircraft to help squelch what was left of the fire; Ashland used a FEMA grant to buy two of its fire trucks; and the Department of Agriculture later provided ranchers with payments of up to a hundred and twenty-five thousand dollars for livestock losses and up to two hundred thousand dollars for fencing. And, of course, Millie and her team, as county officials, were part of "the government" themselves. None of that, however, was to the point, which was the unexpected non-governmental love that the fire sufferers felt coming at them from around the country as if out of nowhere.

A gun-store owner in Oklahoma who raffled off a 9-mm. pistol, an antelope hunt, and a custom rifle explained, "They're all my people." A stranger sent a cash donation and a handwritten note directly to Garth Gardiner after reading about him in the news. She said she wanted to do this for him even though he was a rancher and she was a vegetarian. "Our country's in a pretty turbulent political situation nowadays, but people are still good," Gardiner told me.

One afternoon last summer, I talked with Bernnie Smith in the shade behind the Englewood firehouse. He rolled some office chairs out the back door for us, because the metal building was stifling. Smith is a compact, green-eyed man with a level gaze and a quiet demeanor, and he wore a Western shirt with mother-of-pearl buttons, bluejeans, square-toed cowboy boots, and spurs. My call had interrupted his workday; he'd been moving cattle on his ranch. His horse stood in a stock trailer nearby, behind his club-cab pickup.

Smith said that dealing with the emotions that the firefighters went through after the fire had been kind of a P.T.S.D. experience, because, honestly, they thought the fire had kicked their ass. They held meetings after it and talked about it and sometimes cried, and invited post-stress counsellors a time or two. The guys on his crew range in age from nineteen to over seventy, and they stayed on the fire for, in some cases, two days straight. He worried about one young guy because Smith thought his eyes had been burned, but he turned out to be O.K. The generosity of people in the aftermath had meant an enormous amount. "That hay movement after the fire is the greatest thing that's happened in America in my lifetime," he said. He still loves to look at a pasture full of good grass, but he remarked, "It's pretty until it burns," adding, "That grass is your livelihood and a hazard at the same time." He and I talked for hours, until his horse was whinnying with impatience and kicking the slats of the trailer. Smith never mentioned that while he fought to save lives and houses in Englewood the fire had burned up a third of his cattle herd.

I learned that fact later, from Cara Vanderree, the librarian in Ashland. She is originally from Comanche County, next door, and her family has been in the state for generations. The Ashland Library is the best small library in Kansas, in the judgment of the Kansas Library Association, which recently gave it an award. Vanderree speaks in a sweet, soft voice. A lot of her job, especially in summer, consists of superintending kids who want to use the library's computers. With sweetness and patience, she requires that if a child wants to play on the computer for half an hour the child must read a book for half an hour, and she watches to make sure the child actually reads. Sometimes she decides a difficult child must go, and she says, "Darlin', I can feel your home callin' you."

Before the fire, Vanderree did some archival work for the Kansas Humanities Council, transcribing recorded recollections. After the fire, when everyone was going through a period of talking about it, trying to come to terms with it—a period that has not yet ended—Vanderree got the idea of recording as many survivors of the fire as wanted to participate. Then she would put the recordings on the library's Web site, and make the transcriptions into a book. The Kansas Humanities Council liked the idea and gave her a grant, as did the Kansas Health Foundation. With the assistance of Diana Redger, who has a master's in history and is also the Ashland fire chief's sister, the library has interviewed sixty-nine people.

Vanderree had thought she could ease the chore of transcribing by downloading an app she bought online. She soon discovered that the app couldn't understand a Kansas-Oklahoma accent. When a speaker said the word "town," what the computer somehow heard was "Tehran." The repeated appearance of the Iranian capital in the first-person accounts of a Kansas wildfire gave the project a certain international flavor, but these and many other mistakes became a pain to deal with. She and two assistants went back to doing the transcriptions themselves.

Redger conducted most of the interviews, and at the end of each one she asked the interviewees if anything good had come from the fire. They replied that it made them know their neighbors better, it drew people closer together, and it strengthened the town. I know all these answers are true. If you drive on the plains a lot, you see towns in decline: store windows boarded up on Main Street, houses becoming run-down, local schools closed. Our rural places are emptying out. The frame of a sign with no sign in it could be an emblem for much of small-town America, not only on the plains. Sometimes a town will even lament its fate publicly: "Pray for Fowler," read a sign I drove past in Fowler, Kansas.

Ashland used to have dozens of businesses, a passenger-railroad station on the Santa Fe line, and a movie theatre that showed films every day. All are now gone; but the town nonetheless continues, with the courthouse, an office of the U.S. Department of Agriculture, a public swimming pool, a lumberyard, a good restaurant, the grain elevators, a motel, two bed-and-breakfasts, a public school for K-12, four churches, two banks, the health center, a veterinary clinic, and the library. The March 6th fires affected other towns, but I know of none that did a local-history project about the experience. Ashland's collection of fire stories is like the town's immune system kicking in.

The prairie greened up quickly after the fires, and in some places the grass came back better than before. Then, on March 5, 2018—almost exactly a year after the fires—a wind-driven fire broke out just north of Ashland. Incredulous, the fire department pounced on the blaze, put it out, and watched indignantly lest it make another peep, which it did not. No major fires have threatened Ashland this year. In April, however, big fires burned again in Oklahoma. One that officials called the Rhea Fire started near the town of

Seiling and consumed about three hundred thousand acres, with two deaths. I drove through the area just afterward. Wind turbines in great and stately numbers populate the prairie there—Oklahoma gets more than a quarter of its electricity from wind-generated power. The blades were turning slowly overhead as ashes drifted in the air and made the whole landscape look smudged and blurry. Black skeletons of scrubby trees stretched to the horizon south of Seiling; in other places, the flames had turned open prairie into Sahara-like dunes dotted with spiky black sagebrush stumps.

Since 2005, the prairie states have seen a lot of fires. Several that burned in 2016 now look like preludes to the giants of 2017; other record-setting fires had preceded them. The biggest prairie fire in Texas history burned about a million acres on the Panhandle in 2006. Deke Arndt, a meteorologist and a climate scientist for the National Oceanic and Atmospheric Administration, has written on NOAA's Web site about prairie fires. I called him up and asked if he saw the fires as part of a pattern related to climate change.

Arndt grew up in Oklahoma, and his mother still lives there, so he understands the immediacy of weather in people's lives on the plains. He thinks that may be why Oklahoma produces so many weather scientists.

Arndt explained about the dry line and said that it has been creating wild weather in the region since before humans lived there. But the weather and the climate are two different things, he added (as meteorologists often do). Extreme weather has always occurred on the plains. What is new, and derives from climate change, is that the atmosphere has become hotter and wetter, bringing more rain, causing wetter years (2016, for example), which produce more fuel in the form of grass. As the atmosphere warms, it is also thirstier, so that when dry periods come the air sucks more moisture from the soil and the plants and makes the land more susceptible to fire. We may be witnessing a slow process of desertification in drier parts of the region, but nearly half the people in Arndt's native state doubt that climate change is real.

No one I talked to in Kansas told me that he believed in climate change. Prevailing opinion holds that nothing about the recent extreme weather here is much different from what's always been. People say that Native Americans sometimes used prairie fires as an environmental tool. Some claim that government policy is partly to blame for the recent fires, and often single out the C.R.P.—the Conservation Reserve Program, a federal initiative that, since the nineteen-eighties, has paid farmers to replant and maintain grass cover on their land. The argument resembles the one applied to the long-standing policy of fire suppression in forests: that is, fire suppression and the Conservation Reserve Program both create dangerous buildups of fuel.

As Adam Elliott, who administers the program in Clark County from the U.S.D.A. office in Ashland, told me, the C.R.P. was intended to be a continuation of government efforts to put marginal land back into grass so that it wouldn't blow. A subsequent increase in deer and game-bird populations has become another justification for it. (That increase is evident on summer mornings, even in the middle of Ashland, when you awake to the dulcet "bobwhite" call of the quail.) People say that the fires resulted from a unique combination of a rainy period followed by a drought, with the land made more flammable by the C.R.P. Evidently the government agreed with this analysis, because soon after the 2017 fires the U.S.D.A. expanded the time frame for grazing on otherwise closed C.R.P. land.

When I asked Millie Fudge her opinion about climate change and the fires, she deliberated so long before answering that I thought I'd offended her. Then she said, "I'm not knowledgeable about that. Climate change, if it exists, might have something to do with the fires. But, whether it does or not, I know God is in control. He allows or causes the increase in fires to happen for a reason."

Again she fell silent. "What is the reason?" I asked.

More silence. Then: "The fires are a wake-up call. They will get worse. We humans think we are in charge. We think we are indispensable to God, but he is showing us that he is in control. He is telling us that we need to find God."

When I talked to Mike Harden, the man who plowed and bladed the firebreaks, he also said he did not know if climate change existed. "But I'll tell you, it's never boring trying to raise cattle or crops out here," he went on. "My great-grandparents came out and homesteaded in the eighteen-nineties, and we've seen everything. Floods. Hailstorms. Grasshoppers. Ice storms. Tornadoes. Dust and more dust. And now these fires. Living here takes it out of you, but every year is a little bit different. Whatever the weather is going to do in the future, that's not up to me. All I know is that the Bible says man will eat his bread in the sweat of his face, and that's certainly true if you ranch or farm in western Kansas."

I drove hundreds of miles trying to make geographic sense of the fires. After going back and forth between Oklahoma and Kansas several times, I noticed the name of the river I crossed in Oklahoma near the border: the Cimarron. How could it have taken so long for me to notice the Cimarron River? On the whole Great Plains—in all the West, for my money—no other river name coincides with Western myth so closely or so lyrically. Almost ninety years ago, Edna Ferber wrote a novel, "Cimarron," about the opening of the Oklahoma Territory. The book was made into a movie, starring Richard Dix, which won the Academy Award for Best Picture, in 1931. That movie, in turn, was remade into another "Cimarron," which starred Glenn Ford, in 1960. There've been any number of other movies, TV shows, songs, and albums with "Cimarron" in their titles. It comes from a Spanish word that means, in this context, a runaway horse that lives in wild places.

For years, the Oklahoma Panhandle was No Man's Land, a refuge for outlaws. For complicated reasons, no state had jurisdiction there. A branch of the Chisholm Trail, which the cattle herds followed northward from Texas after the Civil War, crossed the Cimarron River not far from present-day Highway 283. Englewood, just north of the river in Kansas, where law existed, offered entertainment and hotels and a railroad connection for cowboys riding the trail. Outlaws used to commit crimes in Kansas and then escape back across the Cimarron to relative safety. Western-style shootouts, where the bad guy drew on the lawman and the lawman shot the bad guy, occurred on the dusty streets of Englewood. From the even more civilized town of Ashland, marshals set out across the Cimarron to catch and bring back wanted men for trial.

In wet periods, the Cimarron runs at about the volume of a respectable creek back East. I stopped and watched its clear, buckskin-colored water flowing through the willows and the red cedars. A few cottonwoods held up their blackened branches, but otherwise you'd never know that fire had recently raced along this valley. If I were younger, I would have swooned farther back into lonesome-cowboy fantasies, into all the "Cimarron"s of my childhood. But that Wild West past happened to other people. We are of a different time and place—on our own, like the Ashland and Englewood firefighters in the firestorm. As I looked at the Cimarron River, my thoughts were of the present and the soon to come.

Life on a Shrinking Planet

Bill McKibben

November 26, 2018

Thirty years ago, this magazine published "The End of Nature," a long article about what we then called the greenhouse effect. I was in my twenties when I wrote it, and out on an intellectual limb: climate science was still young. But the data were persuasive, and freighted with sadness. We were spewing so much carbon into the atmosphere that nature was no longer a force beyond our influence—and humanity, with its capacity for industry and heedlessness, had come to affect every cubic metre of the planet's air, every inch of its surface, every drop of its water. Scientists underlined this notion a decade later when they began referring to our era as the Anthropocene, the world made by man.

I was frightened by my reporting, but, at the time, it seemed likely that we'd try as a society to prevent the worst from happening. In 1988, George H. W. Bush, running for President, promised that he would fight "the greenhouse effect with the White House effect." He did not, nor did his successors, nor did their peers in seats of power around the world, and so in the intervening decades what was a theoretical threat has become a fierce daily reality. As this essay goes to press, California is ablaze. A big fire near Los Angeles forced the evacuation of Malibu, and an even larger fire, in the Sierra Nevada foothills, has become the most destructive in California's history. After a summer of unprecedented high temperatures and a fall "rainy season" with less than half the usual precipitation, the northern firestorm turned a city called Paradise into an inferno within an hour, razing more than ten thousand buildings and killing at least sixty-three people; more than six hundred others are missing. The authorities brought in cadaver dogs, a lab to match evacuees' DNA with swabs taken from the dead, and anthropologists from California State University at Chico to advise on how to identify bodies from charred bone fragments.

For the past few years, a tide of optimistic thinking has held that conditions for human beings around the globe have been improving. Wars are scarcer, poverty and hunger are less severe, and there are better prospects for wide-scale literacy and education. But there are newer signs that human progress has begun to flag. In the face of our environmental deterioration, it's now reasonable to ask whether the human game has begun to falter—perhaps even to play itself out. Late in 2017, a United Nations agency announced that the number of chronically malnourished people in the world, after a decade of decline, had started to grow again—by thirty-eight million, to a total of eight hundred and fifteen million, "largely due to the proliferation of violent conflicts and climate-related shocks." In June, 2018, the Food and Agriculture Organization of the U.N. found that child labor, after years of falling, was growing, "driven in part by an increase in conflicts and climate-induced disasters."

In 2015, at the U.N. Climate Change Conference in Paris, the world's governments, noting that the earth has so far warmed a little more than one degree Celsius above pre-industrial levels, set a goal of holding the increase this century to 1.5 degrees Celsius (2.7 degrees Fahrenheit), with a fallback target of two degrees (3.6 degrees Fahrenheit). This past October, the U.N.'s Intergovernmental Panel on Climate Change published a special report stating that global warming "is likely to reach 1.5 C between 2030 and 2052 if it continues to increase at the current rate." We will have drawn a line in the sand and then watched a rising tide erase it. The report did not mention that, in Paris, countries' initial pledges would cut emissions only enough to limit warming to 3.5 degrees Celsius (about 6.3 degrees Fahrenheit) by the end of the century, a scale and pace of change so profound as to call into question whether our current societies could survive it.

Scientists have warned for decades that climate change would lead to extreme weather. Shortly before the I.P.C.C. report was published, Hurricane Michael, the strongest hurricane ever to hit the Florida Panhandle, inflicted thirty billion dollars' worth of material damage and killed forty-five people. President Trump, who has argued that global warming is "a total, and very expensive, hoax," visited Florida to survey the wreckage, but told reporters that the storm had not caused him to rethink his decision to withdraw the U.S. from the Paris climate accords. He expressed no interest in the I.P.C.C. report beyond asking "who drew it." (The answer is ninety-one researchers from forty countries.) He later claimed that his "natural instinct" for science made him confident that the climate would soon "change back." A month later, Trump blamed the fires in California on "gross mismanagement of forests."

Human beings have always experienced wars and truces, crashes and recoveries, famines and terrorism. We've endured tyrants and outlasted perverse ideologies. Climate change is different. As a team of scientists recently pointed out in the journal *Nature Climate Change*, the physical shifts we're inflicting on the planet will "extend longer than the entire history of human civilization thus far."

The poorest and most vulnerable will pay the highest price. But already, even in the most affluent areas, many of us hesitate to walk across a grassy meadow because of the proliferation of ticks bearing Lyme disease which have come with the hot weather; we have found ourselves unable to swim off beaches, because jellyfish, which thrive as warming seas kill off other marine life, have taken over the water. The planet's diameter will remain eight thousand miles, and its surface will still cover two hundred million square miles. But the earth, for humans, has begun to shrink, under our feet and in our minds.

"Climate change," like "urban sprawl" or "gun violence," has become such a familiar term that we tend to read past it. But exactly what we've been up to should fill us with awe. During the past two hundred years, we have burned immense quantities of coal and gas and oil—in car motors, basement furnaces, power plants, steel mills—and, as we have done so, carbon atoms have combined with oxygen atoms in the air to produce carbon dioxide. This, along with other gases like methane, has trapped heat that would otherwise have radiated back out to space.

There are at least four other episodes in the earth's half-billion-year history of animal life when CO_2 has poured into the atmosphere in greater volumes, but perhaps never at greater speeds. Even at the end of the Permian Age, when huge injections of CO_2 from volcanoes burning through coal deposits culminated in "The Great Dying," the CO_2 content of the atmosphere grew at perhaps a tenth of the current pace. Two centuries ago, the concentration of CO_2 in the atmosphere was two hundred and seventy-five parts per million; it has now topped four hundred parts per million and is rising more than two parts per million each year. The extra heat that we trap near the planet every day is equivalent to the heat from four hundred thousand bombs the size of the one that was dropped on Hiroshima.

As a result, in the past thirty years we've seen all twenty of the hottest years ever recorded. The melting of ice caps and glaciers and the rising levels of our oceans and seas, initially predicted for the end of the century, have occurred decades early. "I've never been at . . . a climate conference where people say 'that happened slower than I thought it would,'" Christina Hulbe, a New Zealand climatologist, told a reporter for *Grist* last year. This past May, a team of scientists from the University of Illinois reported that there was a thirty-five-per-cent chance that, because of unexpectedly high economic growth rates, the U.N.'s "worst-case scenario" for global warming was too optimistic. "We are now truly in uncharted territory," David Carlson, the former director of the World Meteorological Organization's climate-research division, said in the spring of 2017, after data showed that the previous year had broken global heat records.

We are off the literal charts as well. In August, I visited Greenland, where, one day, with a small group of scientists and activists, I took a boat from the village of Narsaq to a glacier on a nearby fjord. As we made our way across a broad bay, I glanced up at the electronic chart above the captain's wheel, where a blinking icon showed that we were a mile inland. The captain explained that the chart was from five years ago, when the water around us was still ice. The American glaciologist Jason Box, who organized the trip, chose our landing site. "We called this place the Eagle Glacier because of its shape," he said. The name, too, was five years old. "The head and the wings of the bird have melted away. I don't know what we should call it now, but the eagle is dead."

There were two poets among the crew, Aka Niviana, who is Greenlandic, and Kathy Jetnil-Kijiner, from the low-lying Marshall Islands, in the Pacific, where "king tides" recently washed through living rooms and unearthed graveyards. A small lens of fresh water has supported life on the Marshall Islands' atolls for millennia, but, as salt water intrudes, breadfruit trees and banana palms wilt and die. As the Greenlandic ice we were gazing at continues to melt, the water will drown Jetnil-Kijiner's homeland. About a third of the carbon responsible for these changes has come from the United States.

A few days after the boat trip, the two poets and I accompanied the scientists to another fjord, where they needed to change the memory card on a camera that tracks the retreat of the ice sheet. As we took off for the flight home over the snout of a giant glacier, an eight-story chunk calved off the face and crashed into the ocean. I'd never seen anything quite like it for sheer power—the waves rose twenty feet as it plunged into the dark water. You could imagine the same waves washing through the Marshalls. You could almost sense the ice elevating the ocean by a sliver—along the seafront in Mumbai, which already floods on a stormy day, and at the Battery in Manhattan, where the sea-wall rises just a few feet above the water.

When I say the world has begun to shrink, this is what I mean. Until now, human beings have been spreading, from our beginnings in Africa, out across the globe—slowly at first, and then much faster. But a period of contraction is setting in as we lose parts of the habitable earth. Sometimes our retreat will be hasty and violent; the effort to evacuate the blazing California towns along narrow roads was so chaotic that many people died in their cars. But most of the pull-back will be slower, starting along the world's coast-lines. Each year, another twenty-four thousand people abandon Vietnam's sublimely fertile Mekong Delta as crop fields are polluted with salt. As sea ice melts along the Alaskan coast, there is nothing to protect towns, cities, and native villages from the waves. In Mexico Beach, Florida, which was all but eradicated by Hurricane Michael, a resident told the Washington *Post*, "The older people can't rebuild; it's too late in their lives. Who is going to be left? Who is going to care?"

In one week at the end of last year, I read accounts from Louisiana, where government officials were finalizing a plan to relocate thousands of people threatened by the rising Gulf ("Not everybody is going to live where they are now and continue their way of life, and that is a terrible, and emotional, reality to face," one state official said); from Hawaii, where, according to a new study, thirty-eight miles of coastal roads will become impassable in the next few decades; and from Jakarta, a city with a population of ten million, where a rising Java Sea had flooded the streets. In the first days of 2018, a nor'easter flooded downtown Boston; dumpsters and cars floated through the financial district. "If anyone wants to question global warming, just see where the flood zones are," Marty Walsh, the mayor of Boston, told reporters. "Some of those zones did not flood thirty years ago."

According to a study from the United Kingdom's National Oceanography Centre last summer, the damage caused by rising sea levels will cost the world as much as fourteen trillion dollars a year by 2100, if the U.N. targets aren't met. "Like it or not, we will retreat from most of the world's non-urban shorelines in the not very distant future," Orrin Pilkey, an expert on sea levels at Duke University, wrote in his book "Retreat from a Rising Sea." "We can plan now and retreat in a strategic and calculated fashion, or we can worry about it later and retreat in tactical disarray in response to devastating storms. In other words, we can walk away methodically, or we can flee in panic."

But it's not clear where to go. As with the rising seas, rising temperatures have begun to narrow the margins of our inhabitation, this time in the hot continental interiors. Nine of the ten deadliest heat waves in human history have occurred since 2000. In India, the rise in temperature since 1960 (about one degree Fahrenheit) has increased the chance of mass heat-related deaths by a hundred and fifty per cent. The summer of 2018 was the hottest ever measured in certain areas. For a couple of days in June, temperatures in cities in Pakistan and Iran peaked at slightly above a hundred and twenty-nine degrees Fahrenheit, the highest reliably recorded temperatures ever measured. The same heat wave, nearer the shore of the Persian Gulf and the Gulf of Oman, combined triple-digit temperatures with soaring humidity levels to produce a heat index of more than a hundred and forty degrees Fahrenheit. June 26th was the warmest night in history, with the mercury in one Omani city remaining above a hundred and nine degrees Fahrenheit until morning. In July, a heat wave in Montreal killed more than seventy people, and Death Valley, which often sets American records, registered the hottest month ever seen on our planet.

Africa recorded its highest temperature in June, the Korean Peninsula in July, and Europe in August. The *Times* reported that, in Algeria, employees at a petroleum plant walked off the job as the temperature neared a hundred and twenty-four degrees. "We couldn't keep up," one worker told the reporter. "It was impossible to do the work."

This was no illusion; some of the world is becoming too hot for humans. According to the National Oceanic and Atmospheric Administration, increased heat and humidity have reduced the amount of work people can do outdoors by ten per cent, a figure that is predicted to double by 2050. About a decade ago, Australian and American researchers, setting out to determine the highest survivable so-called "wet-bulb" temperature, concluded that when temperatures passed thirty-five degrees Celsius (ninety-five degrees Fahrenheit) and the humidity was higher than ninety per cent, even in "well-ventilated shaded conditions," sweating slows down, and humans can survive only "for a few hours, the exact length of time being determined by individual physiology."

As the planet warms, a crescent-shaped area encompassing parts of India, Pakistan, Bangladesh, and the North China Plain, where about 1.5 billion people (a fifth of humanity) live, is at high risk of such temperatures in the next half century. Across this belt, extreme heat waves that currently happen once every generation could, by the end of the century, become "annual events with temperatures close to the threshold for several weeks each year, which could lead to famine and mass migration." By 2070, tropical regions that now get one day of truly oppressive humid heat a year can expect between a hundred and two hundred and fifty days, if the current levels of greenhouse-gas emissions continue. According to Radley Horton, a climate scientist at the Lamont-Doherty Earth Observatory, most people would "run into terrible problems" before then. The effects, he added, will be "transformative for all areas of human endeavor—economy, agriculture, military, recreation."

Humans share the planet with many other creatures, of course. We have already managed to kill off sixty per cent of the world's wildlife since 1970 by destroying their habitats, and now higher temperatures are starting to take their toll. A new study found that peak-dwelling birds were going extinct; as temperatures climb, the birds can no longer find relief on higher terrain. Coral reefs, rich in biodiversity, may soon be a tenth of their current size.

As some people flee humidity and rising sea levels, others will be forced to relocate in order to find enough water to survive. In late 2017, a study led by Manoj Joshi, of the University of East Anglia, found that, by 2050, if temperatures rise by two degrees a quarter of the earth will experience serious drought and desertification. The early signs are clear: São Paulo came within days of running out of water last year, as did Cape Town this spring. In the fall, a record drought in Germany lowered the level of the Elbe to below twenty inches and reduced the corn harvest by forty per cent. The Potsdam Institute for Climate Impact Research concluded in a recent study that, as the number of days that reach eighty-six degrees Fahrenheit or higher increases, corn and soybean yields across the U.S. grain belt could fall by between twenty-two and forty-nine per cent. We've already overpumped the aquifers that lie beneath most of the world's breadbaskets; without the means to irrigate, we may encounter a repeat of the nineteen-thirties, when droughts and deep plowing led to the Dust Bowl—this time with no way of fixing the problem. Back then, the Okies fled to California, but California is no longer a green oasis. A hundred million trees died in the record drought that gripped the Golden State for much of this decade. The dead limbs helped spread the waves of fire, as scientists earlier this year warned that they could.

Thirty years ago, some believed that warmer temperatures would expand the field of play, turning the Arctic into the new Midwest. As Rex Tillerson, then the C.E.O. of Exxon, cheerfully put it in 2012, "Changes to weather patterns that move crop production areas around—we'll adapt to that." But there is no rich topsoil in the far North; instead, the ground is underlaid with permafrost, which can be found beneath a fifth of the Northern Hemisphere. As the permafrost melts, it releases more carbon into the atmosphere. The thawing layer cracks roads, tilts houses, and uproots trees to create what scientists call "drunken forests." Ninety scientists who released a joint report in 2017 concluded that economic losses from a warming Arctic could approach ninety trillion dollars in the course of the century, considerably outweighing whatever savings may have resulted from shorter shipping routes as the Northwest Passage unfreezes.

Churchill, Manitoba, on the edge of the Hudson Bay, in Canada, is connected to the rest of the country by a single rail line. In the spring of 2017, record floods washed away much of the track. OmniTrax, which owns the line, tried to cancel its contract with the government, declaring what lawyers call a "force majeure," an unforeseen event beyond its responsibility. "To fix things in this era of climate change—

well, it's fixed, but you don't count on it being the fix forever," an engineer for the company explained at a media briefing in July. This summer, the Canadian government reopened the rail at a cost of a hundred and seventeen million dollars—about a hundred and ninety thousand dollars per Churchill resident. There is no reason to think the fix will last, and every reason to believe that our world will keep contracting.

All this has played out more or less as scientists warned, albeit faster. What has defied expectations is the slowness of the response. The climatologist James Hansen testified before Congress about the dangers of human-caused climate change thirty years ago. Since then, carbon emissions have increased with each year except 2009 (the height of the global recession) and the newest data show that 2018 will set another record. Simple inertia and the human tendency to prioritize short-term gains have played a role, but the fossil-fuel industry's contribution has been by far the most damaging. Alex Steffen, an environmental writer, coined the term "predatory delay" to describe "the blocking or slowing of needed change, in order to make money off unsustainable, unjust systems in the meantime." The behavior of the oil companies, which have pulled off perhaps the most consequential deception in mankind's history, is a prime example.

As journalists at InsideClimate News and the Los Angeles *Times* have revealed since 2015, Exxon, the world's largest oil company, understood that its product was contributing to climate change a decade before Hansen testified. In July, 1977, James F. Black, one of Exxon's senior scientists, addressed many of the company's top leaders in New York, explaining the earliest research on the greenhouse effect. "There is general scientific agreement that the most likely manner in which mankind is influencing the global climate is through carbon-dioxide release from the burning of fossil fuels," he said, according to a written version of the speech which was later recorded, and which was obtained by InsideClimate News. In 1978, speaking to the company's executives, Black estimated that a doubling of the carbon-dioxide concentration in the atmosphere would increase average global temperatures by between two and three degrees Celsius (5.4 degrees Fahrenheit), and as much as ten degrees Celsius (eighteen degrees Fahrenheit) at the poles.

Exxon spent millions of dollars researching the problem. It outfitted an oil tanker, the Esso Atlantic, with CO_2 detectors to measure how fast the oceans could absorb excess carbon, and hired mathematicians to build sophisticated climate models. By 1982, they had concluded that even the company's earlier estimates were probably too low. In a private corporate primer, they wrote that heading off global warming and "potentially catastrophic events" would "require major reductions in fossil fuel combustion."

An investigation by the L.A. *Times* revealed that Exxon executives took these warnings seriously. Ken Croasdale, a senior researcher for the company's Canadian subsidiary, led a team that investigated the positive and negative effects of warming on Exxon's Arctic operations. In 1991, he found that greenhouse gases were rising due to the burning of fossil fuels. "Nobody disputes this fact," he said. The following year, he wrote that "global warming can only help lower exploration and development costs" in the Beaufort Sea. Drilling season in the Arctic, he correctly predicted, would increase from two months to as many as five months. At the same time, he said, the rise in the sea level could threaten onshore infrastructure and create bigger waves that would damage offshore drilling structures. Thawing permafrost could make the earth buckle and slide under buildings and pipelines. As a result of these findings, Exxon and other major oil companies began laying plans to move into the Arctic, and started to build their new drilling platforms with higher decks, to compensate for the anticipated rises in sea level.

The implications of the exposés were startling. Not only did Exxon and other companies know that scientists like Hansen were right; they used his NASA climate models to figure out how low their drilling costs in the Arctic would eventually fall. Had Exxon and its peers passed on what they knew to the public, geological history would look very different today. The problem of climate change would not be solved, but the crisis would, most likely, now be receding. In 1989, an international ban on chlorine-containing man-made chemicals that had been eroding the earth's ozone layer went into effect. Last month, researchers reported that the ozone layer was on track to fully heal by 2060. But that was a relatively easy fight, because the chemicals in question were not central to the world's economy, and the manufacturers had readily available substitutes to sell. In the case of global warming, the culprit is fossil fuel, the most lucrative commodity on earth, and so the companies responsible took a different tack.

A document uncovered by the L.A. *Times* showed that, a month after Hansen's testimony, in 1988, an unnamed Exxon "public affairs manager" issued an internal memo recommending that the company "emphasize the uncertainty" in the scientific data about climate change. Within a few years, Exxon, Chevron, Shell, Amoco, and others had joined the Global Climate Coalition, "to coordinate business participation in the international policy debate" on global warming. The G.C.C. coördinated with the National Coal Association and the American Petroleum Institute on a campaign, via letters and telephone calls, to prevent a tax on fossil fuels, and produced a video in which the agency insisted that more carbon dioxide would "end world hunger" by promoting plant growth. With such efforts, it ginned up opposition to the Kyoto Protocol, the first global initiative to address climate change.

In October, 1997, two months before the Kyoto meeting, Lee Raymond, Exxon's president and C.E.O., who had overseen the science department that in the nineteen-eighties produced the findings about climate change, gave a speech in Beijing to the World Petroleum Congress, in which he maintained that the earth was actually cooling. The idea that cutting fossil-fuel emissions could have an effect on the climate, he said, defied common sense. "It is highly unlikely that the temperature in the middle of the next century will be affected whether policies are enacted now, or twenty years from now," he went on. Exxon's own scientists had already shown each of these premises to be wrong.

On a December morning in 1997 at the Kyoto Convention Center, after a long night of negotiation, the developed nations reached a tentative accord on climate change. Exhausted delegates lay slumped on couches in the corridor, or on the floor in their suits, but most of them were grinning. Imperfect and limited though the agreement was, it seemed that momentum had gathered behind fighting climate change. But as I watched the delegates cheering and clapping, an American lobbyist, who had been coördinating much of the opposition to the accord, turned to me and said, "I can't wait to get back to Washington, where we've got this under control."

He was right. On January 29, 2001, nine days after George W. Bush was inaugurated, Lee Raymond visited his old friend Vice-President Dick Cheney, who had just stepped down as the C.E.O. of the oil-drilling giant Halliburton. Cheney helped persuade Bush to abandon his campaign promise to treat carbon dioxide as a pollutant. Within the year, Frank Luntz, a Republican consultant for Bush, had produced an internal memo that made a doctrine of the strategy that the G.C.C. had hit on a decade earlier. "Voters believe that there is no consensus about global warming within the scientific community," Luntz wrote in the memo, which was obtained by the Environmental

Working Group, a Washington-based organization. "Should the public come to believe that the scientific issues are settled, their views about global warming will change accordingly. Therefore, you need to continue to make the lack of scientific certainty a primary issue in the debate."

The strategy of muddling the public's impression of climate science has proved to be highly effective. In 2017, polls found that almost ninety per cent of Americans did not know that there was a scientific consensus on global warming. Raymond retired in 2006, after the company posted the biggest corporate profits in history, and his final annual salary was four hundred million dollars. His successor, Rex Tillerson, signed a five-hundred-billion-dollar deal to explore for oil in the rapidly thawing Russian Arctic, and in 2012 was awarded the Russian Order of Friendship. In 2016, Tillerson, at his last shareholder meeting before he briefly joined the Trump Administration as Secretary of State, said, "The world is going to have to continue using fossil fuels, whether they like it or not."

It's by no means clear whether Exxon's deception and obfuscation are illegal. The company has long maintained that it "has tracked the scientific consensus on climate change, and its research on the issue has been published in publicly available peer-reviewed journals." The First Amendment preserves one's right to lie, although, in October, New York State Attorney General Barbara D. Underwood filed suit against Exxon for lying to investors, which is a crime. What is certain is that the industry's campaign cost us the efforts of the human generation that might have made the crucial difference in the climate fight.

Exxon's behavior is shocking, but not entirely surprising. Philip Morris lied about the effects of cigarette smoking before the government stood up to Big Tobacco. The mystery that historians will have to unravel is what went so wrong in our governance and our culture that we have done, essentially, nothing to stand up to the fossil-fuel industry.

There are undoubtedly myriad intellectual, psychological, and political sources for our inaction, but I cannot help thinking that the influence of Ayn Rand, the Russian émigré novelist, may have played a role. Rand's disquisitions on the "virtue of selfishness" and unbridled capitalism are admired by many American politicians and economists—Paul Ryan, Tillerson, Mike Pompeo, Andrew Puzder, and Donald Trump, among them. Trump, who has called "The Fountainhead" his favorite book, said that the novel "relates to business and beauty and life and inner emotions. That book relates to . . . everything." Long after Rand's death, in 1982, the libertarian gospel of the novel continues to sway our politics: Government is bad. Solidarity is a trap. Taxes are theft. The Koch brothers, whose enormous fortune derives in large part from the mining and refining of oil and gas, have peddled a similar message, broadening the efforts that Exxon-funded groups like the Global Climate Coalition spearheaded in the late nineteen-eighties.

Fossil-fuel companies and electric utilities, often led by Koch-linked groups, have put up fierce resistance to change. In Kansas, Koch allies helped turn mandated targets for renewable energy into voluntary commitments. In Wisconsin, Scott Walker's administration prohibited state land officials from talking about climate change. In North Carolina, the state legislature, in conjunction with real-estate interests, effectively banned policymakers from using scientific estimates of sea-level rise in the coastal-planning process. Earlier this year, Americans for Prosperity, the most important Koch front group, waged a campaign against new bus routes and light-rail service in Tennessee, invoking human liberty. "If someone has the freedom to go where they want, do what they want, they're not going to choose public transit," a spokeswoman for the group explained. In Florida, an anti-renewable-subsidy ballot measure invoked the "Rights of Electricity Consumers Regarding Solar Energy Choice."

Such efforts help explain why, in 2017, the growth of American residential solar installations came to a halt even before March, 2018, when President Trump imposed a thirty-per-cent tariff on solar panels, and why the number of solar jobs fell in the U.S. for the first time since the industry's great expansion began, a decade earlier. In February, at the Department of Energy, Rick Perry—who once skipped his own arraignment on two felony charges, which were eventually dismissed, in order to attend a Koch brothers event—issued a new projection in which he announced that the U.S. would go on emitting carbon at current levels through 2050; this means that our nation would use up all the planet's remaining carbon budget if we plan on meeting the 1.5-degree target. Skepticism about the scientific consensus, Perry told the media in 2017, is a sign of a "wise, intellectually engaged person."

Of all the environmental reversals made by the Trump Administration, the most devastating was its decision, last year, to withdraw from the Paris accords, making the U.S., the largest single historical source of carbon, the only nation not engaged in international efforts to control it. As the Washington *Post* reported, the withdrawal was the result of a collaborative venture. Among the anti-government ideologues and fossil-fuel lobbyists responsible was Myron Ebell, who was at Trump's side in the Rose Garden during the withdrawal announcement, and who, at Frontiers of Freedom, had helped run a "complex influence campaign" in support of the tobacco industry. Ebell is a director of the Competitive Enterprise Institute, which was founded in 1984 to advance "the principles of limited government, free enterprise, and individual liberty," and which funds the Cooler Heads Coalition, "an informal and ad-hoc group focused on dispelling the myths of global warming," of which Ebell is the chairman. Also instrumental were the Heartland Institute and the Koch brothers' Americans for Prosperity. After Trump's election, these groups sent a letter reminding him of his campaign pledge to pull America out. The C.E.I. ran a TV spot: "Mr. President, don't listen to the swamp. Keep your promise." And, despite the objections of most of his advisers, he did. The coalition had used its power to slow us down precisely at the moment when we needed to speed up. As a result, the particular politics of one country for one half-century will have changed the geological history of the earth.

We are on a path to self-destruction, and yet there is nothing inevitable about our fate. Solar panels and wind turbines are now among the least expensive ways to produce energy. Storage batteries are cheaper and more efficient than ever. We could move quickly if we chose to, but we'd need to opt for solidarity and coördination on a global scale. The chances of that look slim. In Russia, the second-largest petrostate after the U.S., Vladimir Putin believes that "climate change could be tied to some global cycles on Earth or even of planetary significance." Saudi Arabia, the third-largest petrostate, tried to water down the recent I.P.C.C. report. Jair Bolsonaro, the newly elected President of Brazil, has vowed to institute policies that would dramatically accelerate the deforestation of the Amazon, the world's largest rain forest. Meanwhile, Exxon recently announced a plan to spend a million dollars—about a hundredth of what the company spends each month in search of new oil and gas—to back the fight for a carbon tax of forty dollars a ton. At a press conference, some of the I.P.C.C.'s authors laughed out loud at the idea that such a tax would, this late in the game, have sufficient impact.

The possibility of swift change lies in people coming together in movements large enough to shift the Zeitgeist. In recent years, despairing at the slow progress, I've been one of many to protest pipelines and to call attention to Big Oil's deceptions. The movement is growing. Since 2015, when four hundred thousand people marched in the streets of New York before the Paris climate talks, activists—often led by indigenous groups and communities living on the front lines of climate change—have blocked pipelines, forced the cancellation of new coal mines, helped keep the major oil companies out of the American Arctic, and persuaded dozens of cities to commit to one-hundred-per-cent renewable energy.

Each of these efforts has played out in the shadow of the industry's unflagging campaign to maximize profits and prevent change. Voters in Washington State were initially supportive of a measure on last month's ballot which would have imposed the nation's first carbon tax—a modest fee that won support from such figures as Bill Gates. But the major oil companies spent record sums to defeat it. In Colorado, a similarly modest referendum that would have forced frackers to move their rigs away from houses and schools went down after the oil industry outspent citizen groups forty to one. This fall, California's legislators committed to using only renewable energy by 2045, which was a great victory in the world's fifth-largest economy. But the governor refused to stop signing new permits for oil wells, even in the middle of the state's largest cities, where asthma rates are high.

New kinds of activism keep springing up. In Sweden this fall, a one-person school boycott by a fifteen-year-old girl named Greta Thunberg helped galvanize attention across Scandinavia. At the end of October, a new British group, Extinction Rebellion—its name both a reflection of the dire science and a potentially feisty response—announced plans for a campaign of civil disobedience. Last week, fifty-one young people were arrested in Nancy Pelosi's office for staging a sit-in, demanding that the Democrats embrace a "Green New Deal" that would address the global climate crisis with policies to create jobs in renewable energy. They may have picked a winning issue: several polls have shown that even Republicans favor more government support for solar panels. This battle is epic and undecided. If we miss the two-degree target, we will fight to prevent a rise of three degrees, and then four. It's a long escalator down to Hell.

Last June, I went to Cape Canaveral to watch Elon Musk's Falcon 9 rocket lift off. When the moment came, it was as I'd always imagined: the clouds of steam venting in the minutes before launch, the immensely bright column of flame erupting. With remarkable slowness, the rocket began to rise, the grip of gravity yielding to the force of its engines. It is the most awesome technological spectacle human beings have produced.

Musk, Jeff Bezos, and Richard Branson are among the billionaires who have spent some of their fortunes on space travel—a last-ditch effort to expand the human zone of habitability. In November, 2016, Stephen Hawking gave humanity a deadline of a thousand years to leave Earth. Six months later, he revised the timetable to a century. In June, 2017, he told an audience that "spreading out may be the only thing that saves us from ourselves." He continued, "Earth is under threat from so many areas that it is difficult for me to be positive."

But escaping the wreckage is, almost certainly, a fantasy. Even if astronauts did cross the thirty-four million miles to Mars, they'd need to go underground to survive there. To what end? The multimillion-dollar attempts at building a "biosphere" in the Southwestern desert in 1991 ended in abject failure. Kim Stanley Robinson, the author of a trilogy of novels about the colonization of Mars, recently called such projects a "moral hazard." "People think if we fuck up here on Earth we can always go to Mars or the stars," he said. "It's pernicious."

The dream of interplanetary colonization also distracts us from acknowledging the unbearable beauty of the planet we already inhabit. The day before the launch, I went on a tour of the vast grounds of the Kennedy Space Center with NASA's public-affairs officer, Greg Harland, and the biologist Don Dankert. I'd been warned beforehand by other NASA officials not to broach the topic of global warming; in any event, NASA's predicament became obvious as soon as we climbed up on a dune overlooking Launch Complex 39, from which the Apollo missions left for the moon, and where any future Mars mission would likely begin. The launchpad is a quarter of a mile from the ocean—a perfect location, in the sense that, if something goes wrong, the rockets will fall into the sea, but not so perfect, since that sea is now rising. NASA started worrying about this sometime after the turn of the century, and formed a Dune Vulnerability Team.

In 2012, Hurricane Sandy, even at a distance of a couple of hundred miles, churned up waves strong enough to break through the barrier of dunes along the Atlantic shoreline of the Space Center and very nearly swamped the launch complexes. Dankert had millions of cubic yards of sand excavated from a nearby Air Force base, and saw to it that a hundred and eighty thousand native shrubs were planted to hold the sand in place. So far, the new dunes have yielded little ground to storms and hurricanes. But what impressed me more than the dunes was the men's deep appreciation of their landscape. "Kennedy Space Center shares real estate with the Merritt Island Wildlife Refuge," Harland said. "We use less than ten per cent for our industrial purposes."

"When you look at the beach, it's like eighteen-seventies Florida—the longest undisturbed stretch on the Atlantic Coast," Dankert said. "We launch people into space from the middle of a wildlife refuge. That's amazing."

The two men talked for a long time about their favorite local species—the brown pelicans that were skimming the ocean, the Florida scrub jays. While rebuilding the dunes, they carefully bucket-trapped and relocated dozens of gopher tortoises. Before I left, they drove me half an hour across the swamp to a pond near the Space Center's headquarters building, just to show me some alligators. Menacing snouts were visible beneath the water, but I was more interested in the sign that had been posted at each corner of the pond explaining that the alligators were native species, not pets. "Putting any food in the water for any reason will cause them to become accustomed to people and possibly dangerous," it went on, adding that, if that should happen, "they must be removed and destroyed."

Something about the sign moved me tremendously. It would have been easy enough to poison the pond, just as it would have been easy enough to bulldoze the dunes without a thought for the tortoises. But NASA hadn't done so, because of a long series of laws that draw on an emerging understanding of who we are. In 1867, John Muir, one of the first Western environmentalists, walked from Louisville, Kentucky, to Florida, a trip that inspired his first heretical thoughts about the meaning of being human. "The world, we are told, was made especially for man—a presumption not supported by all the facts," Muir wrote in his diary. "A numerous class of men are painfully astonished whenever they find anything, living or dead, in all God's universe, which they cannot eat or render in some way what they call useful to themselves." Muir's proof that this self-centeredness was misguided was the alligator, which he could hear roaring in the Florida swamp as he camped nearby, and which clearly caused man mostly trouble. But these animals were wonderful nonetheless, Muir decided—remarkable creatures perfectly adapted to their landscape. "I have better thoughts of those alligators now that I've seen them at home," he wrote. In his diary, he addressed the creatures directly: "Honorable representatives of the great saurian of an older creation, may you long enjoy your lilies and rushes, and be blessed now and then with a mouthful of terror-stricken man by way of dainty."

That evening, Harland and Dankert drew a crude map to help me find the beach, north of Patrick Air Force Base and south of the spot where, in 1965, Barbara Eden emerged from her bottle to greet her astronaut at the start of the TV series "I Dream of Jeannie." There, they said, I could wait out the hours until the pre-dawn rocket launch and perhaps spot a loggerhead sea turtle coming ashore to lay her eggs. And so I sat on the sand. The beach was deserted, and under a near-full moon I watched as a turtle trundled from the sea and lumbered deliberately to a spot near the dune, where she used her powerful legs to excavate a pit. She spent an hour laying eggs, and even from thirty yards away you could hear her heavy breathing in between the whispers of the waves. And then, having covered her clutch, she tracked back to the ocean, in the fashion of others like her for the past hundred and twenty million years.

PART III
Changing the Weather
What We Can Do Now

Green Manhattan

David Owen

October 18, 2004

My wife and I got married right out of college, in 1978. We were young and naïve and unashamedly idealistic, and we decided to make our first home in a utopian environmentalist community in New York State. For seven years, we lived, quite contentedly, in circumstances that would strike most Americans as austere in the extreme: our living space measured just seven hundred square feet, and we didn't have a dishwasher, a garbage disposal, a lawn, or a car. We did our grocery shopping on foot, and when we needed to travel longer distances we used public transportation. Because space at home was scarce, we seldom acquired new possessions of significant size. Our electric bills worked out to about a dollar a day.

The utopian community was Manhattan. (Our apartment was on Sixty-ninth Street, between Second and Third.) Most Americans, including most New Yorkers, think of New York City as an ecological nightmare, a wasteland of concrete and garbage and diesel fumes and traffic jams, but in comparison with the rest of America it's a model of environmental responsibility. By the most significant measures, New York is the greenest community in the United States, and one of the greenest cities in the world. The most devastating damage humans have done to the environment has arisen from the heedless burning of fossil fuels, a category in which New Yorkers are practically prehistoric. The average Manhattanite consumes gasoline at a rate that the country as a whole hasn't matched since the mid-nineteen-twenties, when the most widely owned car in the United States was the Ford Model T. Eighty-two per cent of Manhattan residents travel to work by public transit, by bicycle, or on foot. That's ten times the rate for Americans in general, and eight times the rate for residents of Los Angeles County. New York City is more populous than all but eleven states; if it were granted statehood, it would rank fifty-first in per-capita energy use.

"Anyplace that has such tall buildings and heavy traffic is obviously an environmental disaster—except that it isn't," John Holtzclaw, a transportation consultant for the Sierra Club and the Natural Resources Defense Council, told me. "If New Yorkers lived at the typical American sprawl density of three households per residential acre, they would require many times as much land. They'd be driving cars, and they'd have huge lawns and be using pesticides and fertilizers on them, and then they'd be overwatering their lawns, so that runoff would go into streams." The key to New York's relative environmental benignity is its extreme compactness. Manhattan's population density is more than eight hundred times that of the nation as a whole. Placing one and a half million people on a twenty-three-square-mile island sharply reduces their opportunities to be wasteful, and forces the majority to live in some of the most inherently energy-efficient residential structures in the world: apartment buildings. It also frees huge tracts of land for the rest of America to sprawl into.

My wife and I had our first child in 1984. We had both grown up in suburbs, and we decided that we didn't want to raise our tiny daughter in a huge city. Shortly after she learned to walk, we moved to a small town in northwestern Connecticut, about ninety miles north of midtown Manhattan. Our house, which was built in the late seventeen-hundreds, is across a dirt road from a nature preserve and is shaded by tall white-pine trees. After big rains, we can hear a swollen creek rushing by at the bottom of the hill. Deer, wild turkeys, and the occasional black bear feed themselves in our yard. From the end of our driveway, I can walk several miles through woods to an abandoned nineteenth-century railway tunnel, while crossing only one paved road.

Yet our move was an ecological catastrophe. Our consumption of electricity went from roughly four thousand kilowatt-hours a year, toward the end of our time in New York, to almost thirty thousand kilowatt-hours in 2003—and our house doesn't even have central air-conditioning. We bought a car shortly before we moved, and another one soon after we arrived, and a third one ten years later. (If you live in the country and don't have a second car, you can't retrieve your first car from the mechanic after it's been repaired; the third car was the product of a mild midlife crisis, but soon evolved into a necessity.) My wife and I both work at home, but we manage to drive thirty thousand miles a year between us, mostly doing ordinary errands. Nearly everything we do away from our house requires a car trip. Renting a movie and later returning it, for example, consumes almost two gallons of gasoline, since the nearest Blockbuster is ten miles away and each transaction involves two round trips. When we lived in New York, heat escaping from our apartment helped to heat the apartment above ours; nowadays, many of the Btus produced by our brand-new, extremely efficient oil-burning furnace leak through our two-hundred-year-old roof and into the dazzling star-filled winter sky above.

When most Americans think about environmentalism, they picture wild, unspoiled landscapes—the earth before it was transmogrified by human habitation. New York City is one of the most thoroughly altered landscapes imaginable, an almost wholly artificial environment, in which the terrain's primeval contours have long since been obliterated and most of the parts that resemble nature (the trees on side streets, the rocks in Central Park) are essentially decorations. Ecology-minded discussions of New York City often have a hopeless tone, and focus on ways in which the city might be made to seem somewhat less oppressively man-made: by increasing the area devoted to parks and greenery, by incorporating vegetation into buildings themselves, by reducing traffic congestion, by easing the intensity of development, by creating open space around structures. But most such changes would actually undermine the city's extraordinary energy efficiency, which arises from the characteristics that make it surreally synthetic.

Because densely populated urban centers concentrate human activity, we think of them as pollution crisis zones. Calculated by the square foot, New York City generates more greenhouse gases, uses more energy, and produces more solid waste than most other American regions of comparable size. On a map depicting negative environmental impacts in relation to surface area, therefore, Manhattan would look like an intense hot spot, surrounded, at varying distances, by belts of deepening green.

If you plotted the same negative impacts by resident or by household, however, the color scheme would be reversed. My little town has about four thousand residents, spread over 38.7 thickly wooded square miles, and there are many places within our town limits from which no sign of settlement is visible in any direction. But if you moved eight million people like us, along with our dwellings and possessions and current rates of energy use, into a space the size of New York City, our profligacy would be impossible to miss, because you'd have to stack our houses and cars and garages and lawn tractors and swimming pools and septic tanks higher than skyscrapers. (Conversely, if you made all eight million New Yorkers live at the density of my town, they would require a space equivalent to the land area of the six New England states plus Delaware and New Jersey.) Spreading people out increases the damage they do to the environment, while making the problems harder to see and to address.

Of course, living in densely populated urban centers has many drawbacks. Even wealthy New Yorkers live in spaces that would seem cramped to Americans living almost anywhere else. A well-to-do friend of mine who grew up in a town house in Greenwich Village thought of his upbringing as privileged until, in prep school, he visited a classmate from the suburbs and was staggered by the house, the lawn, the cars, and the swimming pool, and thought, with despair, You mean I could live like this? Manhattan is loud and dirty, and the subway is depressing, and the fumes from the cars and cabs and buses can make people sick. Presumably for environmental reasons, New York City has one of the highest childhood-asthma rates in the country, with an especially alarming concentration in East Harlem.

Nevertheless, barring an almost inconceivable reduction in the earth's population, dense urban centers offer one of the few plausible remedies for some of the world's most discouraging environmental ills. To borrow a term from the jargon of computer systems, dense cities are scalable, while sprawling suburbs are not. The environmental challenge we face, at the current stage of our assault on the world's non-renewable resources, is not how to make our teeming cities more like the pristine countryside. The true challenge is how to make other settled places more like Manhattan. This notion has yet to be widely embraced, partly because it is counterintuitive, and partly because most Americans, including most environmentalists, tend to view cities the way Thomas Jefferson did, as "pestilential to the morals, the health, and the liberties of man." New York is the place that's fun to visit but you wouldn't want to live there. What could it possibly teach anyone about being green?

New York's example, admittedly, is difficult for others to imitate, because the city's remarkable population density is the result not of conscientious planning but of a succession of serendipitous historical accidents. The most important of those accidents was geographic: New York arose on a smallish island rather than on the mainland edge of a river or a bay, and the surrounding water served as a physical constraint to outward expansion. Manhattan is like a typical seaport turned inside out—a city with a harbor around it, rather than a harbor with a city along its edge. Insularity gave Manhattan more shoreline per square mile than other ports, a major advantage in the days when one of the world's main commercial activities was moving cargoes between ships. It also drove early development inward and upward.

A second lucky accident was that Manhattan's street plan was created by merchants who were more interested in economic efficiency than in boulevards, parks, or empty spaces between buildings. The resulting crush of architecture is actually humanizing, because it brings the city's commercial, cultural, and other offerings closer together, thereby increasing their accessibility—a point made forty-three years ago by the brilliantly iconoclastic urban thinker Jane Jacobs, in her landmark book "The Death and Life of Great American Cities."

A third accident was the fact that by the early nineteen-hundreds most of Manhattan's lines had been filled in to the point where not even Robert Moses could easily redraw them to accommodate the great destroyer of American urban life, the automobile. Henry Ford thought of cars as tools for liberating humanity from the wretchedness of cities, which he viewed with as much distaste as Jefferson did. In 1932, John Nolen, a prominent Harvard-educated urban planner and landscape architect, said, "The future city will be spread out, it will be regional, it will be the natural product of the automobile, the good road, electricity, the telephone, and the radio, combined with the growing desire to live a more natural, biological life under pleasanter and more natural conditions." This is the idea behind suburbs, and it's still seductive. But it's also a prescription for sprawl and expressways and tremendous waste.

New York City's obvious urban antithesis, in terms of density and automobile use, is metropolitan Los Angeles, whose metastatic outward growth has been virtually unimpeded by the lay of the land, whose early settlers came to the area partly out of a desire to create space between themselves and others, and whose main development began late enough to be shaped by the needs of cars. But a more telling counterexample is Washington, D.C., whose basic layout was conceived at roughly the same time as Manhattan's, around the turn of the nineteenth century. The District of Columbia's original plan was created by an eccentric French-born engineer and architect named Pierre-Charles L'Enfant, who befriended General Washington during the Revolutionary War and asked to be allowed to design the capital. Many of modern Washington's most striking features are his: the broad, radial avenues; the hublike traffic circles; the sweeping public lawns and ceremonial spaces.

Washington is commonly viewed as the most intelligently beautiful—the most European—of large American cities. Ecologically, though, it's a mess. L'Enfant's expansive avenues were easily adapted to automobiles, and the low, widely separated buildings (whose height is limited by law) stretched the distance between destinations. There are many pleasant places in Washington to go for a walk, but the city is difficult to get around on foot: the wide avenues are hard to cross, the traffic circles are like obstacle courses, and the grandiloquent empty spaces thwart pedestrians, by acting as what Jane Jacobs calls "border vacuums." (One of Jacobs's many arresting observations is that parks and other open spaces can reduce urban vitality, by creating dead ends that prevent people from moving freely between neighborhoods and by decreasing activity along their edges.) Many parts of Washington, furthermore, are relentlessly homogeneous. There are plenty of dignified public buildings on Constitution Avenue, for example, but good luck finding a dry cleaner, a Chinese restaurant, or a grocery store. The city's horizontal, airy design has also pushed development into the surrounding countryside. The fastest-growing county in the United States is Loudoun County, Virginia, at the rapidly receding western edge of the Washington metropolitan area.

The Sierra Club, an environmental organization that advocates the preservation of wilderness and wildlife, has a national campaign called Challenge to Sprawl. The aim of the program is to arrest the mindless conversion of undeveloped countryside into subdivisions, strip malls, and S.U.V.-clogged expressways. The Sierra Club's Web site features a slide-show-like demonstration that illustrates how various sprawling suburban intersections could be transformed into far more appealing and energy-efficient developments by implementing a few modifications, among them widening the sidewalks and narrowing the streets, mixing residential and commercial uses, moving buildings closer together and closer to the edges of sidewalks (to make them more accessible to pedestrians and to increase local density), and adding public transportation—all fundamental elements of the widely touted anti-sprawl strategy known as Smart Growth. In a recent telephone conversation with a Sierra Club representative involved in Challenge to Sprawl, I said that the organization's anti-sprawl suggestions and the modified streetscapes in the slide show shared many significant features with Manhattan—whose most salient characteristics include wide sidewalks, narrow streets, mixed uses, densely packed buildings, and an extensive network of subways and buses. The representative hesitated, then said that I was essentially correct, although he would prefer that the program not be described in such terms, since emulating New York City would not be considered an appealing goal by most of the people whom the Sierra Club is trying to persuade.

An obvious way to reduce consumption of fossil fuels is to shift more people out of cars and into public transit. In many parts of the country, though, public transit has been stagnant or in decline for years. New York City's Metropolitan Transportation Authority and Department of Transportation account for nearly a third of all the transit passenger miles travelled in the United States and for nearly four times as many passenger miles as the Washington Metropolitan Area Transit Authority and the Los Angeles County Metropolitan Transportation Authority combined.

New York City looks so little like other parts of America that urban planners and environmentalists tend to treat it as an exception rather than an example, and to act as though Manhattan occupied an idiosyncratic universe of its own. But the underlying principles apply everywhere. "The basic point," Jeffrey Zupan, an economist with the Regional Planning Association, told me, "is that you need density to support public transit. In all cities, not just in New York, once you get above a certain density two things happen. First, you get less travel by mechanical means, which is another way of saying you get more people walking or biking; and, second, you get a decrease in the trips by auto and an increase in the trips by transit. That threshold tends to be around seven dwellings per acre. Once you cross that line, a bus company can put buses out there, because they know they're going to have enough passengers to support a reasonable frequency of service."

Phoenix is the sixth-largest city in the United States and one of the fastest-growing among the top ten, yet its public transit system accounts for just one per cent of the passenger miles that New York City's does. The reason is that Phoenix's burgeoning population has spread so far across the desert—greater Phoenix, whose population is a little more than twice that of Manhattan, covers more than two hundred times as much land—that no transit system could conceivably serve it. And no amount of browbeating, public-service advertising, or federal spending can change that.

Cities, states, and the federal government often negate their own efforts to nurture public transit by simultaneously spending huge sums to make it easier for people to get around in cars. When a city's automobile traffic becomes congested, the standard response has long been to provide additional capacity by building new roads or widening existing ones. This approach eventually makes the original problem worse, by generating what transportation planners call "induced traffic": every mile of new highway lures passengers from public transit and other more efficient modes of travel, and makes it possible for residential and commercial development to spread even farther from urban centers. And adding public transit in the hope of reducing automobile congestion is as self-defeating as building new highways, because unclogging roads, if successful, just makes driving seem more attractive, and the roads fill up again. A better strategy would be to eliminate existing traffic lanes and parking spaces gradually, thereby forcing more drivers to use less environmentally damaging alternatives—in effect, "induced transit." One reason New Yorkers are the most dedicated transit users in America is that congestion on the city's streets makes driving extraordinarily disagreeable. The average speed of crosstown traffic in Manhattan is little more than that of a brisk walker, and in midtown at certain times of the day the cars on the side streets move so slowly that they appear almost to be parked. Congestion like that urges drivers into the subways, and it makes life easier for pedestrians and bicycle riders by slowing cars to a point where they constitute less of a physical threat.

Even in New York City, the relationship between traffic and transit is not well understood. A number of the city's most popular recent transportation-related projects and policy decisions may in the long run make the city a worse place to live in by luring passengers back into their cars and away from public transportation: the rebuilding and widening of the West Side Highway, the implementation of EZ-Pass on the city's toll bridges, the decision not to impose tolls on the East River bridges, and the current renovation of the F.D.R. Drive (along with the federally funded hundred-and-thirty-nine-million-dollar Outboard Detour Roadway, which is intended to prevent users of the F.D.R. from being inconvenienced while the work is under way).

Public transit itself can be bad for the environment if it facilitates rather than discourages sprawl. The Washington Metropolitan Area Transit Authority is considering extensions to some of the most distant branches of its system, and those extensions, if built, will allow people to live even farther from the city's center, creating new, non-dense suburbs where all other travel will be by automobile, much of it to malls and schools and gas stations that will be built to accommodate them. Transit is best for the environment when it helps to concentrate people in dense urban cores. Building the proposed Second Avenue subway line would be environmentally sound, because it would increase New Yorkers' ability to live without cars; building a bullet train between Penn Station and the Catskills (for example) would not be sound, because it would enable the vast, fuel-squandering apparatus of suburbia to establish itself in a region that couldn't support it otherwise.

On the afternoon of August 14, 2003, I was working in my office, on the third floor of my house, when the lights blinked, my window air conditioner sputtered, and my computer's backup battery kicked in briefly. This was the beginning of the great blackout of 2003, which halted electric service in parts of eight Northeastern and Midwestern states and in southeastern Canada. The immediate cause was eventually traced to Ohio, but public attention often focussed on New York City, which had the largest concentration of affected power customers. Richard B. Miller, who resigned as the senior energy adviser for the city of New York six weeks before the blackout, reportedly over deep disagreements with the city's energy policy, told me, "When I was with the city, I attended a conference on global warming where somebody said, 'We really need to raise energy and electricity prices in New York City, so that people will consume less.' And my response at that conference was 'You know, if you're talking about raising energy prices in New York City only, then you're talking about something that's really bad for the environment. If you make energy prices so expensive in the city that a business relocates from Manhattan to New Jersey, what you're really talking about, in the simplest terms, is a business that's moving from a subway stop to a parking lot. And which of those do you think is worse for the environment?'"

People who live in cities use only about half as much electricity as people who don't, and people who live in New York City generally use less than the urban average. A truly enlightened energy policy would reward city dwellers and encourage others to follow their good example. Yet New York City residents pay more per kilowatt-hour than almost any other American electricity customers; taxes and other government charges, most of which are not enumerated on electricity bills, can constitute close to twenty per cent of the cost of power for residential and commercial users in New York. Richard Miller, after leaving his job with New York City, went to work as a lawyer in Consolidated Edison's regulatory affairs department, spurred by his thinking about the environment. He believes that state and local officials have historically taken unfair advantage of the fact that there is no political cost to attacking a big utility. Con Ed pays more than six hundred million dollars a year in property taxes, making it by far the city's largest property-tax payer, and those charges inflate electric bills. Meanwhile, the cost of driving is kept artificially low. (Fifth Avenue and the West Side Highway don't pay property taxes, for example.) "In addition," Miller said, "the burden of improving the city's air has fallen far more heavily on power plants, which contribute only a small percentage of New York City's air pollution, than it has on cars—even though motor vehicles are a much bigger source."

Last year, the National Building Museum, in Washington, D.C., held a show called "Big & Green: Toward Sustainable Architecture in the 21st Century." A book of the same name was published in conjunction with the show, and on the book's dust jacket was a photograph of 4 Times Square, also known as the Condé Nast Building, a forty-eight-story glass-and-steel tower between Forty-second and Forty-third Streets, a few blocks west of Grand Central Terminal. (*The New Yorker*'s offices occupy two floors in the building.) When 4 Times Square was built, in 1999, it was considered a major breakthrough in urban development. As Daniel Kaplan, a principal of Fox & Fowle Architects, the firm that designed it, wrote in an article in *Environmental Design & Construction* in 1997, "When thinking of green architecture, one usually associates smaller scale," and he cited as an example the headquarters of the Rocky Mountain Institute, a nonprofit environmental research and consulting firm based in Snowmass, Colorado. The R.M.I. building is a four-thousand-square-foot, superinsulated, passive-solar structure with curving sixteen-inch-thick walls, set into a hillside about fifteen miles north of Aspen. It was erected in the early eighties and serves partly as a showcase for green construction technology. (It is also the home of Amory Lovins, who is R.M.I.'s co-founder and chief executive officer.) R.M.I. contributed to the design of 4 Times Square, which has many innovative features, among them collection chutes for recyclable materials, photovoltaic panels incorporated into parts of its skin, and curtain-wall construction with exceptional shading and insulating properties.

These are all important innovations. In terms of the building's true ecological impact, though, they are distinctly secondary. (The power generated by the photovoltaic panels supplies less than one per cent of the building's requirements.) The two greenest features of 4 Times Square are ones that most people never even mention: it is big, and it is situated in Manhattan.

Environmentalists have tended to treat big buildings as intrinsically wasteful, because large amounts of energy are expended in their construction, and because the buildings place intensely localized stresses on sewers, power lines, and water systems. But density can create the same kinds of ecological benefits in individual structures that it does in entire communities. Tall buildings have much less exposed exterior surface per square foot of interior space than smaller buildings do, and that means they present relatively less of themselves to the elements, and their small roofs absorb less heat from the sun during cooling season and radiate less heat from inside during heating season. (The beneficial effects are greater still in Manhattan, where one building often directly abuts another.) A study by Michael Phillips and Robert Gnaizda, published in *CoEvolution Quarterly* in 1980, found that an ordinary apartment in a typical building near downtown San Francisco used just a fifth as much heating fuel as a new tract house in Davis, a little more than seventy miles away. Occupants of tall buildings also do a significant part of their daily coming and going in elevators, which, because they are counterweighted and thus require less motor horsepower, are among the most energy-efficient passenger vehicles in the world.

Bruce Fowle, a founder of Fox & Fowle, told me, "The Condé Nast Building contains 1.6 million square feet of floor space, and it sits on one acre of land. If you divided it into forty-eight one-story suburban office buildings, each averaging thirty-three thousand square feet, and spread those one-story buildings around the countryside, and then added parking and some green space around each one, you'd end up consuming at least a hundred and fifty acres of land. And then you'd have to provide infrastructure, the highways and everything else." Like many other buildings in Manhattan, 4 Times Square doesn't even have a parking lot, because the vast majority of the six thousand people who work inside it don't need one. In most other parts of the country, big parking lots are not only necessary but are required by law. If my town's zoning regulations applied in Manhattan, 4 Times Square would have needed sixteen thousand parking spaces, one for every hundred square feet of office floor space.

The Rocky Mountain Institute's showcase headquarters has double-paned krypton-filled windows, which admit seventy-five per cent as much light as ordinary windows while allowing just ten per cent as much heat to escape in cold weather. That's a wonderful feature, and one of many in the building which people ought to copy. In other ways, though, the R.M.I. building sets a very poor environmental example. It was built in a fragile location, on virgin land more than seven thousand feet above sea level. With just four thousand square feet of interior space, it can hold only six of R.M.I.'s eighteen full-time employees; the rest of them work in a larger building a mile away. Because the two buildings are in a thinly populated area, they force most employees to drive many miles—including trips between the two buildings—and they necessitate extra fuel consumption by delivery trucks, snowplows, and other vehicles. If R.M.I.'s employees worked on a single floor of a big building in Manhattan (or in downtown Denver) and lived in apartments nearby, many of them would be able to give up their cars, and the thousands of visitors who drive to Snowmass each year to learn about environmentally responsible construction could travel by public transit instead.

Picking on R.M.I.—which is one of the world's most farsighted environmental organizations—may seem unfair, but R.M.I., along with many other far-sighted environmental organizations, shares responsibility for perpetuating the powerful anti-city bias of American environmentalism. That bias is evident in the technical term that is widely used for sprawl: "urbanization." Thinking of freeways and strip malls as "urban" phenomena obscures the ecologically monumental difference between Phoenix and Manhattan, and fortifies the perception that population density is an environmental ill. It also prevents most people from recognizing that R.M.I.'s famous headquarters—which sits on an isolated parcel more than a hundred and eighty miles from the nearest significant public transit system—is sprawl.

When I told a friend recently that I thought New York City should be considered the greenest community in America, she looked puzzled, then asked, "Is it because they've started recycling again?" Her question reflected a central failure of the American environmental movement: that too many of us have been made to believe that the most important thing we can do to save the earth and ourselves is to remember each week to set our cans and bottles and newspapers on the curb. Recycling is popular because it enables people to relieve their gathering anxieties about the future without altering the way they live. But most current recycling has, at best, a neutral effect on the environment, and much of it is demonstrably harmful. As William McDonough and Michael Braungart point out in "Cradle to Cradle: Remaking the Way We Make Things," most of the materials we place on our curbs are merely "downcycled"—converted to a lower use, providing a pause in their inevitable journey to a landfill or an incinerator—often with a release of toxins and a net loss of fuel, among other undesirable effects.

By far the worst damage we Americans do to the planet arises not from the newspapers we throw away but from the eight hundred and fifty million or so gallons of oil we consume every day. We all know this at some level, yet we live like alcoholics in denial. How else can we explain that our cars have grown bigger, heavier, and less fuel-efficient at the same time that scientists have become more certain and more specific about the consequences of our addiction to gasoline?

On a shelf in my office is a small pile of recent books about the environment which I plan to reread obsessively if I'm found to have a terminal illness, because they're so unsettling that they may make me less upset about being snatched from life in my prime. At the top of the pile is "Out of Gas: The End of the Age of Oil," by David Goodstein, a professor at the California Institute of Technology, which was published earlier this year. "The world will soon start to run out of conventionally produced, cheap oil," Goodstein begins. In succeeding pages, he lucidly explains that humans have consumed almost a trillion barrels of oil (that's forty-two trillion gallons), or about half of the earth's total supply; that a devastating global petroleum crisis will begin not when we have pumped the last barrel out of the ground but when we have reached the halfway point, because at that moment, for the first time in history, the line representing supply will fall through the line representing demand; that we will probably pass that point within the current decade, if we haven't passed it already; that various well-established laws of economics are about to assert themselves, with disastrous repercussions for almost everything; and that "civilization as we know it will come to an end sometime in this century unless we can find a way to live without fossil fuels."

Standing between us and any conceivable solution to our energy nightmare are our cars and the asphalt-latticed country we have built to oblige them. Those cars have defined our culture and our lives. A car is speed and sex and power and emancipation. It makes its driver a self-sufficient nation of one. It is everything a city is not.

Most of the car's most tantalizing charms are illusory, though. By helping us to live at greater distances from one another, driving has undermined the very benefits that it was meant to bestow. Ignacio San Martín, an architecture professor and the head of the graduate urban-design program at the University of

Arizona, told me, "If you go out to the streets of Phoenix and are able to see anybody walking—which you likely won't—they are going to tell you that they love living in Phoenix because they have a beautiful house and three cars. In reality, though, once the conversation goes a little bit further, they are going to say that they spend most of their time at home watching TV, because there is absolutely nothing to do." One of the main attractions of moving to the suburbs is acquiring ground of your own, yet you can travel for miles through suburbia and see no one doing anything in a yard other than working on the yard itself (often with the help of a riding lawnmower, one of the few four-wheeled passenger vehicles that get worse gas mileage than a Hummer). The modern suburban yard is perfectly, perversely self-justifying: its purpose is to be taken care of.

In 1801, in his first Inaugural address, Thomas Jefferson said that the American wilderness would provide growing room for democracy-sustaining agrarian patriots "to the thousandth and thousandth generation." Jefferson didn't foresee the interstate highway system, and his arithmetic was off, in any case, but he nevertheless anticipated (and, in many ways, embodied) the ethos of suburbia, of anti-urbanism, of sprawl. The standard object of the modern American dream, the single-family home surrounded by grass, is a mini-Monticello. It was the car that put it within our reach. But what a terrible price we have paid—and have yet to pay—for our liberation from the city.

Big Foot

Michael Specter

February 25, 2008

A little more than a year ago, Sir Terry Leahy, who is the chief executive of the Tesco chain of supermarkets, Britain's largest retailer, delivered a speech to a group called the Forum for the Future, about the implications of climate change. Leahy had never before addressed the issue in public, but his remarks left little doubt that he recognized the magnitude of the problem. "I am not a scientist," he said. "But I listen when the scientists say that, if we fail to mitigate climate change, the environmental, social, and economic consequences will be stark and severe. . . . There comes a moment when it is clear what you must do. I am determined that Tesco should be a leader in helping to create a low-carbon economy. In saying this, I do not underestimate the task. It is to take an economy where human comfort, activity, and growth are inextricably linked with emitting carbon and to transform it into one which can only thrive without depending on carbon. This is a monumental challenge. It requires a revolution in technology and a revolution in thinking. We are going to have to rethink the way we live and work."

Tesco sells nearly a quarter of the groceries bought in the United Kingdom, it possesses a growing share of the markets in Asia and Europe, and late last year the chain opened its first stores in America. Few corporations could have a more visible—or forceful—impact on the lives of their customers. In his speech, Leahy, who is fifty-two, laid out a series of measures that he hoped would ignite "a revolution in green consumption." He announced that Tesco would cut its energy use in half by 2010, drastically limit the number of products it transports by air, and place airplane symbols on the packaging of those which it does. More important, in an effort to help consumers understand the environmental impact of the choices they make every day, he told the forum that Tesco would develop a system of carbon labels and put them on each of its seventy thousand products. "Customers want us to develop ways to take complicated carbon calculations and present them simply," he said. "We will therefore begin the search for a universally accepted and commonly understood measure of the carbon footprint of every product we sell—looking at its complete life cycle, from production through distribution to consumption. It will enable us to label all our products so that customers can compare their carbon footprint as easily as they can currently compare their price or their nutritional profile."

Leahy's sincerity was evident, but so was his need to placate his customers. Studies have consistently demonstrated that, given a choice, people prefer to buy products that are environmentally benign. That choice, however, is almost never easy. "A carbon label will put the power in the hands of consumers to choose how they want to be green," Tom Delay, the head of the British government's Carbon Trust, said. "It will empower us all to make informed choices and in turn drive a market for low-carbon products." Tesco was not alone in telling people what it would do to address the collective burden of our greenhouse-gas emissions. Compelled by economic necessity as much as by ecological awareness, many corporations now seem to compete as vigorously to display their environmental credentials as they do to sell their products.

In Britain, Marks & Spencer has set a goal of recycling all its waste, and intends to become carbon-neutral by 2012—the equivalent, it claims, of taking a hundred thousand cars off the road every year. Kraft Foods recently began to power part of a New York plant with methane produced by adding bacteria to whey, a by-product of cream cheese. Not to be outdone, Sara Lee will deploy solar panels to run one of its bakeries, in New Mexico. Many airlines now sell "offsets," which offer passengers a way to invest in projects that reduce CO_2 emissions. In theory, that would compensate for the greenhouse gas caused by their flights. This year's Super Bowl was fuelled by wind turbines. There are carbon-neutral investment banks, carbon-neutral real-estate brokerages, carbon-neutral taxi fleets, and carbon-neutral dental practices. Detroit, arguably America's most vivid symbol of environmental excess, has also staked its claim. ("Our designers know green is the new black," Ford declares on its home page. General Motors makes available hundreds of green pictures, green stories, and green videos to anyone who wants them.)

Possessing an excessive carbon footprint is rapidly becoming the modern equivalent of wearing a scarlet letter. Because neither the goals nor acceptable emissions limits are clear, however, morality is often mistaken for science. A recent article in *New Scientist* suggested that the biggest problem arising from the epidemic of obesity is the additional carbon burden that fat people—who tend to eat a lot of meat and travel mostly in cars—place on the environment. Australia briefly debated imposing a carbon tax on families with more than two children; the environmental benefits of abortion have been discussed widely (and simplistically). Bishops of the Church of England have just launched a "carbon fast," suggesting that during Lent parishioners, rather than giving up chocolate, forgo carbon. (Britons generate an average of a little less than ten tons of carbon per person each year; in the United States, the number is about twice that.)

Greenhouse-gas emissions have risen rapidly in the past two centuries, and levels today are higher than at any time in at least the past six hundred and fifty thousand years. In 1995, each of the six billion people on earth was responsible, on average, for one ton of carbon emissions. Oceans and forests can absorb about half that amount. Although specific estimates vary, scientists and policy officials increasingly agree that allowing emissions to continue at the current rate would induce dramatic changes in the global climate system. To avoid the most catastrophic effects of those changes, we will have to hold emissions steady in the next decade, then reduce them by at least sixty to eighty per cent by the middle of the century. (A delay of just ten years in stopping the increase would require double the reductions.) Yet, even if all carbon emissions stopped today, the earth would continue to warm for at least another century. Facts like these have transformed carbon dioxide into a strange but powerful new currency, difficult to evaluate yet impossible to ignore.

A person's carbon footprint is simply a measure of his contribution to global warming. (CO_2 is the best known of the gases that trap heat in the atmosphere, but others—including water vapor, methane, and nitrous oxide—also play a role.) Virtually every human activity—from watching television to buying a quart of milk—has some carbon cost associated with it. We all consume electricity generated by burning fossil fuels; most people rely on petroleum for transportation and heat. Emissions from those activities are not hard to quantify. Watching a plasma television for three hours every day contributes two hundred and fifty kilograms of carbon to the atmosphere each year; an LCD television is responsible for less than half that number. Yet the calculations required to assess the full environmental impact of how we live can be dazzlingly complex. To sum them up on a label will not be easy. Should the carbon label on a jar of peanut butter include the emissions caused by the fertilizer, calcium, and potassium applied to the original crop of peanuts? What about the energy used to boil the peanuts once they have been harvested, or to mold the jar and print the labels? Seen this way, carbon costs multiply rapidly. A few months ago, scientists at the Stockholm Environment Institute reported that the carbon footprint of Christmas—including food, travel, lighting, and gifts—was six hundred and fifty kilograms per person. That is as much, they estimated, as the weight of "one thousand Christmas puddings" for every resident of England.

As a source of global warming, the food we eat—and how we eat it—is no more significant than the way we make clothes or travel or heat our homes and offices. It certainly doesn't compare to the impact made by tens of thousands of factories scattered throughout the world. Yet food carries enormous symbolic power, so the concept of "food miles"—the distance a product travels from the farm to your home—is often used as a kind of shorthand to talk about climate change in general. "We have to remember our goal: reduce emissions of greenhouse gases," John Murlis told me not long ago when we met in London. "That should be the world's biggest priority." Murlis is the chief scientific adviser to the Carbon Neutral Company, which helps corporations adopt policies to reduce their carbon footprint as well as those of the products they sell. He has also served as the director of strategy and chief scientist for Britain's Environment Agency. Murlis worries that in our collective rush to make choices that display personal virtue we may be losing sight of the larger problem. "Would a carbon label on every product help us?" he asked. "I wonder. You can feel very good about the organic potatoes you buy from a farm near your home, but half the emissions—and half the footprint—from those potatoes could come from the energy you use to cook them. If you leave the lid off, boil them at a high heat, and then mash your potatoes, from a carbon standpoint you might as well drive to McDonald's and spend your money buying an order of French fries."

One particularly gray morning last December, I visited a Tesco store on Warwick Way, in the Pimlico section of London. Several food companies have promised to label their products with the amount of carbon-dioxide emissions associated with making and transporting them. Last spring, Walkers crisps (potato chips) became the first of them to reach British stores, and they are still the only product on the shelves there with a carbon label. I walked over to the crisp aisle, where a young couple had just tossed three bags of Walkers Prawn Cocktail crisps into their shopping cart. The man was wearing fashionable jeans and sneakers without laces. His wife was toting a huge Armani Exchange bag on one arm and dragging their four-year-old daughter with the other. I asked if they paid attention to labels. "Of course," the man said, looking a bit insulted. He was aware that Walkers had placed a carbon label on the back of its crisp packages; he thought it was a good idea. He just wasn't sure what to make of the information.

Few people are. In order to develop the label for Walkers, researchers had to calculate the amount of energy required to plant seeds for the ingredients (sunflower oil and potatoes), as well as to make the fertilizers and pesticides used on those potatoes. Next, they factored in the energy required for diesel tractors to collect the potatoes, then the effects of chopping, cleaning, storing, and bagging them. The packaging and printing processes also emit carbon dioxide and other greenhouse gases, as does the petroleum used to deliver those crisps to stores. Finally, the research team assessed the impact of throwing the empty bags in the trash, collecting the garbage in a truck, driving to a landfill, and burying them. In the end, the researchers—from the Carbon Trust—found that seventy-five grams of greenhouse gases are expended in the production of every individual-size bag of potato chips.

"Crisps are easy," Murlis had told me. "They have only one important ingredient, and the potatoes are often harvested near the factory." We were sitting in a deserted hotel lounge in Central London, and Murlis stirred his tea slowly, then frowned. "Let's just assume every mother cares about the environment—what then?" he asked. "Should the carbon content matter more to her than the fat content or the calories in the products she buys?"

I put that question to the next shopper who walked by, Chantal Levi, a Frenchwoman who has lived in London for thirty-two years. I watched her grab a large bag of Doritos and then, shaking her head, return it to the shelf. "Too many carbohydrates," she said. "I try to watch that, but between the carbs and the fat and the protein it can get to be a bit complicated. I try to buy locally grown, organic food," she continued. "It tastes better, and it's far less harmful to the environment." I asked if she was willing to pay more for products that carried carbon labels. "Of course," she said. "I care about that. I don't want my food flown across the world when I can get it close to home. What a waste."

It is a logical and widely held assumption that the ecological impacts of transporting food—particularly on airplanes over great distances—are far more significant than if that food were grown locally. There are countless books, articles, Web sites, and organizations that promote the idea. There is even a "100-Mile Diet," which encourages participants to think about "local eating for global change." Eating locally produced food has become such a phenomenon, in fact, that the word "locavore" was just named the 2007 word of the year by the New Oxford American Dictionary.

Paying attention to the emissions associated with what we eat makes obvious sense. It is certainly hard to justify importing bottled water from France, Finland, or Fiji to a place like New York, which has perhaps the cleanest tap water of any major American city. Yet, according to one recent study, factories throughout the world are burning eighteen million barrels of oil and consuming forty-one billion gallons of fresh water every day, solely to make bottled water that most people in the U.S. don't need.

"Have a quick rifle through your cupboards and fridge and jot down a note of the countries of origin for each food product," Mark Lynas wrote in his popular handbook "Carbon Counter," published last year by HarperCollins. "The further the distance it has travelled, the bigger the carbon penalty. Each glass of orange juice, for example, contains the equivalent of two glasses of petrol once the transport costs are included. Worse still are highly perishable fresh foods that have been flown in from far away—green beans from Kenya or lettuce from the U.S. They may be worth several times their weight in jet fuel once the transport costs are factored in."

Agricultural researchers at the University of Iowa have reported that the food miles attached to items that one buys in a grocery store are twenty-seven times higher than those for goods bought from local sources. American produce travels an average of nearly fifteen hundred miles before we eat it. Roughly forty per cent of our fruit comes from overseas and, even though broccoli is a vigorous plant grown throughout the country, the broccoli we buy in a supermarket is likely to have been shipped eighteen hundred miles in a refrigerated truck. Although there are vast herds of cattle in the U.S., we import ten per cent of our red meat, often from as far away as Australia or New Zealand.

In his speech last year, Sir Terry Leahy promised to limit to less than one per cent the products that Tesco imports by air. In the United States, many similar efforts are under way. Yet the relationship between food miles and their carbon footprint is not nearly as clear as it might seem. That is often true even when the environmental impact of shipping goods by air is taken into consideration. "People should stop talking about food miles," Adrian Williams told me. "It's a foolish concept: provincial, damaging, and simplistic." Williams is an agricultural researcher in the Natural Resources Department of Cranfield University, in England. He has been commissioned by the British government to analyze the relative environmental impacts of a number of foods. "The idea that a product travels a certain distance and is therefore worse than one you raised nearby—well, it's just idiotic," he said. "It doesn't take into consideration the land use, the type of transportation, the weather, or even the season. Potatoes you buy in winter, of course, have a far higher environmental ticket than if you were to buy them in August." Williams pointed out that when people talk about global warming they usually speak only about carbon dioxide. Making milk or meat contributes less CO_2 to the atmosphere than building a house or making a washing machine. But the animals produce methane and nitrous oxide, and those are greenhouse gases, too. "This is not an equation like the number of calories or even the cost of a product," he said. "There is no one number that works."

Many factors influence the carbon footprint of a product: water use, cultivation and harvesting methods, quantity and type of fertilizer, even the type of fuel used to make the package. Sea-freight emissions are less than a sixtieth of those associated with airplanes, and you don't have to build highways to berth a ship. Last year, a study of the carbon cost of the global wine trade found that it is actually more "green" for New Yorkers to drink wine from Bordeaux, which is shipped by sea, than wine from California, sent by truck. That is largely because shipping wine is mostly shipping glass. The study found that "the efficiencies of shipping drive a 'green line' all the way to Columbus, Ohio, the point where a wine from Bordeaux and Napa has the same carbon intensity."

The environmental burden imposed by importing apples from New Zealand to Northern Europe or New York can be lower than if the apples were raised fifty miles away. "In New Zealand, they have more sunshine than in the U.K., which helps productivity," Williams explained. That means the yield of New Zealand apples far exceeds the yield of those grown in northern climates, so the energy required for farmers to grow the crop is correspondingly lower. It also helps that the electricity in New Zealand is mostly generated by renewable sources, none of which emit large amounts of CO_2. Researchers at Lincoln University, in Christchurch, found that lamb raised in New Zealand and shipped eleven thousand miles by boat to England produced six hundred and eighty-eight kilograms of carbon-dioxide emissions per ton, about a fourth the amount produced by British lamb. In part, that is because pastures in New Zealand need far less fertilizer than most grazing land in Britain (or in many parts of the United States). Similarly, importing beans from Uganda or Kenya—where the farms are small, tractor use is limited, and the fertilizer is almost always manure—tends to be more efficient than growing beans in Europe, with its reliance on energy-dependent irrigation systems.

Williams and his colleagues recently completed a study that examined the environmental costs of buying roses shipped to England from Holland and of those exported (and sent by air) from Kenya. In each case, the team made a complete life-cycle analysis of twelve thousand rose stems for sale in February—in which all the variables, from seeds to store, were taken into consideration. They even multiplied the CO_2 emissions for the air-freighted Kenyan roses by a factor of nearly three, to account for the increased effect of burning fuel at a high altitude. Nonetheless, the carbon footprint of the roses from Holland—which are almost always grown in a heated greenhouse—was six times the footprint of those shipped from Kenya. Even Williams was surprised by the magnitude of the difference. "Everyone always wants to make ethical choices about the food they eat and the things they buy," he told me. "And they should. It's just that what seems obvious often is not. And we need to make sure people understand that before they make decisions on how they ought to live."

How do we alter human behavior significantly enough to limit global warming? Personal choices, no matter how virtuous, cannot do enough. It will also take laws and money. For decades, American utilities built tall smokestacks, hoping to keep the pollutants they emitted away from people who lived nearby. As emissions are forced into the atmosphere, however, they react with water molecules and then are often blown great distances by prevailing winds, which in the United States tend to move from west to east. Those emissions—principally sulfur dioxide produced by coal-burning power plants—are the primary source of acid rain, and by the nineteen-seventies it had become clear that they were causing grave damage to the environment, and to the health of many Americans. Adirondack Park, in upstate New York, suffered more than anywhere else: hundreds of streams, ponds, and lakes there became so acidic that they could no longer support plant life or fish. Members of Congress tried repeatedly to introduce legislation to reduce sulfur-dioxide levels, but the Reagan Administration (as well as many elected officials, both Democratic and Republican, from regions where sulfur-rich coal is mined) opposed any controls, fearing that they would harm the economy. When the cost of polluting is negligible, so are the incentives to reducing emissions.

"We had a complete disaster on our hands," Richard Sandor told me recently, when I met with him at his office at the Chicago Climate Exchange. Sandor, a dapper sixty-six-year-old man in a tan cable-knit cardigan and round, horn-rimmed glasses, is the exchange's chairman and C.E.O. In most respects, the exchange operates like any other market. Instead of pork-belly futures or gold, however, CCX members buy and sell the right to pollute. Each makes a voluntary (but legally binding) commitment to reduce emissions of greenhouse gases—including carbon dioxide, methane, and nitrous oxide—and hydrofluorocarbons. Four hundred corporations now belong to the exchange, including a growing percentage of America's largest manufacturers. The members agree to reduce their emissions by a certain amount every year, a system commonly known as cap and trade. A baseline target, or cap, is established, and companies whose emissions fall below that cap receive allowances, which they can sell (or save to use later). Companies whose emissions exceed the limit are essentially fined and forced to buy credits to compensate for their excess.

Sandor led me to the "trading floor," which, like most others these days, is a virtual market populated solely by computers. "John, can you get the carbon futures up on the big screen?" Sandor yelled to one of his colleagues. Suddenly, a string of blue numbers slid across the monitor. "There is our 2008 price," Sandor said. Somebody had just bid two dollars and fifteen cents per ton for carbon futures.

A former Berkeley economics professor and chief economist at the Chicago Board of Trade, Sandor is known as the "father of financial futures." In the nineteen-seventies, he devised a market in interest rates which, when they started to fluctuate, turned into an immense source of previously untapped wealth. His office is just north of the Board of Trade, where he served for two years as vice-chairman. The walls are filled with interest-rate arcana and mortgage memorabilia; his desk is surrounded by monitors that permit him to track everything from catastrophic-risk portfolios to the price of pollution.

Sandor invents markets to create value for investors where none existed before. He sees himself as "a guy from the sixties"—but one who believes that free markets can make inequality disappear. So, he wondered, why not offer people the right to buy and sell shares in the value of reduced emissions? "At first, people laughed when I suggested the whole future idea," he said. "They didn't see the point of hedging on something like interest rates, and when it came to pollution rights many people just thought it was wrong to take a business approach to environmental protection."

For Sandor, personal factors like food choices and driving habits are small facets of a far larger issue: making pollution so costly that our only rational choice is to stop. When he started, though, the idea behind a sulfur-dioxide-emissions market was radical. It also seemed distasteful; opponents argued that codifying the right to pollute would only remove the stigma from an unacceptable activity. You can't trade something unless you own it; to grant a company the right to trade in emissions is also to give it a property right over the atmosphere. (This effect was noted most prominently when the Reagan Administration deregulated airport landing rights, in 1986. Airlines that already owned the rights to land got to keep those rights, while others had to buy slots at auction; in many cases, that meant that the country's richest airlines were presented with gifts worth millions of dollars.)

Sandor acknowledges the potential for abuse, but he remains convinced that emissions will never fall unless there is a price tag attached to them. "You are really faced with a couple of possibilities when you want to control something," he told me. "You can say, 'Hey, we will allow you to use only x amount of these pollutants.' That is the command approach. Or you can make a market."

In the late nineteen-eighties, Sandor was asked by an Ohio public-interest group if he thought it would be possible to turn air into a commodity. He wrote an essay advocating the creation of an exchange for sulfur-dioxide emissions. The idea attracted a surprising number of environmentalists, because it called for large and specific reductions; conservatives who usually oppose regulation approved of the market-driven solution.

When Congress passed the Clean Air Act, in 1990, the law included a section that mandated annual acid-rain reductions of ten million tons below 1980 levels. Each large smokestack was fitted with a device to measure sulfur-dioxide emissions. As a way to help meet the goals, the act enabled the creation of the market. "Industry lobbyists said it would cost ten billion dollars in electricity increases a year. It cost one billion," Sandor told me. It soon became less expensive to reduce emissions than it was to pollute. Consequently, companies throughout the country suddenly discovered the value of investing millions of dollars in scrubbers, which capture and sequester sulfur dioxide before it can reach the atmosphere.

Sandor still enjoys describing his first sulfur trade. Representatives of a small Midwestern town were seeking a loan to build a scrubber. "They were prepared to borrow millions of dollars and leverage the city to do it," he told me. "We said, 'We have a better idea.'" Sandor arranged to have the scrubber installed with no initial cost, and the apparatus helped the city fall rapidly below its required emissions cap. He then calculated the price of thirty years' worth of that municipality's SO_2 emissions and helped arrange a loan for the town. "We gave it to them at a significantly lower rate than any bank would have done," Sandor said. "It was a fifty-million-dollar deal and they saved seven hundred and fifty thousand dollars a year—and never had to pay a balloon mortgage at the end. I mention this because trading that way not only allows you to comply with the law, but it provides creative financing tools to help structure the way investments are made. It encourages people to comply at lower costs, because then they will make money."

The program has been an undisputed success. Medical and environmental savings associated with reduced levels of lung disease and other conditions have been enormous—more than a hundred billion dollars a year, according to the E.P.A. "When is the last time you heard somebody even talking about acid rain?" Sandor asked. "It was going to ravage the world. Now it is not even mentioned in the popular press. We have reduced emissions from eighteen million tons to nine million, and we are going to halve it again by 2010. That is as good a social policy as you are ever likely to see."

No effort to control greenhouse-gas emissions or to lower the carbon footprint—of an individual, a nation, or even the planet—can succeed unless those emissions are priced properly. There are several ways to do that: they can be taxed heavily, like cigarettes, or regulated, which is the way many countries have established mileage-per-gallon standards for automobiles. Cap and trade is another major approach—although CO_2 emissions are a far more significant problem for the world than those which cause acid rain, and any genuine solution will have to be global.

Higher prices make conservation appealing—and help spark investment in clean technologies. When it costs money to use carbon, people begin to seek profits from selling fuel-efficient products like long-lasting light bulbs, appliances that save energy, hybrid cars, even factories powered by the sun. One need only look at the passage of the Clean Water Act, in 1972, to see that a strategy that combines legal limits with realistic pricing can succeed. Water had always essentially been free in America, and when something is free people don't value it. The act established penalties that made it expensive for factories to continue to pollute water. Industry responded at once, and today the United States (and much of the developed world) manufactures more products with less water than it did fifty years ago. Still, whether you buy a plane ticket, an overcoat, a Happy Meal, a bottle of wine imported from Argentina, or a gallon of gasoline, the value of the carbon used to make those products is not reflected by their prices.

In 2006, Sir Nicholas Stern, a former chief economist of the World Bank, who is now the head of Britain's Economic Service, issued a comprehensive analysis of the implications of global warming, in which he famously referred to climate change as "the greatest market failure the world has ever seen." Sir Nicholas suggested that the carbon emissions embedded in almost every product ought, if priced realistically, to cost about eighty dollars a ton.

Trading schemes have many opponents, some of whom suggest that attaching an acceptable price to carbon will open the door to a new form of colonialism. After all, since 1850, North America and Europe have accounted for seventy per cent of all greenhouse-gas emissions, a trend that is not improving. Stephen Pacala, the director of Princeton University's Environmental Institute, recently estimated that half of the world's carbon-dioxide emissions come from just seven hundred million people, about ten per cent of the population.

If prices were the same for everyone, however, rich countries could adapt more easily than countries in the developing world. "This market driven mechanism subjects the planet's atmosphere to the *legal* emission of greenhouse gases," the anthropologist Heidi Bachram has written. "The arrangement parcels up the atmosphere and establishes the routinized buying and selling of 'permits to pollute' as though they were like any other international commodity." She and others have concluded that such an approach would be a recipe for social injustice.

No one I spoke to for this story believes that climate change can be successfully addressed solely by creating a market. Most agreed that many approaches—legal, technological, and financial—will be necessary to lower our carbon emissions by at least sixty per cent over the next fifty years. "We will have to do it all and more," Simon Thomas told me. He is the chief executive officer of Trucost, a consulting firm that helps gauge the full burden of greenhouse-gas emissions and advises clients on how to address them. Thomas takes a utilitarian approach to the problem, attempting to convince corporations, pension funds, and other investors that the price of continuing to ignore the impact of greenhouse-gas emissions will soon greatly exceed the cost of reducing them.

Thomas thinks that people finally are beginning to get the message. Apple computers certainly has. Two years ago, Greenpeace began a "Green my Apple" campaign, attacking the company for its "iWaste." Then, last spring, not long before Apple launched the iPhone, Greenpeace issued a guide to electronics which ranked major corporations on their tracking, reporting, and reduction of toxic chemicals and electronic waste. Apple came in last. The group's findings were widely reported, and stockholders took notice. (A company that sells itself as one of America's most innovative brands cannot afford to ignore the environmental consequences of its manufacturing processes.) Within a month, Steve Jobs, the company's C.E.O., posted a letter on the Apple Web site promising a "greener Apple." He committed the company to ending the use of arsenic and mercury in monitors and said that the company would shift rapidly to more environmentally friendly LCD displays.

"The success of approaches such as ours relies on the idea that even if polluters are not paying properly now there is some reasonable prospect that they will have to pay in the future," Thomas told me. "If that is true, then we know the likely costs and they are of significant value. If polluters never have to pay, then our approach will fail.

"You have to make it happen, though," he went on. "And that is the job of government. It has to set a level playing field so that a market economy can deliver what it's capable of delivering." Thomas, a former investment banker, started Trucost nearly a decade ago. He mentioned the free-market economist Friedrich von Hayek, who won the Nobel Prize in Economics in 1974. "There is a remarkable essay in which he shows how an explosion, say, in a South American tin mine could work its way through the global supply chain to increase the price of canned goods in Europe," Thomas said. I wondered what the price of tin could have to do with the cost of global warming.

"It is very much to the point," Thomas answered. "Tin became more expensive and the market responded. In London, people bought fewer canned goods. The information travelled all the way from that mine across the world without any person in that supply chain even knowing the reasons for the increase. But there was less tin available and the market responded as you would have hoped it would." To Thomas, the message was simple: "If something is priced accurately, its value will soon be reflected in every area of the economy."

Without legislation, it is hard to imagine that a pricing plan could succeed. (The next Administration is far more likely to act than the Bush Administration has been. The best-known climate-change bill now before

Congress, which would mandate capping carbon limits, was written by Senator Joseph Lieberman. Hillary Clinton, Barack Obama, and John McCain are co-sponsors. Most industrial leaders, whatever their ideological reservations, would prefer a national scheme to a system of rules that vary from state to state.) Even at today's anemic rates, however, the market has begun to function. "We have a price of carbon that ranges from two to five dollars a ton," Sandor told me. "And everyone says that is too cheap. Of course, they are right. But it's not too cheap for people to make money.

"I got a call from a scientist a while ago"—Isaac Berzin, a researcher at M.I.T. "He said, 'Richard, I have a process where I can put an algae farm next to a power plant. I throw some algae in and it becomes a super photosynthesis machine and sucks the carbon dioxide out of the air like a sponge. Then I gather the algae, dry it out, and use it as renewable energy." Berzin asked Sandor whether, if he was able to take fifty million tons of carbon dioxide out of the atmosphere in this way, he could make a hundred million dollars.

"I said, 'Sure,'" Sandor recalled, laughing. "Two dollars a ton, why not? So he sends me a term paper. Not a prospectus, even." Sandor was skeptical, but it didn't take Berzin long to raise twenty million dollars from investors, and he is now working with the Arizona Public Service utility to turn the algae into fuel. Sandor shook his head. "This is at two dollars a ton," he said. "The lesson is important: price stimulates inventive activity. Even if you think the price is too low or ridiculous. Carbon has to be rationed, like water and clean air. But I absolutely promise that if you design a law and a trading scheme properly you are going to find everyone from professors at M.I.T. to the guys in Silicon Valley coming out of the woodwork. That is what we need, and we need it now."

In 1977, Jimmy Carter told the American people that they would have to balance the nation's demand for energy with its "rapidly shrinking resources" or the result "may be a national catastrophe." It was a problem, the President said, "that we will not solve in the next few years, and it is likely to get progressively worse through the rest of this century. We must not be selfish or timid if we hope to have a decent world for our children and grandchildren." Carter referred to the difficult effort as the "moral equivalent of war," a phrase that was widely ridiculed (along with Carter himself, who wore a cardigan while delivering his speech, to underscore the need to turn down the thermostat).

Carter was prescient. We are going to have to reduce our carbon footprint rapidly, and we can do that only by limiting the amount of fossil fuels released into the atmosphere. But what is the most effective—and least painful—way to achieve that goal? Each time we drive a car, use electricity generated by a coal-fired plant, or heat our homes with gas or oil, carbon dioxide and other heat-trapping gases escape into the air. We can use longer-lasting light bulbs, lower the thermostat (and the air-conditioning), drive less, and buy more fuel-efficient cars. That will help, and so will switching to cleaner sources of energy. Flying has also emerged as a major carbon don't—with some reason, since airplanes at high altitudes release at least ten times as many greenhouse gases per mile as trains do. Yet neither transportation—which accounts for fifteen per cent of greenhouse gases—nor industrial activity (another fifteen per cent) presents the most efficient way to shrink the carbon footprint of the globe.

Just two countries—Indonesia and Brazil—account for about ten per cent of the greenhouse gases released into the atmosphere. Neither possesses the type of heavy industry that can be found in the West, or for that matter in Russia or India. Still, only the United States and China are responsible for greater levels of emissions. That is because tropical forests in Indonesia and Brazil are disappearing with incredible speed. "It's really very simple," John O. Niles told me. Niles, the chief science and policy officer for the environmental group Carbon Conservation, argues that spending five billion dollars a year to prevent deforestation in countries like Indonesia would be one of the best investments the world could ever make. "The value of that land is seen as consisting only of the value of its lumber," he said. "A logging company comes along and offers to strip the forest to make some trivial wooden product, or a palm-oil plantation. The governments in these places have no cash. They are sitting on this resource that is doing nothing for their economy. So when a guy says, 'I will give you a few hundred dollars if you let me cut down these trees,' it's not easy to turn your nose up at that. Those are dollars people can spend on schools and hospitals."

The ecological impact of decisions like that are devastating. Decaying trees contribute greatly to increases in the levels of greenhouse gases. Plant life absorbs CO_2. But when forests disappear, the earth loses one of its two essential carbon sponges (the other is the ocean). The results are visible even from space. Satellite photographs taken over Indonesia and Brazil show thick plumes of smoke rising from the forest. According to the latest figures, deforestation pushes nearly six billion tons of CO_2 into the atmosphere every year. That amounts to thirty million acres—an area half the size of the United Kingdom—chopped down each year. Put another way, according to one recent calculation, during the next twenty-four hours the effect of losing forests in Brazil and Indonesia will be the same as if eight million people boarded airplanes at Heathrow Airport and flew en masse to New York.

"This is the greatest remaining opportunity we have to help address global warming," Niles told me. "It's a no-brainer. People are paying money to go in and destroy those forests. We just have to pay more to prevent that from happening." Niles's group has proposed a trade: "If you save your forest and we can independently audit and verify it, we will calculate the emissions you have saved and pay you for that." The easiest way to finance such a plan, he is convinced, would be to use carbon-trading allowances. Anything that prevents carbon dioxide from entering the atmosphere would have value that could be quantified and traded. Since undisturbed farmland has the same effect as not emitting carbon dioxide at all, people could create allowances by leaving their forests untouched or by planting new trees. (Rain forests are essential to planetary vitality in other ways, too, of course. More than a third of all terrestrial species live in forest canopies. Rising levels of CO_2 there alter the way that forests function, threatening to increase flooding and droughts and epidemics of plant disease. Elevated CO_2 in the forest atmosphere also reduces the quality of the wood in the trees, and that in turn has an impact on the reproduction of flowers, as well as that of birds, bees, and anything else that relies on that ecosystem.)

From both a political and an economic perspective, it would be easier and cheaper to reduce the rate of deforestation than to cut back significantly on air travel. It would also have a far greater impact on climate change and on social welfare in the developing world. Possessing rights to carbon would grant new power to farmers who, for the first time, would be paid to preserve their forests rather than destroy them. Unfortunately, such plans are seen by many people as morally unattractive. "The whole issue is tied up with the misconceived notion of 'carbon colonialism,'" Niles told me. "Some activists do not want the Third World to have to alter their behavior, because the problem was largely caused by us in the West."

Environmental organizations like Carbon Trade Watch say that reducing our carbon footprint will require restructuring our lives, and that before we in the West start urging the developing world to do that we ought to make some sacrifices; anything else would be the modern equivalent of the medieval practice of buying indulgences as a way of expiating one's sins.

"You have to realize that, in the end, people are trying to buy their way out of bad behavior," Tony Juniper, the director of Friends of the Earth, told me. "Are we really a society that wants to pay rich people not to fly on private jets or countries not to cut down their trees? Is that what, ultimately, is morally right and equitable?"

Sandor dismisses the question. "Frankly, this debate just makes me want to scream," he told me. "The clock is moving. They are slashing and burning and cutting the forests of the world. It may be a quarter of global warming and we can get the rate to two per cent simply by inventing a preservation credit and making that forest have value in other ways. Who loses when we do that?

"People tell me, well, these are bad guys, and corporate guys who just want to buy the right to pollute are bad, too, and we should not be giving them incentives to stop. But we need to address the problems that exist, not drown in fear or lose ourselves in morality. Behavior changes when you offer incentives. If you want to punish people for being bad corporate citizens, you should go to your local church or synagogue and tell God to punish them. Because that is not our problem. Our problem is global warming, and my job is to reduce greenhouse gases at the lowest possible cost. I say solve the problem and deal with the bad guys somewhere else."

The Tesco corporate headquarters are spread across two low-slung, featureless buildings in an unusually dismal part of Hertfordshire, about half an hour north of London. Having inspired many of the discussions about the meaning of our carbon footprint, the company has been criticized by those who question the emphasis on food. As Adrian Williams, the Cranfield agricultural researcher, put it, the company has been "a little bit shocked" by the discovery that its original goal, to label everything, was naïve.

The process has indeed been arduous. Tesco has undertaken a vast—and at times lonely—attempt to think about global warming in an entirely new way, and the company shows little sign of pulling back. "We are spending more than a hundred million pounds a year trying to increase our energy efficiency and reduce CO_2 emissions," Katherine Symonds told me. A charismatic woman with an abiding belief that global warming can be addressed rationally, Symonds is the corporation's climate-change manager. "We are trying to find a way to help consumers make choices they really want to make—choices that mean something to them. This is not all about food. We just happen to be in the food business.

"One of our real responsibilities is to say to our customers, 'The most important thing you can do to effect climate change is insulate your house properly,'" she went on. "'Next would be to get double-glazed windows,'" which prevent heat from escaping in the winter. "Third, everyone should get a new boiler.' We are trying to put this into context, not to say, 'Buy English potatoes.'" Consumers are unlikely to stop shopping. Economies won't stand still, either; those of China and India are expanding so speedily that people often ask whether sacrifices anywhere else can even matter.

"We have to be careful not to rush from denial to despair," John Elkington told me, when I visited him not long ago at his offices at SustainAbility, the London-based environmental consulting firm he helped found more than two decades ago. He believes there is a danger that people will feel engulfed by the challenge, and ultimately helpless to address it.

"We are in an era of creative destruction," he said. A thin, easygoing man with the look of an Oxford don, Elkington has long been one of the most articulate of those who seek to marry economic prosperity with environmental protection. "What happens when you go into one of these periods is that before you get to the point of reconstruction things have to fall apart. Detroit will fall apart. I think Ford"—a company that Elkington has advised for years—"will fall apart. They have just made too many bets on the wrong things. A bunch of the institutions that we rely on currently will, to some degree, decompose. I believe that much of what we count as democratic politics today will fall apart, because we are simply not going to be able to deal with the scale of change that we are about to face. It will profoundly disable much of the current political class."

He sat back and smiled softly. He didn't look worried. "I wrote my first report on climate change in 1978, for Herman Kahn, at the Hudson Institute," he explained. "He did not at all like what I was saying, and he told me, 'The trouble with you environmentalists is that you see a problem coming and you slam your foot on the brakes and try and steer away from the chasm. The problem is that it often doesn't work. Maybe the thing to do is jam your foot on the pedal and see if you can just jump across.' At the time, I thought he was crazy, but as I get older I realize what he was talking about. The whole green movement in technology is in that space. It is an attempt to jump across the chasm."

The Great Oasis

Burkhard Bilger

December 19, 2011

In the Al Hajar Mountains of northern Oman, at the eastern edge of the Arabian Desert, high above the white terraces and minarets of Muscat, rain comes rarely and then in floods. *Hajar* means "rock" in Arabic, and the mountains are made of little else—a fractal landscape of umber and dusty limestone, thrust from the sea more than sixty-five million years ago and still shaped more by salt water than by sweet. When the clouds burst, as they do a few times a year, the rain skitters from the slopes like oil from a griddle, gathers into rivulets and swiftly moving sheets, and tumbles into the wadies that wind between peaks. The ancient Omanis built networks of aqueducts and underground *falajes* to funnel the water to their crops. Oases of mango, date palm, sweet lemon, and lime still survive on this system, their fruit knuckled in on itself against the heat, smaller and more pungent than their Indian ancestors. But on most slopes the only traces of green are a few umbrella-thorn trees, *Acacia tortilis*, anchored to the bare rock. Their roots can descend more than a hundred feet in search of groundwater.

"It used to be much wetter here when I was a boy," Hamad Reesi said, as our S.U.V. lurched up a gravel switchback in the foothills. "You never had to buy fodder for your goats." Ali al-Abdullatif nodded, then yanked the steering wheel to one side to avoid a dropoff. Next to him, Pieter Hoff dozed in the passenger seat. Abdullatif is the chairman of the Horticultural Association of Oman, a slender, cultivated man more comfortable potting plants than going on desert excursions. Hoff is a Dutch inventor and former tulip and lily grower who had come to Oman to test an experimental tree-planting device. We'd spent the past few hours bumping over back roads, stopping every few minutes to look at trees that might be good for Hoff's project: hardy natives like *Ziziphus spina-christi*, said to have provided the thorns for Jesus' crown, and *Salvadora persica*, the toothbrush tree. Its fibrous twigs were laced with fluoride and antiseptics. Word had it that this area was home to one of the last baobabs in northern Oman, but we'd got lost trying to find it and had picked up Reesi, a local farmer, as a guide. The great tree was deep in the mountains, he said. We would never reach it on our own.

Abdullatif and Reesi wore the traditional white robes and embroidered prayer caps of Omani Muslims. They were born and raised here—although Abdullatif had done his horticultural training in England, at Canterbury College, in Kent—and had seen the country transformed, in forty years, from a near-medieval land of warring tribes to a unified and oil-rich sultanate. When Abdullatif was a boy, firewood was still gathered by Bedouin nomads and brought in by camel; water arrived by donkey in goatskin bags and barrels once filled with ghee. Now the coast was dotted with desalinization plants and it was sometimes hard to tell that Oman was a desert nation. Along the boulevards and highways of Muscat, the medians were as lush as croquet lawns. Weeping casuarina trees lined the shoulders between beds of petunia, bougainvillea, and topiary trimmed like battlements. The sultan was said to be an environmentalist—he'd recently decreed the construction of the country's first botanical garden—and he wanted his capital green.

The illusion didn't last beyond the city limits. Most of Oman averages less than six inches of rain a year—barely enough to sustain native plants, much less thirsty exotics like the petunias. On the northern coast, long known as the country's fertile crescent, so much groundwater has been tapped for farms, orchards, and date-palm plantations that salt water has seeped into the aquifers. "Look at this," Abdullatif told us at one point, gesturing at a line of dead palms along the road, their fronds decaying to dust on the ground. "Complete destruction."

When we reached the top of the first pass, Abdullatif pulled onto an overlook and killed the engine. The sun was setting, the road getting harder to follow, and he seemed ready to turn back. "I don't like the drops on the sides here," he said, as he got out of the car. "These sheer drops. They do not make me feel very secure." After a while, Hoff shook himself awake and joined Abdullatif outside. Tall and pale, with a bladelike nose and a thinning crown of blond hair, he was built for cooler climes. Perched beside the dusky, heavy-lidded Abdullatif, he looked like an egret about to snack on a lizard.

"What if the car stops?" he said. "Is there a hotel here?"

"Yes, a very big, open hotel. We have one banana left. We share it."

Hoff laughed. To the west, the high peaks of the Al Hajar rose rank upon rank into the coppery sky, the empty plains half in shadow below them. "We have a saying in Holland," he said. "'If you call out in the desert, no one will hear you.'" But I knew what he was thinking: not so long ago, these mountains were covered with desert junipers and groves of bitter olive. What would it take to bring them back?

The desert is a good place for visionaries. It can flower in the mind even as it withers at your feet. About a third of all land on the planet has been claimed by it—almost twenty million square miles—and the percentage increases every year. Where rain is scarce and the ground is stripped of trees, where soil is eroded by the steady beat of sun, hooves, and seasonal farming, a landscape can turn to dust in a generation. "These are real deserts that are being born today, under our eyes," the French botanist André Aubréville warned in 1949, when he popularized the term "desertification." "The desert always menaces." In the past century, over most of the globe, the amount of dust in the air has doubled.

It's an old story in some ways. Deserts have been advancing and retreating for much of the earth's history, driven by tectonic shifts and planetary wobbles beyond our control. The Sahara and the Arabian Peninsula haven't been green for thousands of years.

What has changed is the fact that global warming is making climates more extreme. Regional rainfall is hard to predict in the long term, but most models agree on the over-all pattern. "The wet will get wetter and the dry will get dryer," Isaac Held, a research scientist with the National Oceanic and Atmospheric Administration, told me. By the end of the century, according to the Intergovernmental Panel on Climate Change, rainfall could decrease by fifteen to twenty per cent in the Middle East and by twenty-five per cent in North Africa. "That's a lot," Held said. The recent drought and famine in Somalia, which has killed tens of thousands of people and driven many more into Kenya and Ethiopia, is a preview of things to come.

To Hoff, the solution seems straightforward. If we can replant the forests lost to desertification, he says, we can provide food, fuel, shade, and shelter on an enormous scale. We can conserve water, fertilize the soil, protect wildlife, and cool the atmosphere. Every year, human industry sends about nine billion tons of carbon into the air. An acre of trees, planted in a desert, could pull two to three tons of that carbon back down. "Multiplied by five billion, we have solved the problem," Hoff says.

The math is a little fuzzy, admittedly. Five billion acres is an area twice the size of Europe. Even if all of it could be reforested, the trees would gradually stop sequestering carbon as they matured. Still, the benefits would be dramatic and the idea isn't as far-fetched as it seems. Since the mid-nineteen-sixties, Israel has forested tens of thousands of acres of the Negev Desert, using simple irrigation systems to collect and distribute the rainfall. In Kenya, the Green Belt Movement founded by the late Wangari Maathai, who won the Nobel Peace Prize in 2004, has planted more than forty-five million trees. And the Chinese have outdone everyone. Since 1982, they have planted more than forty billion trees, many of them in a nearly three-thousand-mile strip along the southern edge of the Gobi Desert. Forests that were clear-cut for agriculture during the Great Leap Forward, fed into furnaces for the ironworks of the Cultural Revolution, or sawed up for housing and other needs (chopsticks alone consume nearly an acre of trees a day) have been replanted on an equally epic scale. By 2020, the Chinese plan to add another hundred million acres of trees—an area larger than Germany.

As global temperatures rise, reforestation schemes seem to grow ever more extravagant—fever dreams of the desert's future. One project, proposed three years ago by a group of British and Norwegian designers, would consist of long chains of greenhouses and orchards, running for miles across the Sahara. The trees and crops would subsist on seawater pumped from the coast and desalinated using heat and power from huge solar arrays. Another proposal, from the Swedish architect Magnus Larsson, would make use of an organism called *Bacillus pasteurii*, which can turn sand into sandstone. In Larsson's scheme, great masses of the bacteria would be injected into dunes across the breadth of the Sahara, creating a bulwark against the sand and solid footing for a shelterbelt of trees. Water would collect in the sandstone's cool, porous substructure, sustaining the trees' roots and any settlers who wished to move inside. In Larsson's drawings, the underground rooms have the groovy, biomorphic look of an old Yes album cover.

George Taylor, a former agriculture and environment officer for U.S.A.I.D., remembers fielding a number of such proposals when he worked in Africa in the nineteen-eighties and nineties. "Delusional development," he calls them. Yet a variation on Larsson's idea has attracted substantial political and financial backing. The Great Green Wall, as it's known, was first proposed in the mid-eighties and finally approved by the African Union in 2007. The exact shape it will take is still a matter for debate, but the idea, in its original form, is thrillingly simple. To halt the spread of the

Sahara, eleven African nations have agreed to erect a wall of trees across the dusty shoulders of the continent. It will stretch from the Atlantic coast of Senegal to the east coast of Djibouti, across sand and scrub and desiccated grassland, in a column nine miles wide and almost five thousand miles long. If and when it's completed, it will be the largest feat of horticulture in human history. "The desert is a spreading cancer," Abdoulaye Wade, the President of Senegal, declared at a summit in Chad in 2010. "We must fight it. That is why we have decided to join in this titanic battle."

The Great Green Wall is a collective effort, which is to say, a patchwork affair. Each country along its path will reforest its segment after its own fashion, coördinated by a pan-African agency in Chad. The Global Environmental Facility has allocated a hundred and nineteen million dollars for the countries involved in the project, but that barely counts as seed money. A desert, once established, is hard to push back. African leaders have neither the means to mobilize a billion tree-planting farmers nor the money for irrigation systems like those in Israel. How can they grow a forest in the desert?

When I asked foresters and environmentalists that question, they tended to give conflicting answers: the solution lay in technology or grass-roots activism, they said, land reform or carbon-credit financing, drought-tolerant trees or water-retaining gels. And so, this past year, I went to Oman and then to sub-Saharan Africa to look at two of the most promising, albeit contradictory, approaches. Both of them, as it happened, were espoused by Dutchmen.

Hoff's invention, which he calls the Waterboxx, was inspired by a trip to Italy in 1994. He was driving past a barren mountain range near Naples, and he began to wonder what it would take to grow trees there. The local climate was fairly dry, but the problem was less a matter of moisture than of timing. Even deserts can get as much as twenty inches of rain a year, but it all comes down at once. The plants that survive tend to rely on condensation—"They drink from the air," as Hoff puts it. In Africa's Namib Desert, Welwitschia plants have been known to live for more than a thousand years on the dew that they absorb through their long, porous leaves. What if a device could be built on the same principle? Hoff thought. It could collect rain and dew, then release it to a seed or sapling one drop at a time.

Hoff had never lived in a desert. He didn't know much about tree planting or industrial design. But he did have a good business sense and a lifetime's experience with plants. Born in 1953, he grew up in a small farming village in West Friesland, the eldest of nine children. His father was a tulip and lily grower with a modernizing bent—he was one of the first farmers in northern Holland to own a tractor—and Pieter showed an early gift for breeding new varieties. (Most of his lilies were named after Santana songs: Moonflower, Black Magic Woman, and his best-seller, a canary-and-aubergine number called Festival.) In 1976, Hoff and two of his brothers bought the farm from their father and began to expand it. By 2003, when they sold the place, it was Holland's largest grower of lilies.

Hoff went on to devote most of his time and the greater part of his fortune—some thirteen million dollars, at last count—to developing the Waterboxx. By the time I met him, he had spent five years shuttling from desert to desert, testing prototypes with local agronomists. That month alone, he'd been in Kenya, Kuwait, and Bahrain, with Spain and India still to come. "I always search for the most extreme places, where no one expects anything to grow," he said. All told, he had planted some sixty thousand trees in twenty countries, with a few vegetable patches and vineyards thrown in. (Robert Mondavi Winery was testing Waterboxxes in the Napa Valley.) Oman was his most challenging site yet. If he could grow trees there, Hoff figured, he could grow them anywhere.

On the morning before our trip to find the baobab, Hoff took Abdullatif and me to see his latest plantings, in the port city of Sohar. The site was a former camel racetrack, levelled to make way for an industrial park. The soil was gray and gravelly, compacted into hardpan by bulldozers and steamrollers. A viewing platform had been erected nearby—what exactly was on view wasn't clear—surrounded by a hurricane fence topped by razor wire. Beyond it, the land lay table flat in every direction, punctuated only by some container cranes along the coast and the distant flares of an oil refinery. Hoff knelt down and poured a handful of dust into his hand. "It's like the moon," he said.

The Waterboxxes were arrayed in a circle around the platform. There were forty in all—a gift from Queen Beatrix of the Netherlands to Sultan Qaboos bin Sa'id, commemorating his forty years in power. Like most good tools, they weren't much to look at: a set of simple ideas combined to surprising effect. Each box had a round, four-gallon tank molded out of polypropylene, with an open-ended shaft in the middle where the seedlings grew. The lid was modelled on a lotus leaf, with radiating folds that collected the rain and the dew and sluiced them into a pair of drains. "If you have a rain shower of only four inches, then this is full," Hoff said. A wick at the bottom of the tank carried the water to the root at a rate of about four tablespoons a day—a single tank could sustain a seedling for about a year without a refill. The whole box functioned as a temperature regulator, Hoff said. The water absorbed heat by day and released it by night. The shaft was shaped to let in the morning and evening sun but throw shade over the seedling at midday.

Hoff reached down and pried the lid off one of the boxes. "Mother Nature plants trees differently than people do," he said. "We buy a very big plant and then we dig a hole for it. Mother Nature starts with a seed." Saplings from nurseries have well-developed secondary roots, which spread laterally through the ground. Short and densely woven, they draw maximum sustenance from the soil, but they need water right away and by the bucketful. A seed can afford to wait. Encased in dung from a passing bird or other animal, it can survive for months without rain. If the soil is dry, it can put all its energy into sending a single taproot in search of groundwater.

Hoff liked to collect pictures of trees in improbable places: ponderosa pines sprouting from boulders in the Sierras, holly oaks clinging to ledges of Extremadura granite in Spain. A taproot is one of nature's astonishments, he said. It can worm itself into the tiniest crack, then expand a few cells at a time, generating pressures of up to seven hundred and twenty-five pounds per square inch—enough to split paving stones or punch holes through brick walls.

The Waterboxx is designed to encourage such persistence. Like nature, it begins with a seed, then gives it just enough sustenance to survive until it finds water. After a year, when the root reaches wetter soil, the box can be lifted away and reused. "See these young leaves?" Hoff said, pointing to a cluster of heart-shaped buds, bright green against the darkened soil. "They show that the root is already tapping deeper."

This was a Moringa seedling—one of a number of species that Hoff thought might do well here and in Africa. In its native India, it was known as the drumstick tree for its bulbous seedpods, but a generation of tree-planting N.G.O.s had given it a new name: the Miracle Tree. Moringas are among the world's most nutritious plants. Their leaves can be eaten raw, cooked, or ground into baby formula. They contain four times the calcium of milk, three times the potassium of bananas, four times the Vitamin A of carrots, seven times the Vitamin C of oranges, and about half again the protein of soybeans. The seeds can be pressed for an unsaturated fat like olive oil or crushed into a powder that purifies water: its electrolytes attract impurities and precipitate them out of the fluid. Best of all, Moringas are fast-growing and extremely drought-tolerant. "They will not die," Abdullatif said. "A Moringa you cannot kill."

When Hoff planted these seedlings a month earlier, he'd watered the soil and filled the tanks. Now they had to fend for themselves. Soon the heat would rise above a hundred and twenty degrees, with no rain for six months. Yet Hoff could count on close to ninety per cent of his seedlings surviving into their second year. (A few months earlier, *Popular Science* had named the Waterboxx its Innovation of the Year.) All of these Moringas were alive, as were about half the tamarisks that he'd planted in other boxes. Since he'd planted the seedlings in pairs, most of the boxes held at least one living tree. "Next year, it will be one and a half metres," he said, pointing to a survivor.

He stood up and swept his gaze around the site. If this project succeeded, he hoped to plant six thousand more trees in Sohar, and still more in the surrounding desert—a forest of Moringas surging up from the sand, marching west across the Arab Peninsula toward Africa. "Within forty years, if the world wants it, everywhere is covered with trees again," he said.

The Great Green Wall alone could require five billion seedlings or more, not counting those which would die along the way. But Hoff wasn't intimidated. "My country has sixty thousand miles of canals," he said. "We made them over the last two thousand years, all by hand. We have eight thousand miles of dikes, all made by hand. That is one and a half times the length of the Chinese Wall. If tomorrow there is no electricity, we are beneath the water. That is Holland. It's not a country; it's the largest art work in the world. That is why I tell people, if we really want to replant five billion acres, we can do it. It is a matter of determined will."

The cost of planting trees is hard to gauge, given the wildly varying results. The Chinese spend less than fifty cents a tree—just enough to plant a seedling and pray that it will live. The Israelis invest closer to forty dollars, with a success rate of around ninety per cent. Hoff is somewhere in between. Until recently, he gave his Waterboxxes away, chalking them up as a research expense. But, as his costs have mounted, his strategy has changed. "I have to live," he told me. "And you can't reforest deserts if there is no money-making business model." Depending on the size of the order, he now sells his boxes for around fifteen to twenty-five dollars. In a project like the Great Green Wall, he says, the box could be reused up to ten times, bringing the final cost down to as little as a dollar-fifty a tree.

The one thing missing from this equation is local people. When Hoff talks about the world wanting to reforest five billion acres, he mostly means Western donors and environmentalists. But even if he can persuade them to send a billion Waterboxxes to Africa, the seedlings will have to be planted and cared for by local farmers. And they'll have to be convinced that it's worth it.

China is a cautionary example. Most of its reforestation has been done involuntarily, by villagers obliged to meet national quotas. "If you add up all the acres, more trees have probably been planted there than anywhere else," Nick Menzies, the executive director of the Asia Institute at U.C.L.A., told me. "But the survival rates have been dismal." Farmers have set seedlings in the poorest soil, to keep their crops and pastures clear, or planted them upside down, to spite the authorities. They've laid them in identical grids, regardless of the terrain, and never bothered to water or thin them out. They've plowed up tenacious old prairies, where trees rarely grew, then left the seedlings and the topsoil to blow away. Decades after the reforestation along the Gobi began, the desert still claims more than a thousand square miles of land annually. "Every year, we plant trees," one popular saying goes. "But we never see a forest."

The Great Green Wall will face even tougher conditions. The Sahara is an ungovernable landscape—more than three million square miles of sand and stone, advancing and retreating with every change in the weather. Between 1980 and 1984, the desert's southern border moved a hundred and forty-five miles south; by 1990, it had shifted sixty-six miles north. Even if the Great Green Wall survives, the land to the south of it can still turn to desert from drought or overgrazing. The Sahara is bordered by a semi-arid savanna known as the Sahel. (In Arabic, *sahel* means "edge of the desert.") To the north, it gets as little as eight inches of rain a year. To the south, rainfall rises, but so do demands on farmers. In the past sixty years, populations in the Sahel have more than quintupled.

"The idea that there will be an uninterrupted green wall from Dakar to Djibouti—I think it's a dream," Chris Reij, an agroforestry specialist at the Free University of Amsterdam, told me. "If you want to maximize your difficulties, go there." Reij, who is sixty-two, has worked in the Sahel for more than half his life, first for Oxfam International, then for the Centre for International Cooperation, based at his university. Like Hoff, he grew up in Holland and speaks impeccable English—the crisp, almost jaunty sort common among northern Europeans. But if Hoff is a businessman turned environmentalist, as trusting of the free market as he is of photosynthesis, Reij is a radical at heart. He was brought up as a devout Calvinist and studied in Amsterdam in the nineteen-sixties. While Hoff was breeding lilies on his father's farm, Reij protested the war in Vietnam and debated the Club of Rome's limits to growth. "You develop a certain sense about the world and what needs to be done," he told me. "I had a nice picture of Che Guevara in my room."

Reij and Hoff have never met, though not for lack of trying on Hoff's part. Reij's long history in the Sahel, and his ties to academia and to private foundations, have made him one of the gatekeepers for funding in the region. Last year, a Dutch businessman sent Reij an e-mail on Hoff's behalf. "He was suggesting a meeting to discuss the large-scale inclusion of the Waterboxx in the Great Green Wall," Reij told me. "My reaction was that I had no intention whatsoever to participate." The Waterboxx is much too costly for most farmers, Reij told me, and it fails to address the fundamental issue of desertification: what sort of agriculture makes sense in this place? "Tree planting is bloody difficult in dry lands," he said. "When I come across people who say it's possible to re-green two billion hectares, I start observing the clouds."

And yet Reij's own vision is no less ambitious than Hoff's. For the past two years, Reij and Tony Rinaudo, an Australian agronomist and former missionary, have organized what they call the African Re-Greening Initiatives. Instead of trying to erect a wall of trees across the Sahara, Reij told me, African leaders should look fifty miles south, where a green revolution is already under way. In the past twenty-five years, farmers in Burkina Faso, Mali, and Niger have reforested vast stretches of the Sahel. They've done so with little money and no modern equipment, in some of the world's most politically volatile regions. Theirs may be the greatest environmental success in African history, Reij said, and it violates almost every rule of reforestation.

Burkina Faso lies just above the equator, between the sands of Mali to the north and the tropical forests of Ghana and Ivory Coast to the south. When I met Reij there in the spring, it was the middle of the dry season. The noonday sun could send temperatures soaring to a hundred and fifteen degrees and a hot harmattan wind blew down from the desert. On the drive north from the capital, Ouagadougou, a faint haze hung over the countryside, more dust than condensation. To either side of the road, the grasses were parched brittle and sere, the red soil baked hard beneath them. The only signs of life were a few Senegalese fire finches, darting like sparks among the shea trees.

"These are the really difficult months," Reij said. In the open country, the farmers and their livestock had retreated to their mud-brick compounds—miniature fortresses with low walls and circular towers—or to the mosques that spired above the treetops. But in the cities the escalating heat and rumors of distant revolution had begun to inflame local tempers. When we arrived in the city of Ouahigouya, we found most of its municipal buildings burned. A few days earlier, police officers in a nearby city had beaten a student to death, sparking a riot. "The officers are still in hiding," our driver said, pointing gleefully to the blackened shell of the police station. "They're afraid to come out!" Within days, soldiers would be rioting in the capital as well, looting the houses of government officials and injuring the mayor.

To keep tabs on the situation, Reij chatted with local bartenders or with a Syrian family that owned his favorite hotel in town, but nothing seemed to faze him. Going from village to village in his S.U.V., he looked like a retired professor on holiday: floppy hat, comfortable belly, high domed forehead fringed with gray. An avid bird-watcher, he had a beautiful, warbling whistle with which he often accompanied himself as he drove. "I'm a born optimist," he said. Or perhaps he just had a longer memory than most.

When Reij first came to Burkina Faso with his wife, in 1978, he had been hired by Oxfam as a regional planner. It was hard to know where to begin. The Sahel was in the midst of a devastating long-term drought: close to a million people died of famine between 1972 and 1984, prompting huge shipments of Western aid. In Burkina Faso's Central Plateau, some villages lost as many as a quarter of their families, as they were driven into exile, and water levels fell between twenty and forty inches a year. "I was gripped by a certain fear," Reij recalled. "Am I going to work here? Am I going to live here?"

Much of the conservation work in the area was a disaster of good intentions. Two years before Reij arrived, Erik Eckholm, of the *Times*, had warned that the Sahel faced a desperate fuelwood shortage—"the other energy crisis," he called it. Almost every tree within forty miles of Ouagadougou had been cut down. To meet the demands of multiplying populations, one study estimated, reforestation would have to increase fiftyfold. "It was 'Oh, my God, the desert is expanding!

Let's plant some trees!'" George Taylor, of U.S.A.I.D., recalled. "So they did these industrial-scale plantings in every country across the Sahel. Big Caterpillars bulldozing down what they thought was useless brush. And then putting in just huge fields of eucalyptus."

The Sahel had inherited from its French colonists a strictly compartmentalized approach to agriculture: crops over here, trees over there. Farmers were told to clear and plow their fields, enrich them with chemicals, and plant them with improved species. Foresters would oversee the tree plantations. It was a tidy, seemingly scientific method, but it often failed miserably. Without trees to shelter the fields, the topsoil dried up and blew away. Without farmers to tend to the trees, the seedlings died. By the early eighties, crop yields were down to less than four hundred pounds per acre—in the United States, the average cornfield produced fourteen times that amount—and the fuelwood crisis was worse than ever.

In the spring of 1984, Reij drove across the border into southeastern Niger to design a soil-and-water-conservation project. He'd grown to love the Sahel by then—"It felt like a second mother country," he told me—but he despaired of its future. "It was a drought year, a very bad drought year," he said. "There was so much sand and dust in the air that you couldn't see a hundred metres. Even at noon, we had to drive with our headlights on. I thought Niger was being blown off the map." That same year, Reij recalled, the French environmentalist and former Presidential candidate René Dumont visited the region and reached the same conclusion. "Burkina Faso isn't a developing country," he declared. "It's a disappearing country."

Tony Rinaudo, Reij's future partner in the Re-Greening Initiatives, was working in the city of Maradi in those days, near Niger's southern border. He had come to the Sahel from Australia four years earlier with his wife and infant son (they would have three more children in Africa). He was in charge of organizing tree-planting and farming projects for a Christian aid group, Serving in Mission, but was soon doing famine-relief work as well. "I was in shock," he told me. "We had windstorms that would bury the seed or carry it away. We had a mouse plague. We had locust swarms—hatchlings moving across the ground like a carpet. We had crows who knew where the drill holes were. For a young agricultural adviser—I was born in 1957—it was just mind-boggling."

One afternoon, Rinaudo was driving into the country with a trailer-load of tree seedlings when his truck began to get mired in deep sand. He got out to let some air out of the tires, for better traction, and looked around at the barren scrub that encircled him. "I'm wasting my time," he remembers thinking. "It wouldn't matter if I had millions of dollars and dozens of staff. I wouldn't have a chance. The water table was forty to sixty metres deep. I'd plant the trees and watch them die."

It was in that moment of surrender, he says, that an altogether different thought struck him. What if he had things backward? Every year, the villagers cleared the brush to make room for crops, and planted trees around them. And every year the plantings failed and the brush resprouted from its old rootstocks. What if they just let it grow? What if they cut back only a portion of the native trees, let the rest mature, and planted crops between them?

The idea went against conventional wisdom in almost every way. In temperate areas like Europe, the United States, and southeastern Australia, the growing season is short and sunlight is a limited resource. The best way to take advantage of it is to plant your crops beneath an open sky. Most of modern agriculture—its rolling plains and mechanized harvesters, monocultures and center-pivot irrigation systems—is predicated on that assumption. Nothing is allowed to block the sun or get in the way of the equipment. And that means getting rid of the trees.

In the Sahel, the situation is reversed. The sun beats ceaselessly down. There's too little water for irrigation, too little money for mechanized harvesting. The trick isn't to maximize your crops' exposure, Rinaudo realized, but to minimize it—to provide shelter and shade from the wind and the withering heat. It's not enough to plant a row of trees around your field. You have to grow them side by side with your other plants, as a secondary crop. That way, you can harvest grains and vegetables on the ground, fruits and nuts in the trees.

Farmers in the tropics have been doing this since agriculture began, Dennis Garrity, of the World Agroforestry Centre, told me. But the tradition was largely quashed by colonization. In the French colonial system, trees were the property of the state—even those which grew on a farmer's land. Pruning without a permit could earn you a fine; felling a tree could get you jail time. "The laws were put in place to protect the forests," Garrity said. "They did the opposite." Caught between agronomists who insisted on cleared fields and foresters who claimed any new seedlings as state property, farmers in Niger had learned to avoid growing trees altogether. If a seedling sprouted in a field, the farmer dug it up before an *agent forestier* noticed. If an N.G.O. paid for a communal woodlot, it was dutifully planted, fenced in, and left to die. In most cases, the trees cut down to make the fences outnumbered the ones inside.

"So I went to the forestry department," Rinaudo recalled. "And I said, 'Look, what you're doing isn't working. You have a forest guard at the entrance to every city, but the wood comes in anyway. Would you let us try something? If you give people permission to harvest trees, we will teach them to take care of them.'" As it turned out, the foresters were easier to convince than the farmers. "We had ten or twelve in as many villages who agreed to try it out," Rinaudo said. "They were ridiculed and laughed at. As soon as it looked like the idea was taking root, people would come in at night and cut the trees back down." The real battle wasn't against the Sahara, he discovered, but against people's ideas. "If we could change their minds, we could change everything."

In the end, the farmers were given an ultimatum: unless they protected their trees, they would get no food from Rinaudo's famine-relief program. "A lot of them hated me," he said. "They protected roughly half a million trees, but, when the famine was over, two-thirds of them chopped down their trees again."

It was the other third that made all the difference.

Rinaudo and his family went home to Australia in 1999, but Reij continued to work on conservation projects in the Sahel. In 2001, he co-edited a book called "Farmer Innovation in Africa," but most of what he described were small local changes. Nothing that could be called a movement. Then, in the summer of 2004, Reij returned to Niger for the first time in a decade. "There were villages that I used to be able to see at distances of several kilometres, the land was so barren," he told me. "Now I couldn't see them. There was too much vegetation. I thought, Huh. Something is happening here."

A few days later, Reij sent a note to Gray Tappan, a geographer at the U.S. Geological Survey. Tappan had spent more than twenty years documenting land use and vegetation in the Sahel with satellite images and aerial photographs. Like Reij and Rinaudo, he had worked in the area during the terrible droughts of the early eighties, but he hadn't been back in years; and he hadn't heard any talk of reforestation. "A lot of the satellite images were medium or coarse resolution, where you couldn't see the trees," he told me. Even if you could, the area that Reij had seen was dominated by winter-thorn trees, which lose their leaves in the rainy season. In the satellite images, they didn't appear green. "The whole phenomenon flew under the radar," Tappan said.

Tappan promised Reij that he'd do an aerial survey. Then he began to dig into earlier records of the country's vegetation. Beginning in 1955, when Niger was still a colony, French cartographers had taken tens of thousands of aerial photographs. Tappan found copies of these at the official mapping agency in Niamey, then he compared them with his own survey results, as well as with photographs from 1975. "I was blown away," he told me. "I'd never seen something like that." There weren't just a few more trees along the roadsides. Entire stretches of the country had been reforested—more than twelve million acres in all. What's more, the most densely forested areas weren't in parks or nature preserves. They were on farms.

"Twelve million acres equals about two hundred million trees," Reij told me. "No other country in the Sahel, or even in Africa, has managed to do that." We were sitting in a bar in Ouahigouya, looking at aerial photographs on Reij's laptop. One set was from 1975, the other from 2005. In the older pictures, the skin of the continent lay pale and featureless below, creased only by a streambed or two and by the faint crosshatchings of roads. Thirty years later, it was covered with trees. It's tempting to credit the climate, Reij said, since rainfall has increased somewhat in that same period. But rainy periods haven't had this effect on Niger in the past, or on its neighbors today. He pulled up a photograph of the border between Niger and Nigeria. "Same landscape, same people, same culture," he said. Yet Nigeria, which gets more rain, was nearly barren. Niger, to the north, was cloaked in forests. "You can literally see the border from space," he said.

After Reij and Tappan made their discovery, they travelled through southern Niger together to try to figure out what had happened. "Of all the places for trees to return, why there?" Tappan wondered. It was the poorest country in the world, according to the United Nations Development Programme, and it barely had a rainy season. The two men eventually traced the story back to Rinaudo and his epiphany that day outside Maradi. Farmers who still had trees standing after the drought in 1984 found that their harvests increased. Women who used to spend an average of two and a half hours a day collecting wood now spent half an hour. And their neighbors, seeing this, began to protect seedlings on their own land. "The penny dropped," Rinaudo told me. "Trees stopped being a weed."

Until Reij called to give him Tappan's results, in 2004, Rinaudo had no idea how far things had come.

But he has since returned to Niger three times as an adviser for World Vision, a Christian relief and development organization. Since 2009, he has helped lead the African Re-Greening Initiatives. "I don't think that I'm creating a green movement or turning all these villagers into environmentalists," he told me. "The bottom line is just this: growing trees puts food on the table and money in your pocket."

To Reij, what he's seen has been a kind of regional awakening—an outpouring of new ideas from farmers and aid workers left with no other choice. Sometimes, he said, it takes a near-death experience for a person or a people to truly change their behavior—to pull themselves out of the sand by their own hair, as people in Mali like to say. "*Il faut reculer pour mieux sauter*," Reij said, quoting Montaigne. You have to step back in order to jump.

Worldwide, more than a billion acres of forests could be regenerated naturally, according to the World Resources Institute. But the re-greening of Niger won't always be easy to replicate. In some places, like Burkina Faso's Central Plateau, the rootstocks of native trees died off long ago. The soil on the Central Plateau is lateritic—laced with aluminum and iron oxides, which give it its rusty red color. Mixed with water and baked in the sun, it hardens into a crust that's nearly impermeable to the rain. *Zipele*, the locals call it.

One morning, Reij took me to see Ousseni Kindo, one of his favorite farmers in the area. Kindo owns eleven parched acres on a hillside north of Ouahigouya. Lean and grizzled, he wore a midnight-blue robe and a cream-colored prayer cap, and kept up a high-pitched patter as he showed us around. (He claimed to be fifty-four and brought out a birth certificate to prove it; when we pointed out that it said he was sixty-nine, he laughed and admitted that he couldn't read.) His compound was home to two wives, fourteen children, a pair of zebu cattle, and an unruly mob of sheep and goats. To keep the latter away from his seedlings, he carried a homemade slingshot, armed with a sharp rock.

When Kindo first began farming, in 1985, the land was barren. In some spots, the topsoil had washed away, exposing the rock below. During the next twenty-five years, Kindo tried every trick he could to improve things—some picked up by trial and error, some from other farmers, some from N.G.O.s like Oxfam. He made compost piles and girded the slopes with low stone ridges to prevent erosion; he set up termite colonies to churn the soil, and hung feeders to attract birds, for their droppings. On the neighbor's property, the soil was as hard and flat as a clay tennis court; on Kindo's side, it was dimpled with circular pits. This was an ancient technique called *zai* agriculture, Reij said—a kind of low-tech version of Hoff's Waterboxx. The pits, each about a foot wide, collected the rain and funnelled it toward the plants at the bottom. Kindo added a little dung to each one, to fertilize it and seed it with native species. The rest was up to natural selection. "If the seed dies, it doesn't cost you anything," Dennis Garrity, of the World Agroforestry Centre, told me. "If it survives, you know it'll be hardy."

Kindo's land had the scrubby look of savanna, yet every inch of it was cultivated. The trees came in dozens of species and were spaced about twenty feet apart, as in an arboretum. Their branches formed a canopy above the rows of millet, sorghum, cowpeas, and dryland rice at their feet. Some trees provided fruit: tamarind, marula, raisinier, custard apple. Others, like the winter thorns, fertilized the soil by fixing nitrogen with their roots. Shea seeds could be pressed to make butter for cooking or skin care; neem leaves made an excellent mosquito repellent. And almost every tree had some medicinal value: marula bark for malaria and rheumatism; neem oil for acne and contraception; tamarind leaves for jaundice, boils, dysentery, and hemorrhoids.

Farming this way is hard work, Kindo admitted: "I'm always here, even during the dry season. Even if a baby is being baptized, I'm here." The *zais* had to be chopped out with a pickaxe or a hoe, the trees continually pruned and protected from goats and poachers. (The scarecrow on Kindo's farm wasn't meant for birds; it helped ward off local gold miners, who came scavenging for wood.) Still, if the Sahel has a surplus of anything, it's manual laborers. And agroforestry more than repays the effort. On average, *zais* increase yields between thirty and eighty per cent, Reij said—"It should really be a hundred, because the starting point is often *zipele*." The thirteen hundred pounds of grain that Kindo harvested per acre, and the food, fuel, fodder, and medicine that he got from his trees, had carried him from subsistence to surplus. "With my cowpeas alone," he said, "I can pay for food to keep my family for a year."

At the top of the hill, where Kindo's property ended, one of his neighbors came over to greet us. Small, gaunt, and well into his fifties, the man had lost his right arm in a car accident. His left arm was still strong enough to heft an axe, however, and he was busy digging a *zai*. Thirty-five years after Erik Eckholm warned of "the other energy crisis," farmers like these were turning his predictions upside down. Instead of stripping the land of forest, they'd added three-quarters of a million acres to the Central Plateau, and another million acres to southern Mali. "It's counterintuitive, but it's true," Garrity told me. "The more people, the more trees."

Within a generation or two, Reij believes, agroforestry could spread across the southern Sahel, forming a green zone many times wider than the Great Green Wall. If so, it will arrive just in time. Even if farmers manage to double their harvests, they'll barely be able to feed the doubling population. The current famine in Somalia, which could affect twelve million people or more before it's over, has shown how precarious the situation is. "All those children between five and fifteen—bloody hell! What will they be living off in ten years?" Reij said. "We should act as if the Devil is knocking at the door—and he *is* knocking."

Reij's call to action can seem contradictory. The best way to help local farmers, recent history suggests, is to leave well enough alone. Yet groups like Oxfam and Rinaudo's Serving in Mission have played a crucial role in the regreening—if only, at times, by undoing decades of bad advice from former colonists. Even in areas where forests can resprout naturally, there are wells to be dug, roads to be built, and land reforms to be encouraged. In parts of Niger, Mali, and Burkina

Faso, the authorities now let farmers trim or cut down their trees as they see fit. But in most of the Sahel the old colonial system of government control continues.

Last January, Reij began a project with the World Wide Web Foundation to broadcast the latest innovations over cell phones and radio. But the Re-Greening Initiatives' most effective tactic, by far, has been much more direct: take the poorest farmers, pile them into a van, and show them what agroforestry has done for their neighbors. "This isn't a big project with a cloud of money," Reij said. "It's a movement with a bunch of people with their noses in the same direction."

Whether Reij can persuade others to join him remains to be seen. African leaders are loath to give up their literal vision of a Great Green Wall, the journalist Mark Hertsgaard recently wrote in *The Nation.* Yet most Western donors are convinced that this vision is doomed. Hertsgaard, who wrote about the regreening of the Sahel in his 2011 book, "Hot: Living Through the Next Fifty Years on Earth," believes that the Global Environmental Facility will push for a grass-roots approach to the Wall, and that it's unlikely to fund any proposals that are scientifically suspect. But when I spoke to Mohamed Bakarr, a senior specialist with the G.E.F., he wasn't eager to force the issue. "It's the country's prerogative," he told me. "One cannot rule out the idea that some will want to have tree plantations. Those are things that we cannot control."

When it comes to planting trees in the desert, unfortunately, mad ambition is often its own reward. The more titanic the struggle, in the words of the Senegalese President Abdoulaye Wade, the greater the appeal. Wade turned eighty-five in May, and is said to think of the Great Green Wall as a valedictory gesture—an arboreal Arc de Triomphe. Last year, when an official promotional video for the Wall was released, it looked as if it could have been made three decades ago. It showed bulldozers clearing away brush in Senegal while happy volunteers and soldiers planted nursery seedlings all in a row. "When I saw that, I just started to tear my remaining hair out," Reij told me.

Near the end of our trip, Reij and I drove north into Mali, up a winding road that led to the cliffs of Bandiagara. The cliffs are the leading edge of a sandstone escarpment that juts a thousand feet above the Seno Plains. Centuries ago, people built mud-brick structures in the shadow of the rock, like those at Mesa Verde, that are still used as granaries by local villagers. "This is all farmland now," our guide told us, as we looked out from the top. "Twenty years ago, people left this place to go elsewhere. Now they are coming back." I could see the thatched roofs of a village tucked among some mango trees below. Beyond them, to the south and west, airy groves of winter thorn and acacia stretched to the horizon. The wind whipped across the plains so steady and sharp that it made my eyes water. But there was no sand in it. "If you come here on a clear day, you can see all the way to Bankass," the guide said. "Green! Green! Green!"

It was a spectacular view, but not so different, I thought, from the one that I'd seen a month earlier, with Pieter Hoff, in the Al Hajar Mountains. In both places, people were fighting to stave off an encroaching desert; they just had different means at their disposal. The Sahel had manpower but no money; Oman had money but little manpower. Hoff's twenty-dollar Waterboxx might not interest the subsistence farmers of Mali, but then Sultan Qaboos and his citizens weren't likely to pick up axes and dig *zais*. There is a place for technical ingenuity and epic reforestation schemes—for turning the surface of the earth into an art work. It might be in Holland or Israel, southern Spain or someday, perhaps, the Sahel. But you have to start somewhere.

In Oman, on the evening that Hoff and Abdullatif went searching for the baobab, the light had begun to fail when we finally climbed back into the car. The road had narrowed to an old camel path and the dropoffs were more precipitous than ever, but we decided to forge ahead. "I will never forget this baobab," Abdullatif muttered. "But first I have to pass through this in one piece." Soon we were off-road entirely, the undercarriage of the S.U.V. clanging like a gong, a cloud of dust billowing around us. At one point, we stopped to ask directions from two small boys standing outside a mud hut. And then, suddenly, there it was, next to a small stream at the bottom of a hill: the great tree.

In the deepening dusk, it looked like an apparition out of "The Arabian Nights"—a fat caliph surrounded by his fan-fluttering harem. It was wider than our vehicle and at least forty feet tall, with knobby gray branches that hung low over the water. Its wrinkled hide was embedded with thousands of pebbles and nails—offerings, Abdullatif said, from villagers who hoped they would rid them of tooth decay. Like the Moringas in Sohar, this baobab had almost certainly come from someplace else. Half a millennium ago, perhaps, a settler from Zanzibar or East Africa had brought a seedpod here and planted it in a rare patch of wet soil. And then, for hundreds of years, his or her descendants had kept the water flowing past, chasing away goats and sheep and hungry cows until the tree's roots grew deep, its branches so strong and thick that they could fend for themselves.

"If we could plant a tree like this in Sohar, all of Oman would come to see it," Hoff said, gazing up at the branches. In the meadow behind him, Abdullatif and our guide were kneeling on the ground for their evening prayers, bowing low toward Mecca and the setting sun. Their voices came to us as a steady murmur mixed with the rustlings of leaves. When they were done, Hoff bent down and picked something off the ground, then walked contentedly back to the car. His pockets were full of seeds.

The Climate Fixers

Michael Specter

May 14, 2012

Late in the afternoon on April 2, 1991, Mt. Pinatubo, a volcano on the Philippine island of Luzon, began to rumble with a series of the powerful steam explosions that typically precede an eruption. Pinatubo had been dormant for more than four centuries, and in the volcanological world the mountain had become little more than a footnote. The tremors continued in a steady crescendo for the next two months, until June 15th, when the mountain exploded with enough force to expel molten lava at the speed of six hundred miles an hour. The lava flooded a two-hundred-and-fifty-square-mile area, requiring the evacuation of two hundred thousand people.

Within hours, the plume of gas and ash had penetrated the stratosphere, eventually reaching an altitude of twenty-one miles. Three weeks later, an aerosol cloud had encircled the earth, and it remained for nearly two years. Twenty million metric tons of sulfur dioxide mixed with droplets of water, creating a kind of gaseous mirror, which reflected solar rays back into the sky. Throughout 1992 and 1993, the amount of sunlight that reached the surface of the earth was reduced by more than ten per cent.

The heavy industrial activity of the previous hundred years had caused the earth's climate to warm by roughly three-quarters of a degree Celsius, helping to make the twentieth century the hottest in at least a thousand years. The eruption of Mt. Pinatubo, however, reduced global temperatures by nearly that much in a single year. It also disrupted patterns of precipitation throughout the planet. It is believed to have influenced events as varied as floods along the Mississippi River in 1993 and, later that year, the drought that devastated the African Sahel. Most people considered the eruption a calamity.

For geophysical scientists, though, Mt. Pinatubo provided the best model in at least a century to help us understand what might happen if humans attempted to ameliorate global warming by deliberately altering the climate of the earth.

For years, even to entertain the possibility of human intervention on such a scale—geoengineering, as the practice is known—has been denounced as hubris. Predicting long-term climatic behavior by using computer models has proved difficult, and the notion of fiddling with the planet's climate based on the results generated by those models worries even scientists who are fully engaged in the research. "There will be no easy victories, but at some point we are going to have to take the facts seriously," David Keith, a professor of engineering and public policy at Harvard and one of geoengineering's most thoughtful supporters, told me. "Nonetheless," he added, "it is hyperbolic to say this, but no less true: when you start to reflect light away from the planet, you can easily imagine a chain of events that would extinguish life on earth."

There is only one reason to consider deploying a scheme with even a tiny chance of causing such a catastrophe: if the risks of not deploying it were clearly higher. No one is yet prepared to make such a calculation, but researchers are moving in that direction. To offer guidance, the Intergovernmental Panel on Climate Change (I.P.C.C.) has developed a series of scenarios on global warming. The cheeriest assessment predicts that by the end of the century the earth's average temperature will rise between 1.1 and 2.9 degrees Celsius. A more pessimistic projection envisages a rise of between 2.4 and 6.4 degrees—far higher than at any time in recorded history. (There are nearly two degrees Fahrenheit in one degree Celsius. A rise of 2.4 to 6.4 degrees Celsius would equal 4.3 to 11.5 degrees Fahrenheit.) Until recently, climate scientists believed that a six-degree rise, the effects of which would be an undeniable disaster, was unlikely. But new data have changed the minds of many. Late last year, Fatih Birol, the chief economist for the International Energy Agency, said that current levels of consumption "put the world perfectly on track for a six-degree Celsius rise in temperature. . . . Everybody, even schoolchildren, knows this will have catastrophic implications for all of us."

Tens of thousands of wildfires have already been attributed to warming, as have melting glaciers and rising seas. (The warming of the oceans is particularly worrisome; as Arctic ice melts, water that was below the surface becomes exposed to the sun and absorbs more solar energy, which leads to warmer oceans—a loop that could rapidly spin out of control.) Even a two-degree climb in average global temperatures could cause crop failures in parts of the world that can least afford to lose the nourishment. The size of deserts would increase, along with the frequency and intensity of wildfires. Deliberately modifying the earth's atmosphere would be a desperate gamble with significant risks. Yet the more likely climate change is to cause devastation, the more attractive even the most perilous attempts to mitigate those changes will become.

"We don't know how bad this is going to be, and we don't know when it is going to get bad," Ken Caldeira, a climate scientist with the Carnegie Institution, told me. In 2007, Caldeira was a principal contributor to an I.P.C.C. team that won a Nobel Peace Prize. "There are wide variations within the models," he said. "But we had better get ready, because we are running rapidly toward a minefield. We just don't know where the minefield starts, or how long it will be before we find ourselves in the middle of it."

The Maldives, a string of islands off the coast of India whose highest point above sea level is eight feet, may be the first nation to drown. In Alaska, entire towns have begun to shift in the loosening permafrost. The Florida economy is highly dependent upon coastal weather patterns; the tide station at Miami Beach has registered an increase of seven inches since 1935, according to the National Oceanic and Atmospheric

Administration. One Australian study, published this year in the journal *Nature Climate Change*, found that a two-degree Celsius rise in the earth's temperature would be accompanied by a significant spike in the number of lives lost just in Brisbane. Many climate scientists say their biggest fear is that warming could melt the Arctic permafrost—which stretches for thousands of miles across Alaska, Canada, and Siberia. There is twice as much CO_2 locked beneath the tundra as there is in the earth's atmosphere. Melting would release enormous stores of methane, a greenhouse gas nearly thirty times more potent than carbon dioxide. If that happens, as the hydrologist Jane C. S. Long told me when we met recently in her office at the Lawrence Livermore National Laboratory, "it's game over."

The Stratospheric Particle Injection for Climate Engineering project, or SPICE, is a British academic consortium that seeks to mimic the actions of volcanoes like Pinatubo by pumping particles of sulfur dioxide, or similar reflective chemicals, into the stratosphere through a twelve-mile-long pipe held aloft by a balloon at one end and tethered, at the other, to a boat anchored at sea.

The consortium consists of three groups. At Bristol University, researchers led by Matt Watson, a professor of geophysics, are trying to determine which particles would have the maximum desired impact with the smallest likelihood of unwanted side effects. Sulfur dioxide produces sulfuric acid, which destroys the ozone layer of the atmosphere; there are similar compounds that might work while proving less environmentally toxic—including synthetic particles that could be created specifically for this purpose. At Cambridge, Hugh Hunt and his team are trying to determine the best way to get those particles into the stratosphere. A third group, at Oxford, has been focussing on the effect such an intervention would likely have on the earth's climate.

Hunt and I spoke in Cambridge, at Trinity College, where he is a professor of engineering and the Keeper of the Trinity College clock, a renowned timepiece that gains or loses less than a second a month. In his office, dozens of boomerangs dangle from the wall. When I asked about them, he grabbed one and hurled it at my head. "I teach three-dimensional dynamics," he said, flicking his hand in the air to grab it as it returned. Hunt has devoted his intellectual life to the study of mechanical vibration. His Web page is filled with instructive videos about gyroscopes, rings wobbling down rods, and boomerangs.

"I like to demonstrate the way things spin," he said, as he put the boomerang down and picked up an inflated pink balloon attached to a string. "The principle is pretty simple." Holding the string, Hunt began to bobble the balloon as if it were being tossed by foul weather. "Everything is fine if it is sitting still," he continued, holding the balloon steady. Then he began to wave his arm erratically. "One of the problems is that nothing is going to be still up there. It is going to be moving around. And the question we've got is . . . this pipe"—the industrial hose that will convey the particles into the sky—"is going to be under huge stressors." He snapped the string connected to the balloon. "How do you know it's not going to break? We are really pushing things to the limit in terms of their strength, so it is essential that we get the dynamics of motion right."

Most scientists, even those with no interest in personal publicity, are vigorous advocates for their own work. Not this group. "I don't know how many times I have said this, but the last thing I would ever want is for the project I have been working on to be implemented," Hunt said. "If we have to use these tools, it means something on this planet has gone seriously wrong."

Last fall, the SPICE team decided to conduct a brief and uncontroversial pilot study. At least they thought it would be uncontroversial. To demonstrate how they would disperse the sulfur dioxide, they had planned to float a balloon over Norfolk, at an altitude of a kilometre, and send a hundred and fifty litres of water into the air through a hose. After the date and time of the test was announced, in the middle of September, more than fifty organizations signed a petition objecting to the experiment, in part because they fear that even to consider engineering the climate would provide politicians with an excuse for avoiding tough decisions on reducing greenhouse-gas emissions. Opponents of the water test pointed out the many uncertainties in the research (which is precisely why the team wanted to do the experiment). The British government decided to put it off for at least six months.

"When people say we shouldn't even explore this issue, it scares me," Hunt said. He pointed out that carbon emissions are heavy, and finding a place to deposit them will not be easy. "Roughly speaking, the CO_2 we generate weighs three or four times as much as the fuel it comes from." That means that a short round-trip journey—say, eight hundred miles—by car, using two tanks of gas, produces three hundred kilograms of CO_2. "This is ten heavy suitcases from one short trip," Hunt said. "And you have to store it where it can't evaporate.

"So I have three questions, Where are you going to put it? Who are you going to ask to dispose of this for you? And how much are you reasonably willing to pay them to do it?" he continued. "There is nobody on this planet who can answer any of those questions. There is no established place or technique, and nobody has any idea what it would cost. And we need the answers now."

Hunt stood up, walked slowly to the window, and gazed at the manicured Trinity College green. "I know this is all unpleasant," he said. "Nobody wants it, but nobody wants to put high doses of poisonous chemicals into their body, either. That is what chemotherapy is, though, and for people suffering from cancer those poisons are often their only hope. Every day, tens of thousands of people take them willingly—because they are very sick or dying. This is how I prefer to look at the possibility of engineering the climate. It isn't a cure for anything. But it could very well turn out to be the least bad option we are going to have."

The notion of modifying the weather dates back at least to the eighteen-thirties, when the American meteorologist James Pollard Espy became known as the Storm King, for his (prescient but widely ridiculed) proposals to stimulate rain by selectively burning forests. More recently, the U.S. government project Stormfury attempted for decades to lessen the force of hurricanes by seeding them with silver iodide. And in 2008 Chinese soldiers fired more than a thousand rockets filled with chemicals at clouds over Beijing to prevent them from raining on the Olympics. The relationship between carbon emissions and the earth's temperature has been clear for more than a century: in 1908, the Swedish scientist Svante Arrhenius suggested that burning fossil fuels might help prevent the coming ice age. In 1965, President Lyndon Johnson received a report from his Science Advisory Committee, titled "Restoring the Quality of Our Environment," that noted for the first time the potential need to balance increased greenhouse-gas emissions by "raising the albedo, or the reflectivity, of the earth." The report suggested that such a change could be achieved by spreading small reflective particles over large parts of the ocean.

While such tactics could clearly fail, perhaps the greater concern is what might happen if they succeeded in ways nobody had envisioned. Injecting sulfur dioxide, or particles that perform a similar function, would rapidly lower the temperature of the earth, at relatively little expense—most estimates put the cost at less than ten billion dollars a year. But it would do nothing to halt ocean acidification, which threatens to destroy coral reefs and wipe out an enormous number of aquatic species. The risks of reducing the amount of sunlight that reaches the atmosphere on that scale would be as obvious—and immediate—as the benefits. If such a program were suddenly to fall apart, the earth would be subjected to extremely rapid warming, with nothing to stop it. And while such an effort would cool the globe, it might do so in ways that disrupt the behavior of the Asian and African monsoons, which provide the water that billions of people need to drink and to grow their food.

"Geoengineering" actually refers to two distinct ideas about how to cool the planet. The first, solar-radiation management, focusses on reducing the impact of the sun. Whether by seeding clouds, spreading giant mirrors in the desert, or injecting sulfates into the stratosphere, most such plans seek to replicate the effects of eruptions like Mt. Pinatubo's. The other approach is less risky, and involves removing carbon directly from the atmosphere and burying it in vast ocean storage beds or deep inside the earth. But without a significant technological advance such projects will be expensive and may take many years to have any significant effect.

There are dozens of versions of each scheme, and they range from plausible to absurd. There have been proposals to send mirrors, sunshades, and parasols into space. Recently, the scientific entrepreneur Nathan

Myhrvold, whose company Intellectual Ventures has invested in several geoengineering ideas, said that we could cool the earth by stirring the seas. He has proposed deploying a million plastic tubes, each about a hundred metres long, to roil the water, which would help it trap more CO_2. "The ocean is this giant heat sink," he told me. "But it is very cold. The bottom is nearly freezing. If you just stirred the ocean more, you could absorb the excess CO_2 and keep the planet cold." (This is not as crazy as it sounds. In the center of the ocean, wind-driven currents bring fresh water to the surface, so stirring the ocean could transform it into a well-organized storage depot. The new water would absorb more carbon while the old water carried the carbon it has already captured into the deep.)

The Harvard physicist Russell Seitz wants to create what amounts to a giant oceanic bubble bath: bubbles trap air, which brightens them enough to reflect sunlight away from the surface of the earth. Another tactic would require maintaining a fine spray of seawater—the world's biggest fountain—which would mix with salt to help clouds block sunlight.

The best solution, nearly all scientists agree, would be the simplest: stop burning fossil fuels, which would reduce the amount of carbon we dump into the atmosphere. That fact has been emphasized in virtually every study that addresses the potential effect of climate change on the earth—and there have been many—but none have had a discernible impact on human behavior or government policy. Some climate scientists believe we can accommodate an atmosphere with concentrations of carbon dioxide that are twice the levels of the preindustrial era—about five hundred and fifty parts per million. Others have long claimed that global warming would become dangerous when atmospheric concentrations of carbon rose above three hundred and fifty parts per million. We passed that number years ago. After a decline in 2009, which coincided with the harsh global recession, carbon emissions soared by six per cent in 2010—the largest increase ever recorded. On average, in the past decade, fossil-fuel emissions grew at about three times the rate of growth in the nineteen-nineties.

Although the I.P.C.C., along with scores of other scientific bodies, has declared that the warming of the earth is unequivocal, few countries have demonstrated the political will required to act—perhaps least of all the United States, which consumes more energy than any nation other than China, and, last year, more than it ever had before. The Obama Administration has failed to pass any meaningful climate legislation. Mitt Romney, the presumptive Republican nominee, has yet to settle on a clear position. Last year, he said he believed the world was getting warmer—and humans were a cause. By October, he had retreated. "My view is that we don't know what is causing climate change on this planet," he said, adding that spending huge sums to try to reduce CO_2 emissions "is not the right course for us." China, which became the world's largest emitter of greenhouse gases several years ago, constructs a new coal-burning power plant nearly every week. With each passing year, goals become exponentially harder to reach, and global reductions along the lines suggested by the I.P.C.C. seem more like a "pious wish," to use the words of the Dutch chemist Paul Crutzen, who in 1995 received a Nobel Prize for his work on ozone depletion.

"Most nations now recognize the need to shift to a low-carbon economy, and nothing should divert us from the main priority of reducing global greenhouse gas emissions," Lord Rees of Ludlow wrote in his 2009 forward to a highly influential report on geoengineering released by the Royal Society, Britain's national academy of sciences. "But if such reductions achieve too little, too late, there will surely be pressure to consider a 'plan B'—to seek ways to counteract climatic effects of green-house gas emissions."

While that pressure is building rapidly, some climate activists oppose even holding discussions about a possible Plan B, arguing, as the Norfolk protesters did in September, that it would be perceived as indirect permission to abandon serious efforts to cut emissions. Many people see geoengineering as a false solution to an existential crisis—akin to encouraging a heart-attack patient to avoid exercise and continue to gobble fatty food while simply doubling his dose of Lipitor. "The scientist's focus on tinkering with our entire planetary system is not a dynamic new technological and scientific frontier, but an expression of political despair," Doug Parr, the chief scientist at Greenpeace UK, has written.

During the 1974 Mideast oil crisis, the American engineer Hewitt Crane, then working at S.R.I. International, realized that standard measurements for sources of energy—barrels of oil, tons of coal, gallons of gas, British thermal units—were nearly impossible to compare. At a time when these commodities were being rationed, Crane wondered how people could conserve resources if they couldn't even measure them. The world was burning through twenty-three thousand gallons of oil every second. It was an astonishing figure, but one that Crane had trouble placing into any useful context.

Crane devised a new measure of energy consumption: a three-dimensional unit he called a cubic mile of oil. One cubic mile of oil would fill a pool that was a mile long, a mile wide, and a mile deep. Today, it takes three cubic miles' worth of fossil fuels to power the world for a year. That's a trillion gallons of gas. To replace just one of those cubic miles with a source of energy that will not add carbon dioxide to the atmosphere—nuclear power, for instance—would require the construction of a new atomic plant every week for fifty years; to switch to wind power would mean erecting thousands of windmills each month. It is hard to conceive of a way to replace that much energy with less dramatic alternatives. It is also impossible to talk seriously about climate change without talking about economic development. Climate experts have argued that we ought to stop emitting greenhouse gases within fifty years, but by then the demand for energy could easily be three times what it is today: nine cubic miles of oil.

The planet is getting richer as well as more crowded, and the pressure to produce more energy will become acute long before the end of the century. Predilections of the rich world—constant travel, industrial activity, increasing reliance on meat for protein—require enormous physical resources. Yet many people still hope to solve the problem of climate change just by eliminating greenhouse-gas emissions. "When people talk about bringing emissions to zero, they are talking about something that will never happen," Ken Caldeira told me. "Because that would require a complete alteration in the way humans are built."

Caldeira began researching geoengineering almost by accident. For much of his career, he has focussed on the implications of ocean acidification. During the nineteen-nineties, he spent a year in the Soviet Union, at the Leningrad lab of Mikhail Budyko, who is considered the founder of physical climatology. It was Budyko, in the nineteen-sixties, who first suggested cooling the earth by putting sulfur particles in the sky.

"In the nineteen-nineties, when I was working at Livermore, we had a meeting in Aspen to discuss the scale of the energy-system transformation needed in order to address the climate problem," Caldeira said. "Among the people who attended was Lowell Wood, a protégé of Edward Teller. Wood is a brilliant but sometimes erratic man . . . lots of ideas, some better than others." At Aspen, Wood delivered a talk on geoengineering. In the presentation, he explained, as he has many times since, that shielding the earth properly could deflect one or two per cent of the sunlight that reaches the atmosphere. That, he said, would be all it would take to counter the worst effects of warming.

David Keith was in the audience with Caldeira that day in Aspen. Keith now splits his time between Harvard and Calgary, where he runs Carbon Engineering, a company that is developing new technology to capture CO_2 from the atmosphere—at a cost that he believes would make it sensible to do so. At the time, though, both men considered Wood's idea ridiculous. "We said this will never happen," Caldeira recalled. "We were so certain Wood was nuts, because we assumed you can change the global mean temperature, but you will still get seasonal and regional patterns you can't correct. We were in the back of the room, and neither of us could believe it."

Caldeira decided to prove his point by running a computer simulation of Wood's approach. Scenarios for future climate change are almost always developed using powerful three-dimensional models of the earth and its atmosphere. They tend to be most accurate when estimating large numbers, like average global temperatures. Local and regional weather patterns are more difficult to predict, as anyone who has relied on a five-day weather forecast can understand. Still, in 1998 Caldeira tested the idea, and, "much to my surprise, it seemed to work and work well," he told me. It turned out that reducing sunlight offset the effect of CO_2 both regionally and seasonally. Since then, his results have been confirmed by several other groups.

Recently, Caldeira and colleagues at Carnegie and Stanford set out to examine whether the techniques of solar-radiation management would disrupt the sensitive agricultural balance on which the earth depends. Using two models, they simulated climates with carbon-dioxide levels similar to those which exist today. They then doubled those concentrations to reflect levels that would be likely in several decades if current trends continue unabated. Finally, in a third set of simulations, they doubled the CO_2 in the atmosphere, but added a layer of sulfate aerosols to the stratosphere, which would deflect about two per cent of incoming sunlight from the earth. The data were then applied to crop models that are commonly used to project future yields. Again, the results were unexpected.

Farm productivity, on average, went up. The models suggested that precipitation would increase in the northern and middle latitudes, and crop yields would grow. In the tropics, though, the results were significantly different. There heat stress would increase, and yields would decline. "Climate change is not so much a reduction in productivity as a redistribution," Caldeira said. "And it is one in which the poorest people on earth get hit the hardest and the rich world benefits"—a phenomenon, he added, that is not new.

"I have two perspectives on what this might mean," he said. "One says: humans are like rats or cockroaches. We are already living from the equator to the Arctic Circle. The weather has already become .7 degrees warmer, and barely anyone has noticed or cares. And, yes, the coral reefs might become extinct, and people from the Seychelles might go hungry. But they have gone hungry in the past, and nobody cared. So basically we will live in our gated communities, and we will have our TV shows and Chicken McNuggets, and we will be O.K. The people who would suffer are the people who always suffer.

"There is another way to look at this, though," he said. "And that is to compare it to the subprime-mortgage crisis, where you saw that a few million bad mortgages led to a five-per-cent drop in gross domestic product throughout the world. Something that was a relatively small knock to the financial system led to a global crisis. And that could certainly be the case with climate change. But five per cent is an interesting figure, because in the Stern Report"—an often cited review led by the British economist Nicholas Stern, which signalled the alarm about greenhouse-gas emissions by focussing on economics—"they estimated climate change would cost the world five per cent of its G.D.P. Most economists say that solving this problem is one or two per cent of G.D.P. The Clean Water and Clean Air Acts each cost about one per cent of G.D.P.," Caldeira continued. "We just had a much worse shock to our banking system. And it didn't even get us to reform the economy in any significant way. So why is the threat of a five-per-cent hit from climate change going to get us to transform the energy system?"

Solar-radiation management, which most reports have agreed is technologically feasible, would provide, at best, a temporary solution to rapid warming—a treatment but not a cure. There are only two ways to genuinely solve the problem: by drastically reducing emissions or by removing the CO_2 from the atmosphere. Trees do that every day. They "capture" carbon dioxide in their leaves, metabolize it in the branch system, and store it in their roots. But to do so on a global scale would require turning trillions of tons of greenhouse-gas emissions into a substance that could be stored cheaply and easily underground or in ocean beds.

Until recently, the costs of removing carbon from the atmosphere on that scale have been regarded by economists as prohibitive. CO_2 needs to be heated in order to be separated out; using current technology, the expense would rival that of creating an entirely new energy system. Typically, power plants release CO_2 into the atmosphere through exhaust systems referred to as flues. The most efficient way we have now to capture CO_2 is to remove it from flue gas as the emissions escape. Over the past five years, several research groups—one of which includes David Keith's company, Carbon Engineering, in Calgary—have developed new techniques to extract carbon from the atmosphere, at costs that may make it economically feasible on a larger scale.

Early this winter, I visited a demonstration project on the campus of S.R.I. International, the Menlo Park institution that is a combination think tank and technological incubator. The project, built by Global Thermostat, looked like a very high-tech elevator or an awfully expensive math problem. "When I called chemical engineers and said I want to do this on a planetary scale, they laughed," Peter Eisenberger, Global Thermostat's president, told me. In 1996, Eisenberger was appointed the founding director of the Earth Institute, at Columbia University, where he remains a professor of earth and environmental sciences. Before that, he spent a decade running the materials research institute at Princeton University, and nearly as much time at Exxon, in charge of research and development.

He believes he has developed a system to capture CO_2 from the atmosphere at low heat and potentially at low cost.

The trial project is essentially a five-story brick edifice specially constructed to function like a honeycomb. Global Thermostat coats the bricks with chemicals called amines to draw CO_2 from the air and bind with it. The carbon dioxide is then separated with a proprietary method that uses low-temperature heat—something readily available for free, since it is a waste product of many power plants. "Using low-temperature heat changes the equation," Eisenberger said. He is an excitable man with the enthusiasm of a graduate student and the manic gestures of an orchestra conductor. He went on to explain that the amine coating on the bricks binds the CO_2 at the molecular level, and the amount it can capture depends on the surface area; honeycombs provide the most surface space possible per square metre.

There are two groups of honeycombs that sit on top of each other. As Eisenberger pointed out, "You can only absorb so much CO_2 at once, so when the honeycomb is full it drops into a lower section." Steam heats and releases the CO_2—and the honeycomb rises again. (Currently, carbon dioxide is used commercially in carbonated beverages, brewing, and pneumatic drying systems for packaged food. It is also used in welding. Eisenberger argues that, ideally, carbon waste would be recycled to create an industrial form of photosynthesis, which would help reduce our dependence on fossil fuels.)

Unlike some other scientists engaged in geoengineering, Eisenberger is not bothered by the notion of tinkering with nature. "We have devised a system that introduces no additional threats into the environment," he told me. "And the idea of interfering with benign nature is ridiculous. The Bambi view of nature is totally false. Nature is violent, amoral, and nihilistic. If you look at the history of this planet, you will see cycles of creation and destruction that would offend our morality as human beings. But somehow, because it's 'nature,' it's supposed to be fine." Eisenberger founded and runs Global Thermostat with Graciela Chichilnisky, an Argentine economist who wrote the plan, adopted in 2005, for the international carbon market that emerged from the Kyoto Climate talks. Edgar Bronfman, Jr., an heir to the Seagram fortune, is Global Thermostat's biggest investor. (The company is one of the finalists for Richard Branson's Virgin Earth Challenge prize. In 2007, Branson offered a cash prize of twenty-five million dollars to anyone who could devise a process that would drain large quantities of greenhouse gases from the atmosphere.)

"What is fascinating for me is the way the innovation process has changed," Eisenberger said. "In the past, somebody would make a discovery in a laboratory and say, 'What can I do with this?' And now we ask, 'What do we want to design?,' because we believe there is powerful enough knowledge to do it. That is what my partner and I did." The pilot, which began running last year, works on a very small scale, capturing about seven hundred tons of CO_2 a year. (By comparison, an automobile puts out about six tons a year.) Eisenberger says that it is important to remember that it took more than a century to assemble the current energy system: coal and gas plants, factories, and the worldwide transportation network that has been responsible for depositing trillions of tons of CO_2 into the atmosphere. "We are not going to get it all out of the atmosphere in twenty years," he said. "It will take at least thirty years to do this, but if we start now that is plenty of time. You would just need a source of low-temperature heat—factories anywhere in the world are ideal." He envisions a network of twenty thousand such devices scattered across the planet. Each would cost about a hundred million dollars—a two-trillion-dollar investment spread out over three decades.

"There is a strong history of the system refusing to accept something new," Eisenberger said. "People say I am nuts. But it would be surprising if people didn't call me crazy. Look at the history of innovation! If people don't call you nuts, then you are doing something wrong."

After leaving Eisenberger's demonstration project, I spoke with Curtis Carlson, who, for more than a decade, has been the chairman and chief executive officer of S.R.I. and a leading voice on the future of American innovation. "These geoengineering methods will not be implemented for decades—or ever," he said. Nonetheless, scientists worry that if methane emissions from the Arctic increase as rapidly as some of the data now suggest, climate intervention isn't going to be an option. It's going to be a requirement. "When and where do we have the serious discussion about how to intervene?" Carlson asked. "There are no agreed-upon rules or criteria. There isn't even a body that could create the rules."

Over the past three years, a series of increasingly urgent reports—from the Royal Society, in the U.K., the Washington-based Bipartisan Policy Center, and the Government Accountability Office, among other places—have practically begged decision-makers to begin planning for a world in which geoengineering might be their only recourse. As one recent study from the Wilson International Center for Scholars concluded, "At the very least, we need to learn what approaches to avoid even if desperate."

The most environmentally sound approach to geoengineering is the least palatable politically. "If it becomes necessary to ring the planet with sulfates, why would you do that all at once?" Ken Caldeira asked. "If the total amount of climate change that occurs could be neutralized by one Mt. Pinatubo, then doesn't it make sense to add one per cent this year, two per cent next year, and three per cent the year after that?" he said. "Ramp it up slowly, throughout the century, and that way we can monitor what is happening. If we see something at one per cent that seems dangerous, we can easily dial it back. But who is going to do that when we don't have a visible crisis? Which politician in which country?"

Unfortunately, the least risky approach politically is also the most dangerous: do nothing until the world is faced with a cataclysm and then slip into a frenzied crisis mode. The political implications of any such action would be impossible to overstate. What would happen, for example, if one country decided to embark on such a program without the agreement of other countries? Or if industrialized nations agreed to inject sulfur particles into the stratosphere and accidentally set off a climate emergency that caused drought in China, India, or Africa?

"Let's say the Chinese government decides their monsoon strength, upon which hundreds of millions of people rely for sustenance, is weakening," Caldeira said. "They have reason to believe that making clouds right near the ocean might help, and they started to do that, and the Indians found out and believed—justifiably or not—that it would make their monsoon worse. What happens then? Where do we go to discuss that? We have no mechanism to settle that dispute."

Most estimates suggest that it could cost a few billion dollars a year to scatter enough sulfur particles in the atmosphere to change the weather patterns of the planet. At that price, any country, most groups, and even some individuals could afford to do it. The technology is open and available—and that makes it more like the Internet than like a national weapons program. The basic principles are widely published; the intellectual property behind nearly every technique lies in the public domain. If the Maldives wanted to send airplanes into the stratosphere to scatter sulfates, who could stop them?

"The odd thing here is that this is a democratizing technology," Nathan Myhrvold told me. "Rich, powerful countries might have invented much of it, but it will be there for anyone to use. People get themselves all balled up into knots over whether this can be done unilaterally or by one group or one nation. Well, guess what. We decide to do much worse than this every day, and we decide unilaterally. We are polluting the earth unilaterally. Whether it's life-taking decisions, like wars, or something like a trade embargo, the world is about people taking action, not agreeing to take action. And, frankly, the Maldives could say, 'Fuck you all—we want to stay alive.' Would you blame them? Wouldn't any reasonable country do the same?"

Adaptation

Eric Klinenberg

January 7, 2013

In July, 1995, a scorching heat wave hit Chicago, killing seven hundred and thirty-nine people, roughly seven times as many as died in Superstorm Sandy. Soon after the heat abated, social scientists began to look for patterns behind the deaths. Some of the results were unsurprising: having a working air-conditioner reduced the risk of death by eighty per cent. But fascinating patterns did emerge. For the most part, the geography of heat-wave mortality was consistent with the city's geography of segregation and inequality: eight of the ten community areas with the highest death rates were virtually all African-American, with pockets of concentrated poverty and violent crime, places where old people were at risk of hunkering down at home and dying alone during the heat wave. At the same time, three of the ten neighborhoods with the *lowest* heat-wave death rates were also poor, violent, and predominantly African-American.

Englewood and Auburn Gresham, two adjacent neighborhoods on the hyper-segregated South Side of Chicago, were both ninety-nine per cent African-American, with similar proportions of elderly residents. Both had high rates of poverty, unemployment, and violent crime. Englewood proved to be one of the most perilous places during the disaster, with thirty-three deaths per hundred thousand residents. But Auburn Gresham's death rate was only three per hundred thousand, making it far safer than many of the most affluent neighborhoods on the North Side. Identifying the sources of such resilience is important, and not for merely academic reasons. The changing climate is likely to deliver more severe weather more often, and understanding why some neighborhoods fare better in a crisis than others that resemble them can help us prepare for the next disaster.

For the past decade and a half, governments around the world have been investing in elaborate plans to "climate-proof" their cities—protecting people, businesses, and critical infrastructure against weather-related calamities. Much of this work involves upgrading what engineers call "lifeline systems": the network infrastructure for power, transit, and communications, which is crucial in the immediate aftermath of a disaster. Some of the solutions are capital-intensive and high-tech; some are low- or no-tech approaches, such as organizing communities so that residents know which of their neighbors are vulnerable and how to assist them. The fundamental threat to the human species is, of course, our collective inability to reduce our carbon emissions and slow the pace of climate change. Yet, even if we managed to stop increasing global carbon emissions tomorrow, we would probably experience several centuries of additional warming, rising sea levels, and more frequent dangerous weather events. If our cities are to survive, we have no choice but to adapt.

Klaus Jacob is a geophysicist at Columbia University whose 2009 report on climate risks to New York City contains eerily accurate predictions about what would happen to the city's infrastructure during a major storm surge. He works at the university's Lamont-Doherty Earth Observatory, a sprawling research campus for earth sciences perched above the Hudson River, on the Palisades. Its drab, boxy buildings make it look more like a military base than like a collegiate Arcadia. Jacob's office has the familiar academic freight of books, journals, and papers, but there's also a Day-Glo hard hat that has had some use over the years. When I asked him how he got interested in urban security, he told me about his childhood. "I was born in Stuttgart in 1936, and when the war started my parents moved us to a small village in Bavaria, because they knew we would be safer there," he said. "The family that moved into our home was killed two years later. The building was bombed." Jacob has a thick silver beard and pale-blue eyes, although his foreboding manner makes the effect less Kris Kringle than Old Testament prophet. He gave me an intent look. "I grew up in a war environment. And what I learned is that you can plan your fate, at least to some degree, if you assess your risks and do something about it."

Jacob's early research, which was funded by the United States Air Force, focussed on underground nuclear explosions, and he later studied how structures such as bridges and high-rise buildings could survive seismic shocks. In the nineteen-nineties, Jacob was asked to help New York City evaluate its capacity to withstand storms, and since then he has conducted similar studies for New York State and the M.T.A. His findings were sobering. "Much of the subway system is below sea level already," Jacob explained. During Sandy, several stations and lines filled up like bathtubs. Elevating and redesigning access points and the ventilation system would be an immense undertaking. "It will probably cost billions, maybe tens of billions, to protect it."

Jacob's computer screen displayed two maps of New York City neighborhoods, one color-coded by elevation and the other by population growth since 2000. "Look at the blue zones, which show where we've been developing real estate, and the pink ones, which show where the population is dropping," Jacob said. Downtown neighborhoods on the Hudson—Battery Park, Tribeca, the West Village, West Chelsea, and Hell's Kitchen—were solidly blue; neighborhoods uptown, on higher ground, were pink. "Think about all the projects we conceived more than a decade ago, before we knew about rising sea levels, in the name of waterfront revitalization," he went on. "They've been quite successful, but they've also placed a lot of people at risk."

Genuine adaptation, Jacob believes, means preparing for the inevitable deluge. "The ocean is going to reclaim what we took from it," he said. He thinks that New York can learn from Rotterdam, which has a long history of flooding. After enduring a devastating storm surge in 1953, Rotterdam began building a series of dams, barriers, and seawalls as part of a national project called Delta Works, and five years ago the Dutch government provided funds for an upgrade, the Rotterdam Climate Proof Program. Arnoud Molenaar, who manages it, says his team realized that they could convert the water that comes into the city from the skies and the sea into "blue gold." "Before, we saw the water as a problem," Molenaar told me. "In the Netherlands, we focussed on how to prevent it from coming in. New York City focussed on evacuation, how to get people out of the way. The most interesting thing is figuring out what's between these approaches: what to do with the water once it's there."

In 2005, Rotterdam hosted the Second International Architecture Biennale. The theme was "The Flood." Designers from around the world presented plans for how cities could cope with water in the future, and when the exhibition ended Molenaar's team set out to implement those that would have immediate practical value. Rotterdam is now experimenting with an architecture of accommodation: it has a floating pavilion in the city center, made of three silver half spheres with an exhibition space that's equivalent to four tennis courts; a water plaza that serves as a playground most of the year but is converted into a water-storage facility on days of heavy rainfall; a floodable terrace and sculpture garden along the city's canal; and buildings whose façades, garages, and ground-level spaces have been engineered to be waterproof.

Smart designs have improved other parts of the Netherlands' critical infrastructure. Its communications network features the fastest Internet speed in Europe, and, with I.B.M., it has built a system for water and energy management. It also has a resilient power grid, designed to withstand strong winds and heavy rain. In the United States, most distribution lines are elevated on wooden poles and exposed to falling tree branches; in the Netherlands, the lines are primarily underground and encased in water-resistant pipes. The Dutch grid is circular, rather than being a system of hub and spokes, so that, if a line goes out in one direction, operators can restore power by bringing it in from another source. And it's interconnected to the grids in neighboring countries, which gives the system additional capacity when there are local problems. This network architecture is more resilient in ordinary times, too. In Holland, the average duration of total annual power outages is twenty-three minutes, compared with two hundred and fourteen minutes in New Jersey, Pennsylvania, and New York—not including outages from disasters.

After Sandy, there was a five-day blackout in lower Manhattan, because the walls protecting Con Ed's substation along the East River, at twelve and a half feet above the ground, were eighteen inches too low to stop the storm surge and prevent the consequent equipment explosions. When I asked Jacob about this, he threw up his hands in exasperation. "Just put it on a high platform and use more underwater cable," he said. "We've had it available for a long time now. These are just moderate investments, in the millions of dollars. It's a small price to pay for more resilience."

The island nation of Singapore—where 5.2 million people are packed into seven hundred and ten square kilometres of land, much of which is perilously close to sea level—offers other lessons. Singapore began adapting to dangerous weather thirty years ago, after a series of heavy rains during monsoon seasons caused repeated flooding in the low-lying city center. The country has always had a difficult relationship with water. Its geography makes it vulnerable to heavy seasonal rains and frequent flooding but there is never a sufficient supply of usable water, and in recent years Singapore's dependency on Malaysian water sources has led to political conflicts. Climate change, with its rising sea levels and increase in heavy rains, threatens the city-state's stability. But Singapore's government also sees this as an opportunity.

The Marina Barrage and Reservoir, which opened in 2008, is at the heart of Singapore's two-billion-dollar campaign to improve drainage infrastructure, reduce the size of flood-prone areas, and enhance the quality of city life. It has nine operable crest gates, a series of enormous pumps, and a ten-thousand-hectare catchment area that is roughly one-seventh the size of the country. The system not only protects low-lying urban neighborhoods from flooding during heavy rains; it also eliminates the tidal influence of the surrounding seawater, creating a rain-fed supply of freshwater that currently meets ten per cent of Singapore's demand. Moreover, by stabilizing water levels in the Marina basin the barriers have produced better conditions for water sports. The Marina's public areas, which include a sculpture garden, a water-play space, a green roof with dramatic skyline vistas, and the Sustainable Singapore Gallery, bolster the city's tourist economy as well.

The Marina is just one of Singapore's adaptation projects. The Mass Rapid Transit system has elevated the access points for the underground rail system to at least a metre above the highest recorded flood levels. To minimize damage, the Public Utility Board has improved its drainage systems. In the nineteen-seventies, thirty-two hundred hectares of land were flood-prone; today, only forty-nine hectares are. Singapore is further reducing its dependence on imported water by building new facilities for desalinating seawater, and developing technology for using reclaimed and treated wastewater in industrial settings. To reduce its energy consumption, the Building and Construction Authority requires that all new structures be insulated with materials designed to retain cool temperatures. Today, Singapore is better prepared not only for extreme weather but also for meeting future demands for power and water as its population grows.

Jacob doesn't think Rotterdam's or Singapore's arrangements can simply be replicated elsewhere, but he's impressed by their ambition and foresight. After Sandy, New York paid the price for its lack of preparation. In recent decades, American utility companies have spent relatively little on research and development. One industry report estimates that, in 2009, research-and-development investments made by all U.S. electrical-power utilities amounted to at most $700 million, compared with $6.3 billion by I.B.M. and $9.1 billion by Pfizer. In 2009, however, the Department of Energy issued $3.4 billion in stimulus grants to a hundred smart-grid projects across the United

States, including many in areas that are prone to heat waves and hurricanes. The previous year, Hurricane Ike had knocked out power to two million customers in Houston, and full restoration took nearly a month. When the city received $200 million in federal funds to install smart-grid technology, it quickly put crews to work. Nearly all Houston households have been upgraded to the new network, one that should be more reliable when the next storm arrives.

Smart grids are in the early stages, but already they have several advantages over the old power systems. Digital meters, which are installed in households and at key transmission points, automatically generate real-time information about both consumers and suppliers, allowing utility providers to detect failures immediately, and sometimes also to identify the cause. This means that, after an outage, operators don't need to wait for calls from angry customers or field reports from crews. Moreover, smart grids are flexible, capable of being fed by disparate sources of energy, including systems powered by the sun and the wind. When the energy industry develops better technologies for storing power from these renewable resources, the new networks should be capable of integrating them.

Reëngineered grids will ultimately offer other benefits. "The situational awareness of the system might allow operators to reconfigure the system, either before or after the event, to maintain service," Leonardo Dueñas-Osorio, an engineering professor at Rice University who is developing resilience metrics for critical infrastructure systems, told me. "As a hurricane approaches, operators could 'island' areas that look like they will get the most damage. This breaks the system into small clusters and prevents cascading failures. It gives the operators more control, more capacity to keep the power going or get it back." Smart meters also enable consumers to go online anytime to learn when and how they use energy and how much they're spending. Already there's evidence that customers with this information are adjusting their behavior accordingly: easing off on air-conditioning, drying their clothes at night. Creating a smarter, more resilient grid for New York will be expensive, but not as expensive as a future filled with recurring outages during ordinary times and long-lasting failures when the weather turns menacing.

The communications system, too, is vulnerable to weather extremes; America's mobile-phone networks have always been less reliable than those in Europe, and regularly fail in catastrophes. During Sandy, emergency workers in New York and New Jersey were unable to communicate with colleagues who came from other states, because there's no nationwide network for first responders, and those from outside the region depended on cellular networks that were down. "Good public policies could potentially make these new networks much more resilient," Harold Feld, the senior vice-president of the digital-rights advocacy group Public Knowledge, says. The networks he envisages are flexible and have redundancies: "They can back each other up." Smart phones give the networks additional capacities for emergency communications, such as reverse-911 messaging that can be sent from government agencies to all customers in Zip Codes where dangerous weather is approaching, with geographically specific instructions on whether to evacuate or how to stay safe.

Unfortunately, the cellular industry has resisted efforts to regulate it, as the old telephone network is regulated, and there are no federal laws establishing minimum requirements for backup power during emergencies, no standards for how and when providers will share networks or drop roaming charges to give more people access to information, and no rules for reporting what caused extended outages. "We have a public interest in building robust networks," Feld says. "And by now it's clear that we're not going to get them by letting industry regulate itself."

New York City will inevitably explore ways to reduce flooding. There are relatively inexpensive measures, such as restoring wetlands and planting oyster beds, and then there are more ambitious, capital-intensive approaches: engineers at the Dutch firm Arcadis have designed a $6.5-billion barrier that would go just north of the Verrazano-Narrows Bridge. Others have proposed a five-mile gate that would stretch from Sandy Hook, New Jersey, to Rockaway, New York. Malcolm Bowman, who runs the Storm Surge Research Group at SUNY Stony Brook, is among the leading advocates for such a barrier. "We can't just sit around waiting for the next catastrophe," he argues. "The time is now, and it's really just a matter of political will."

But there are debates about the long-term efficacy of these barriers. "Barriers are at best an intermediate solution," Jacob told me in his office. "They will require at least twenty years to build, because we'd need environmental-impact reports, and buy-in from the federal government, the state governments of New York, New Jersey, and Connecticut, and probably also about three hundred municipalities. If all that happens, we'd get protection for perhaps a few decades. Walls will keep out storm surges, but not the rising ocean, and they could cause a sense of false security that prevents us from finding real solutions."

Jacob's office lacks the high-powered computing equipment that one finds in the labs where engineers are designing sophisticated models for sea barriers. "I'm a conceptual thinker," Jacob told me. "I do the modelling in my head. And if we spend all our resources on expensive safety systems that are not sustainable we're not going to solve the problem. Sometimes engineers don't see things holistically. Earth science helps us see the bigger picture." Eventually, Jacob believes, the city will need to make a "managed retreat" to higher ground. "We have a lot of high areas that we're not using, or that we've used for cemeteries, in Queens. I think we need to switch the living and dead, and I think the dead would understand."

He turned again to his monitor. "Look at the map," he said, tracing the coastline in Brooklyn, Queens, and Staten Island, then running his fingers along the Hudson and the East River. "This is where the rising water will hit."

Still, a strategy of resilience will involve more than changes to our physical infrastructure. Increasingly, governments and disaster planners are recognizing the importance of social infrastructure: the people, places, and institutions that foster cohesion and support. "There's a lot of social-science research showing how much better people do in disasters, how much longer they live, when they have good social networks and connections," says Nicole Lurie, a former professor of health policy at RAND's graduate school and at the University of Minnesota, who has been President Obama's assistant secretary for preparedness and response since 2009. "And we've had a pretty big evolution in our thinking, so promoting community resilience is now front and center in our approach."

The Chicago heat wave proved to be a case study in this respect. Researchers who sifted through the data (I was among them) noticed that women fared far better than men, because they have stronger ties to friends and family and are less prone to isolation. Latinos, who had high levels of poverty, had an easier time than other ethnic groups in Chicago, simply because in Chicago they tend to live in crowded apartments and densely packed neighborhoods, places where dying alone is nearly impossible.

The key difference between neighborhoods like Auburn Gresham and others that are demographically similar turned out to be the sidewalks, stores, restaurants, and community organizations that bring people into contact with friends and neighbors. The people of Englewood were vulnerable not just because they were black and poor but also because their community had been abandoned. Between 1960 and 1990, Englewood lost fifty per cent of its residents and most of its commercial outlets, as well as its social cohesion. "We used to be much closer, more tight-knit," says Hal Baskin, who has lived in Englewood for fifty-two years and currently leads a campaign against neighborhood violence. "Now we don't know who lives across the street or around the corner. And old folks are apprehensive about leaving their homes." Auburn Gresham, by contrast, experienced no population loss during that period. In 1995, residents walked to diners and grocery stores. They knew their neighbors. They participated in block clubs and church groups. "During the heat wave, we were doing wellness checks, asking neighbors to knock on each other's doors," Betty Swanson, who has lived in Auburn Gresham for nearly fifty years, says. "The presidents of our block clubs usually know who's alone, who's aging, who's sick. It's what we always do when it's very hot or very cold here."

Robert J. Sampson, a sociologist at Harvard University, has been measuring the strength of social ties, mutual assistance, and nonprofit organizations in Chicago communities for nearly two decades. He has found that the benefits of living in a neighborhood with a robust social infrastructure are significant during ordinary times as well as during disasters. In 1990, life expectancy in Auburn Gresham was five years higher than it was in Englewood. And, during the severe heat waves that are likely to hit Chicago and other cities in the near future, living in a neighborhood like Auburn Gresham is the rough equivalent of having a working air-conditioner in each room.

Since 1995, officials in Chicago have begun to take these factors into account. During times of intense heat, the city has urged the local media to advise neighbors, friends, and family to check in on one another. City agencies have maintained a database that lists the names, addresses, and phone numbers of old, chronically ill, and otherwise vulnerable people, and city workers call or visit to make sure they're safe. Churches and civic organizations have encouraged neighbors to look out for one another, as have family and friends. In Englewood, meanwhile, residents and community organizations have invented their own version of the Rotterdam strategy, turning their main problem, abandonment, into an advantage. Their goal is to transform Englewood into a hub of urban farming, with gardens that create stronger community ties as well as fresh produce and shade.

Englewood, as it happens, is just a few miles from the neighborhoods where Barack Obama worked as a community organizer during the late nineteen-eighties, learning first hand why social ties matter. Obama must have been thinking of places like Englewood when, as a U.S. senator in 2005, he connected the effects of Hurricane Katrina to the slow-motion disaster that New Orleans's vulnerable neighborhoods endured every day. "I hope we realize that the people of New Orleans weren't just abandoned during the hurricane," he said. "They were abandoned long ago." Katrina, Obama continued, should "awaken us to the great divide that continues to fester in our midst" and inspire us to "prevent such a failure from ever occurring again."

Obama was one of many members of Congress who believed that Katrina exposed the shortcomings of a national-security strategy that marginalized non-terrorist threats. The following year, Congress passed the Post-Katrina Emergency Management Reform Act, which expanded FEMA's authority, and the Pandemic and All Hazards Preparedness Act, which authorized new programs to improve public-health responses, ranging from risk communications to targeted support for vulnerable populations. During his first term, President Obama introduced a new National Health Security Strategy that emphasized preparedness and resilience, calling for the participation of the "whole community"—government agencies, civic organizations, corporations, and citizens—in all aspects of the security plan. "It was a pretty big evolution in our own thinking, to be able to put community resilience front and center," Nicole Lurie says.

Since March, 2011, when Obama issued a directive on national preparedness, FEMA has embraced a similar approach to community resilience. "Community-engagement" pilot programs funded by the Centers for Disease Control and Prevention have been launched in Los Angeles, Chicago, New York City, and Washington, D.C. "There's always been a big focus on classic infrastructure in mitigation," Alonzo Plough, the director of emergency preparedness and response for the County of Los Angeles, says. "But it's not just engineering that matters. It's social capital. And what this movement is bringing to the fore is that the social infrastructure matters, too."

Sandy revealed serious flaws in all forms of infrastructure in New York and New Jersey. But it also turned up surprising reserves of strength. When I visited Rockaway Beach in mid-November, residents complained about the slow pace of recovery. The power was out. The gas was off. Phone service was spotty. Trains weren't running. Sewage water from the flooding covered the streets. Still, there were some bright spots. The Rockaway Beach Surf Club, which opened in March, in a converted auto-repair shop beneath the El on Beach Eighty-seventh Street, transformed itself into a temporary relief agency when two of its founders returned there after the storm, posted Facebook updates inviting friends to join them, and watched more than five thousand volunteers come to help. It became the main community organization, providing food, cleaning supplies, camaraderie, and manual labor for nearby residents. The surf club's neighbors, including blue-collar families and poor African-Americans who, months before, had worried about how the club would fit into the community, joined in and benefitted from the organization.

Ofelia Mangen, a thirty-year-old who lives with her younger brother in a row house on Beach Ninety-second Street, joined the surf club last summer and spent many nights there volunteering with neighbors and friends. "I brought flowers, bartended, worked the door—whatever was needed," she told me, as we walked down Rockaway Freeway on a mild day in mid-November, past sanitation trucks, police cars, and sidewalks cluttered with debris. "I've just kept doing whatever is needed since the storm hit. But now the needs have changed, and there's obviously a lot more of them." Two weeks had passed since the superstorm, and residents had no power, gas, heat, or hot water for bathing. Stores, restaurants, pharmacies, and gas stations were closed. Trains were inoperable.

Mangen, a graduate student in educational design at New York University, has a steady disposition. She's slim but sturdy, with curly brown hair, thick glasses, and a strong, deliberate voice. She had a plastic crate for carrying things and was dressed for manual labor: black ski cap, black puffer jacket, black arm warmers, cargo pants, and hiking boots. "The sidewalks are still coated with sewage, and there's dangerous shit everywhere," she warned.

We began at her place, which, like all the houses in the area, sustained major flood damage. (The garage door showed a watermark about four feet above ground.) Several boxes of family photographs and a recently restored 1965 Ford Mustang belonging to the homeowner's son were among the casualties. We took supplies from Manhattan to a café where the staff was feeding relief workers, and then walked to the compost garden that Mangen helps run and inspected the soil, which had been drenched by contaminated floodwater. She instructed two arborists, who had arrived from Tennessee the night before, on how to prepare a fallen tree so that neighbors could use it as firewood, and offered respirator masks and large garbage bags to a worker from Illinois who was removing water-damaged drywall next door. Then we headed back to her street. A fallen tree rested ominously on the roof of a two-story apartment building on the corner.

Mangen spotted Junior, a neighbor who works as a contractor, parking his van across the street from us, and she led us to the apartment building so that we could see if anyone was there. "Hello," she shouted, and then banged on the door a few times. "Anybody home?" A young brown-skinned woman peeked out from the second-floor window, beneath the fallen tree. "Are you O.K. up there?" Mangen asked. "Do you need anything?"

"Juice," she answered. "For the baby."

We returned to the surf club, where residents, many with shopping carts, lined the sidewalk, putting in requests for food and supplies and waiting while volunteers fetched them. Mangen introduced me to Brandon d'Leo, a sculptor, and Bradach Walsh, a firefighter, who were directing the club's relief efforts, and they enlisted me in the search for volunteers to meet their neighbors' most pressing needs. "You're from a school," d'Leo noted. "Do you know anyone who teaches plumbers or electricians? We can't get power restored in our homes until we've passed an inspection by someone with certification."

When Mangen returned to her neighbor's apartment to deliver the juice, the woman came down to greet us. She held an infant, and appeared to have a painful sore on her lip. "Can we get you anything else?" Mangen asked. The woman shook her head, but hesitantly. "Listen, there's food at the surf club down on Beach Eighty-seventh," Mangen told her. "It's just a few minutes from here. And there are places giving out medicine now, too." The woman smiled shyly, thanked her, and returned to her dark apartment. A few minutes later, walking back to her place, Mangen had a thought. She called the arborists from Tennessee and asked them if they could take on another job. Soon afterward, the tree resting on her neighbor's rooftop was removed.

Thousands of people whose homes were damaged by Sandy live in neighborhoods that lack strong support networks or community organizations capable of mounting a large relief effort. They tend to be poorer and less educated than typical New Yorkers, with weaker ties to their neighbors as well as to political power brokers. Since Sandy, Michael McDonald, who heads Global Health Initiatives, in Washington, D.C., and worked in Haiti after the 2010 earthquake, has been coördinating relief efforts by volunteer groups, government agencies, corporate consultants, health workers, and residents in vulnerable areas, particularly in the Rockaways. McDonald calls the network the New York Resilience System, and he's convinced that civil society will ultimately determine which people and places will withstand the emerging threats from climate change. In December, I watched him chair a meeting of network participants—they included representatives from New York Cares (the city's largest volunteer organization), the accounting and consulting firm PricewaterhouseCoopers, the New York City Department of Health, and the state Attorney General's Office. "What's actually happening on the ground is not under an incident command system," he told me. "It's the fragile, agile networks that make a difference in situations like these. It's the horizontal relationships like the ones we're building that create security on the ground, not the hierarchical institutions. We're here to unify the effort."

Whether they come from governments or from civil society, the best techniques for safeguarding cities don't just mitigate disaster damage; they also strengthen the networks that promote health and prosperity during ordinary times. Contrast this with our approach to homeland security since 9/11: the checkpoints, the bollards, the surveillance cameras, the no-entry zones. We do not know whether these devices have prevented an attack on an American city, but, as the sociologist Harvey Molotch argues in "Against Security," they have certainly made daily life less pleasant and efficient, imposing costs that are difficult to measure while yielding "almost nothing of value" in the normal course of things.

"We were making some progress on climate-change adaptation in the late nineteen-nineties," Klaus Jacob observed. "But September 11th set us back a decade on extreme-weather hazards, because we started focussing on a completely different set of threats." Effective climate-proofing demands more intelligent design. It should provide benefits not just when disaster strikes but day to day, like Singapore's Marina Barrage, which created new waterfront, parkland, and exhibition spaces, or like a smarter power grid, which helps reduce energy consumption in all weather. That's true of the low-tech and the no-tech measures. Auburn Gresham's advantages over Englewood aren't restricted to mortality rates during a heat wave.

It's a cause for regret that we're not responding to the challenges of climate change with the same resources we've devoted to the war on terror. As long as the threat from global warming seemed remote and abstract, it was easier to ignore. Now climate change is coming to mean something specific, and scary. "Even on a clear day a hundred years from now, the water will be where it is today under storm-surge conditions," Jacob said. More heat waves, wildfires, hurricanes, and floods are to be expected. We are entering an age of extremes. "We can't just rebuild after every disaster," Jacob continued. "We need to pro-build, with a future of climate change in mind."

Power Brokers

Bill McKibben

June 26, 2017

The cacao-farming community of Daban, in Ghana, is seven degrees north of the equator, and it's always hot. In May, I met with several elders there to talk about the electricity that had come to the town a few months earlier, when an American startup installed a solar microgrid nearby. Daban could now safely store the vaccine for yellow fever; residents could charge their cell phones at home rather than walking to a bigger town to do it. As we talked, one of the old men handed me a small plastic bag of water, the kind street vendors sell across West Africa—you just bite off a corner and drink. The water was ice-cold and refreshing, but it took me an embarrassingly long moment to understand the pleasure with which he offered it: cold water was now available in this hot place. There was enough power to run a couple of refrigerators, and so *coldness* was, for the first time, a possibility.

I'd come to Daban to learn about the boom in solar power in sub-Saharan Africa. The spread of cell phones in the region has made it possible for residents to pay daily or weekly bills using mobile money, and now the hope is that, just as cell phones bypassed the network of telephone lines, solar panels will enable many rural consumers to bypass the electric grid. From Ghana, I travelled to Ivory Coast, and then to Tanzania, and along the way I encountered a variety of new solar ventures, most of them American-led. Some, such as Ghana's Black Star Energy, which had electrified Daban, install solar microgrids, small-scale versions of the giant grid Americans are familiar with. Others, such as Off-Grid Electric, in Tanzania and Ivory Coast, market home-based solar systems that run on a panel installed on each individual house. These home-based systems can't produce enough current for a fridge, but they can supply each home with a few lights, a mobile-phone charger, and, if the household can afford it, a small, super-efficient flat-screen TV.

In another farming town, in Ivory Coast, I talked to a man named Abou Traoré, who put his television out in a courtyard most nights, so that neighbors could come by to watch. He said that they tuned in for soccer matches—the village tilts Liverpool, but has a large pocket of Manchester United supporters. What else did he watch? Traoré considered. "I like the National Geographic channel," he replied—that is, the broadcast arm of the institution that became famous showing Westerners pictures of remote parts of Africa.

There are about as many people living without electricity today as there were when Thomas Edison lit his first light bulb. More than half are in sub-Saharan Africa. Europe and the Americas are almost fully electrified, and Asia is quickly catching up, but the absolute number of Africans without power remains steady. A World Bank report, released in May, predicted that, given current trends, there could still be half a billion people in sub-Saharan Africa without power by 2040. Even those with electricity can't rely on it: the report noted that in Tanzania power outages were so common in 2013 that they cost businesses fifteen per cent of their annual sales. Ghanaians call their flickering power *dum/sor*, or "off/on." Vivian Tsadzi, a businesswoman who lives not far from the Akosombo Dam, which provides about a third of the nation's power, said that most of the time "it's *dum dum dum dum*." The dam's head of hydropower generation, Kwesi Amoako, who retired last year, told me that he is proud of the structure, which created the world's largest man-made lake. But there isn't an easy way to increase the country's hydropower capacity, and drought, caused by climate change, has made the system inconsistent, meaning that Ghana will have to look elsewhere for electricity. "I've always had the feeling that one of the main thrusts should be domestic solar," Amoako said. "And I think we should put the off-grid stuff first, because the consumer wants it so badly."

Electrifying Africa is one of the largest development challenges on earth. Until recently, most people assumed that the continent would electrify in the same manner as the rest of the globe. "The belief was, you'd eventually build the U.S. grid here," Xavier Helgesen, the American co-founder and C.E.O. of Off-Grid Electric, told me. "But the U.S. is the richest country on earth, and it wasn't fully electrified until the nineteen-forties, and that was in an era of cheap copper for wires, cheap timber for poles, cheap coal, and cheap capital. None of that is so cheap anymore, at least not over here."

Solar electricity, on the other hand, has become inexpensive, in part because the price of solar panels has fallen at the same time that the efficiency of light bulbs and appliances has dramatically increased. In 2009, a single compact fluorescent bulb and a lead-acid battery cost about forty dollars; now, using L.E.D. bulbs and lithium-ion batteries, you can get four times as much light for the same price. In 2009, a radio, a mobile-phone charger, and a solar system big enough to provide four hours of light and television a day would have cost a Kenyan a thousand dollars; now it's three hundred and fifty dollars.

President Trump has derided renewable energy as "really just an expensive way of making the tree huggers feel good about themselves." But many Western entrepreneurs see solar power in Africa as a chance to reach a large market and make a substantial profit. This is a nascent industry, which, at the moment, represents a small percentage of the electrification in the region, and is mostly in rural areas. There's plenty of uncertainty about its future, and no guarantee that it will spread at the pace of cell phones. Still, in the past eighteen months, these businesses have brought electricity to hundreds of thousands of consumers—many of them in places that the grid failed to reach, despite a hundred-year head start. Funding, much of it from private investors based in Silicon Valley or Europe, is flowing into this sector—more than two hundred million dollars in venture capital last year, up from nineteen million in 2013—and companies are rapidly expanding their operations with the new money. M-Kopa, an American startup that launched in Kenya, in 2011, now has half a million pay-as-you-go solar customers; d.light, a competitor with offices in California, Kenya, China, and India, says that it is adding eight hundred new households a day. Nicole Poindexter, the founder and C.E.O. of Black Star, told me that every million dollars the company raises in venture capital delivers power to seven thousand people. She expects Black Star to be profitable within the next three years.

Like many of the American entrepreneurs I met in Africa, Poindexter has a background in finance. A graduate of Harvard Business School, she worked as a derivatives trader before leading business development at Opower, a software platform for utilities customers that was acquired by Oracle last year. (Unlike many of these entrepreneurs, who tend to skew white and male, Poindexter is African-American.) She decided to start the company in 2015, after she began to learn about energy poverty. She recalled watching TV coverage of the Ebola epidemic in Liberia. "There was a lot of coughing in the background, and I was thinking, That's someone with Ebola," she said. "But it wasn't. It was from the smoke in the room from the fire." Last year, in the Ghanaian community of Kofihuikrom, one of the first towns that Black Star served, the company erected twenty-two solar panels. Today, the local clinic no longer has to deliver babies by flashlight. The town chief, Nana Kwaku Appiah, said that he was so excited that he initially left his lights on inside all night. "Our relatives from the city used to not come here to visit," he said. "Now they do."

When I visited the Tanzanian headquarters of Off-Grid Electric, in the city of Arusha, the atmosphere was reminiscent of Palo Alto or Mountain View, with standing desks and glassed-in conference rooms for impromptu meetings. Erick Donasian, the company's head of service in Tanzania, grew up in a powerless house three miles from the office and joined the company in 2013; he said that, along with his enthusiasm for the company's goals, one attraction of working there is that it is far less formal than many Tanzanian businesses, where "you have to tuck your shirt in, which I hate the most." Off-Grid's Silicon Valley influence was clearest in the T-shirt Helgesen wore. It read "Make something people want," and sported the logo for Y Combinator, Silicon Valley's most famous incubator, where Helgesen's wife had recently developed a bartering app.

Helgesen, who is thirty-eight years old and lanky, with hair that he regularly brushes out of his eyes, grew up in Silver Bay, Minnesota, a small town on the shore of Lake Superior. At fourteen, he came up with the idea of leasing the municipal mini-golf course for a summer, and tripled revenues by offering season passes and putting on special promotions for visiting hockey teams. As a sophomore at Notre Dame, in 1999, he set up a Web site that posted the college's freshman register online, so that, as he put it, "you'd actually know who that cute girl you saw in anthro class was." Helgesen started similar sites at other colleges, but, he told me, "I wasn't as good a programmer as Zuckerberg. Even if I'd gotten it completely right, it would have been more Friendster than Facebook." His first major company, Better World Books, founded in 2002, took the model of charity used-book drives and moved it online. It's now one of the biggest sellers of used books on Amazon, and has helped raise twenty-five million dollars for literacy organizations, including Books for Africa.

Helgesen made his first trip to Tanzania in 2006, to visit recipients of Better World's funding and to go on safari. "I was staying at a fancy lodge near Kilimanjaro, and I remember thinking, How do things really work around here?" Helgesen said. He paid a local man to take him to the nearest village. "I was peppering him with questions: 'Do young people go to the city?' 'How much does coffee sell for?'" The experience, he said, "flipped my mind-set from 'People in Africa are poor and they need our help and our donated books' to 'This is what an emerging economy looks like. This is young people, this is entrepreneurialism, this is where growth will be.'" During a second trip to Africa, he went scuba diving in Lake Malawi ("to see the cichlid fish, which keep their babies in their mouths"), and was invited to dinner by his scuba instructor. "It was a decent-sized town, maybe twenty thousand people, but absolutely no electricity," Helgesen said. "It was all narrow alleys—they were bustling, but they were pitch-black."

In 2010, Helgesen won a Skoll Scholarship to Oxford, for M.B.A. students seeking "entrepreneurial solutions for urgent social and environmental challenges," and spent the year researching the renewables market. He found two like-minded business partners, and, in 2012, they set up shop in Arusha. At first, they planned to build solar microgrids to power cell-phone towers and sell the excess electricity to locals, but, Helgesen said, "it became clear that that was a pretty expensive way to go." So they visited customers in their homes to ask them what they wanted. "Those conversations were the smartest thing we ever did," Helgesen said. "I remember this one customer, she had a baby, and she would keep the kerosene lamp on low all night, as a night-light. It was costing thirty dollars a month in kerosene. And I was, like, Wow, for thirty dollars a month I could do a lot better."

Helgesen decided to "start with the customer, and the price point they could pay, and build the business behind that." Matt Schiller, the thirty-two-year-old vice-president of business operations, said that, in some ways, it is an easy sell. "If we talk to a hundred customers, not one says, 'I'd rather have kerosene,'" he told me. "Not one says, 'I'd like the warm glow of the kerosene lights.' In fact, when we were designing the L.E.D.s, we focus-grouped lights. And the engineers assumed they'd want a warmer light, because that's what they were used to. But, no, they picked the bluest, hardest light you can imagine. That's modernity. That's clean."

There were solar panels in sub-Saharan Africa before companies like Off-Grid arrived, but customers generally had to pay for them up front, a forbidding prospect for many. "Cost is important to the customer at the bottom, but risk is even more important," Helgesen told me. "A bad decision when you're that poor can mean your kids don't eat or go to school, which is why people tend to be conservative. And which is why kerosene was winning. There was no risk. You could buy it a tiny bit at a time."

Off-Grid, like several of its competitors, finances the panels, so that people can pay the same small monthly amounts they were paying for kerosene. Customers in Tanzania put down about thirteen dollars to buy Off-Grid's cheapest starter kit: a panel, a battery, a few L.E.D. lights, a phone charger, and a radio. Then they pay about eight dollars a month for three years, after which they own the products outright. The most popular system adds a few more lights and a flat-screen TV, for a higher down payment and about twice the monthly price. Customers pay their bill by phone; if they don't pay, the system stops working, and after a while it is repossessed. That scenario, it turns out, is uncommon: less than two per cent of the loans in Tanzania have gone bad.

Despite Off-Grid's Silicon Valley vibe, it faces challenges unfamiliar to software companies. Aidan Leonard, Off-Grid's Arusha-based general counsel, told me that the company "requires a lot of people walking around selling things and installing things and fixing things. There's a lot of hardware—someone's got a physical box in their house, and a panel on the roof, and they have to pay for it on a monthly basis." Poindexter, of Black Star, put the problem more bluntly. "We're a utility company," she told me, and utilities are a difficult business.

In America, utilities are burdened with infrastructure, such as the endless poles and wires that come down in storms. Off-Grid doesn't have to worry about poles, and the wires only run a few feet, from panel to battery to appliance. Still, the company is working with technology that is brand-new and needs to be made cheaply in order to be affordable. When solar energy first came to Africa, it was expensive and unreliable. Arne Jacobson, a professor of environmental-resources engineering at Humboldt State University, in California, is a couple of decades older than most of the entrepreneurs I met in Africa. He got his doctorate studying the first generation of home solar in Kenya, in the late nineteen-nineties. "In Kenya, I was trying to understand the quality of the panels that had started to flood the market," he said. Much of the technology had "big troubles. Chinese panels, panels from the U.K., all this low-quality junk coming in. Later, L.E.D.s that failed in hours or days instead of lasting thousands of hours, as they should. People's first experiences were often really bad."

Jacobson has spent his career in renewable energy; he helped build the world's first street-legal hydrogen-fuel-cell vehicle, in 1998. He now runs Humboldt's Schatz Energy Research Center. ("You want to know why a lot of early solar research happened in Humboldt?" he asked me. "Because there were a lot of back-to-the-land types here, and they had cash because they were growing dope.") After seeing the unpredictability of solar technology, he created, in 2007, what he calls a "de facto consumer-protection bureau for this nascent industry." The program, Lighting Global, which is run under the umbrella of the World Bank Group, tests and certifies panels, bulbs, and appliances to make sure that they work as promised. Jacobson credits this innovation with making investors more willing to put their money into companies such as Off-Grid, which has now raised more than fifty-five million dollars. His main testing lab is in Shenzhen, China, near most of the solar-panel manufacturers. He also has facilities in Nairobi, New Delhi, and Addis Ababa, and some of the work is still done in the basement of his building at Humboldt, where there's an "integrating sphere" for measuring light output from a bulb, and a machine that switches radios on and off to see if they'll eventually break.

Because many of Off-Grid's potential customers have experience with bad products, or know someone who has, the company takes extra steps to build trust with its clients. After an Off-Grid installer shows up on his motorbike, he opens the product carton with great solemnity; in an Ivorian village, I watched along with seventeen neighbors, who nodded as the young man held up each component, one by one. He then climbed onto the roof of the house, nailed on a solar panel about the size of a placemat, and used a crowbar to lift up the corrugated-tin roof to run the wire inside. He screwed the battery box to the cement-block wall and walked the customer through the process of switching lights on and off several times, something the man had never done before. The company also offers a service guarantee: as long as customers are making their payments, they can call a number on the box and a repairman will arrive within three days. These LightRiders, as the company calls them, are trained to trouble-shoot small problems. They travel by motorcycle, and if they can't make repairs easily they replace the system with a new one and haul the old unit back to headquarters.

This sales-and-installation system presents some engineering challenges. When the company expanded into Ivory Coast, last year, it had to redesign its packaging to fit on the smaller motorcycles used there. It also runs into problems coördinating coverage across a vast area where most houses don't have conventional addresses. "We had to build our own internal software to make it possible," Kim Schreiber, who runs Off-Grid's marketing operations in Africa, said. "We optimize, via G.P.S. coördinates, the best routes for our riders to take. The LightRider turns on his phone every morning, and he has a list of his tasks for the day, so he knows what parts to take with him."

Solar companies also contend with the complexity of the mobile-payment systems. In Ghana, where many customers don't use mobile money, Poindexter's Black Star team instead sells scratch cards from kiosks, which give customers a code they need to enter on their meter box to top up their account. Off-Grid delivers these codes over the phone, but the company still needs a call center, manned by fifteen people, to help customers with the mechanics of paying. Nena Sanderson, who runs Off-Grid's Tanzanian operation, showed me the steps entailed in paying a bill through a ubiquitous mobile-money system called M-Pesa. There are ten screens, and the process ends with the input of a sixteen-digit code. "And I have a smartphone," she said. "Now, imagine a feature phone, and imagine you may not know how to read, and the screen is a lot smaller, and it's probably scratched up. Mobile money is a great enabler, but it's not frictionless." One of Off-Grid's competitors, PEGAfrica, has printed the whole sequence on a wristband, which it gives to customers.

Because one of the biggest obstacles to the growth of solar power in the region is the lack of available cash, many of these companies are essentially banks as well as utilities, providing loans to customers who may have no credit history. That can make it hard to figure out what to charge people. "What you see in this space is at least eight to ten decent-sized pay-as-you-go solar companies, all trying to parse through what the actual end price to the customer really is," Peter Bladin, who spent many years in leadership roles at Microsoft and now invests in several of these firms, told me. Bladin first started studying distributed solar—solar electricity produced near where it is used—in Bangladesh, where the Nobel Prize winner Muhammad Yunus used his Grameen microcredit network to finance and distribute panels and batteries. Lacking that established financial architecture, companies in sub-Saharan Africa are constantly experimenting with different plans: Off-Grid began by offering ten-year leases, but found that customers wanted to own their systems more quickly, and so the payments are now spread out over three years. PEGAfrica customers buy their system in twelve months, but the company gives them hospitalization insurance as a bonus. Black Star is a true utility: the customers in the communities where it builds microgrids will always pay bills, but the charges start at only two dollars a month. (The business model depends on customers steadily increasing the amount of energy they buy, as they move from powering televisions to powering small businesses.) Companies like Burro—a Ghanaian outfit launched by Whit Alexander, the Seattle entrepreneur who founded Cranium games—sell lamps and chargers and panels outright, saving customers credit fees but limiting the number of people who can afford the products.

This uncertainty about the most practical financial model reflects the fact that in sub-Saharan Africa there is a great deal of economic diversity, both between countries and within them. One morning, I found myself walking down a line of houses in the Arushan suburb of Morombo. At the first house, a two-room cinder-block structure with a broken piece of mirror on one wall, a woman talked with me as we sat on the floor. The home represented a big step up for her, she said—she and her husband had rented a place for years, until they were able to buy this plot of land and build this house. She had a solar lantern the size of a hockey puck in her courtyard, soaking up rays. (Aid groups have distributed more than a million of these little lamps across the continent.) She assured me that she planned to get a larger solar system soon, but, for many of Africa's poorest people, buying a lantern is the only possible step toward electrification.

Next door, a twenty-six-year-old student named Nehemiah Klimba shared a more solidly built house with his mother. It had a corrugated-iron roof on a truss that let hot air escape, and we sat on a sofa. Klimba said that, as soon as he finished paying off the windows, he was going to electrify. He and his mother were already spending fifteen dollars a month on kerosene and another four dollars charging their cell phones at a local store, so they knew they'd be able to afford the twenty dollars a month for a solar system with a TV.

One door down was the fanciest house I'd seen in weeks. It belonged to a soldier who worked as a U.N. peacekeeper, and the floors were made of polished stone. There was an Off-Grid solar system on the roof, but it was providing only backup power. The owner had paid a hefty fee to connect to the local electric grid, so he faced none of the limitations of a battery replenished by the sun. In his living room, he had a huge TV and speakers; a stainless-steel Samsung refrigerator gleamed in the kitchen.

"This is how the solar revolution happens—one hot sales meeting at a time," Off-Grid's Kim Schreiber whispered to me as we watched one of the company's salesmen, an Ivorian named Seko Serge Lewis, at work. We were visiting the village of Grand Zattry with Off-Grid's Ivory Coast sales director, Max-Marc Fossouo. A couple of dogs tussled nearby; a motorbike rolled past with six people on board. In the courtyard next to us, a woman was doing the day's laundry in a bucket with a washboard. Her husband listened to the sales pitch from Lewis, who was showing him pictures on his cell phone of other customers in the village.

"That's to build up trust," Fossouo said. He'd been providing a play-by-play throughout the hour-long sales call. "This customer is on a big fence," he said. "He's stuck in the trust place. And I'm pretty sure the decision-maker is over there washing the clothes anyway." Fossouo was born in Cameroon and went to school in Paris. In his twenties, he spent seven summers in the U.S., selling books for Southwestern Publishing, a Nashville-based titan of door-to-door marketing. (Rick Perry is another company alum; so is Kenneth Starr.) "I did L.A. for years," he told me. "'Hi, my name is Max. I'm a crazy college student from France, and I'm helping families with their kids' education. I've been talking to your neighbors A, B, and C, and I'd like to talk to you. Do you have a place where I can come in and sit down?'" All selling, he said, is the same: "It starts with a person understanding they have a problem. Someone might live in the dark but not understand that it's a problem. So you have to show them. And then you have to create a sense of urgency to spend the money to solve the problem now."

The man turned down Lewis's pitch. He was worried that he wouldn't be able to make the monthly payments in the lean stretch before the next cacao harvest. "That's crap," Fossouo whispered, pointing again to the man's wife. "He loves this woman, he can move the world for her." When we went to the next house, Fossouo took over. This prospect was a farmer and schoolteacher, and they talked in his classroom, which had a few low desks with shards of slate on top. Fossouo had the man catalogue everything that he was spending on energy: money for kerosene, flashlight batteries, even the gas for the scooter that he borrowed when he needed to charge his phone. Then Fossouo showed him what he had to offer: a radio and four lights, each with a dimmer switch. "Where would you put the lamp?" he asked. "In front of the door? Of course! And the big light in the middle of the room, so when you have a party everyone could see. Now, tell me, if you went to the market to buy all of this, how much would it cost?" Fossouo tried angle after angle. "You have to think big here," he said. "When I talked to your chief, he said, 'Don't think small.' If your kid could see the news on TV, he might say, 'I, too, could be President.'"

"This is great," the man said. "I know you're trying to help us. I just don't have the money. Life is hard, things are expensive. Sometimes we're hungry."

Fossouo nodded. "What if I gave you a way to pay for it?" he asked. "So the dollar wouldn't even come from your pocket? If you get a system, people will pay you to charge their phones. Or, if you had a TV, you could charge people to come watch the football games."

"I couldn't charge a person for coming in to watch a game," the man said. "We're all one big family. If someone is wealthy enough to have a TV, everyone is welcome to it."

The hour ended without a sale, but Fossouo wasn't worried. "It takes two or three approaches on average," he said. "You always have to leave the person in a good place, where he loves you stopping by. This guy wants to finish building his house right now—his house is heavy on him—but it won't be long." As we talked, the first prospect came over, asking for a leaflet and a phone number. His wife, he said, was very interested.

The arrival of electricity is hard for today's Westerners to imagine. Light means differences in sleeping and eating patterns and an increased sense of safety. I talked with one Tanzanian near Arusha who had traded in a kerosene lamp for five Off-Grid bulbs, including a security light outside his door that went on automatically when it got dark. "Crime is here," he said, "but also dangerous animals. Especially snakes.

So it's good to have lights." Everywhere I went, I met parents who said that their children could study at night. "You can feel the effects with their grades now at school," one Ivorian father said. Several town chiefs told me that they hoped to get classroom computers, and one planned to mechanize the well so that townspeople would no longer need to pump water by hand. Farmers in West Africa were getting daily weather reports from Farmerline, a Ghanaian information service that uses G.P.S. to customize the forecasts. "If a farmer puts fertilizer on the field and then it rains, he loses the fertilizer—it washes away," Aloysius Attah, a young Ghanaian entrepreneur who co-founded the service, told me. "And the farmers say they can't tell the rain anymore. My auntie could read the clouds, the birds flying by, but the usual rainfall pattern has shifted."

"Our killer app is definitely the television," Off-Grid's Schreiber said. "If the twenty-four-inch is out of stock, lots of people won't buy." Wandering through newly electrified towns, I saw teen-agers watching action movies. Black Star's Poindexter told me, "There was a kid in town that I liked, Samuel, and when I came back after the power was turned on his arm was in a cast. He'd watched a karate show on TV, and he and his friends were playing it, and he broke his arm. I was horrified—I was, like, society is not prepared for this. And then I remembered that I did the same thing after I watched 'Popeye' as a kid. I ran right into the hedge and had to get twenty stitches. That's kids and TV."

In Daban, after I asked what the most popular program was, everyone began laughing and nodding. "'Kumkum'!" people shouted. "Kumkum Bhagya," an Indian soap opera set in a marriage hall and loosely based on Jane Austen's "Sense and Sensibility," airs every night from seven-thirty to eight-thirty, during which time village life comes to a standstill. "All the chiefs have advocated for everyone to watch, because it's about how relationships are built," the local chief, Nana Oti Awere, said. Of course, the changes brought about by electrification will affect local communities in unpredictable ways that will play out over many years. One mother I spoke to explained that the TV "keeps the children at home at night, instead of roaming around." The Ivorian farmer who told me about the effects on his children's grades went on to say, "In the old time, you had to go outside and talk. Now my neighbor has his TV, I have my TV, and we stay inside."

A decade ago, most experts would have predicted that foreign aid, rather than venture capital, would play a central role in bringing power to sub-Saharan Africa.

Off-Grid Electric has been funded by sources including Tesla and Paul Allen's venture fund, Vulcan. Allen, one of the world's richest men, is worth twenty billion dollars, or roughly half of the G.D.P. of Tanzania, a country of almost fifty-four million people. Should he be able to make yet more money off the electrification of African huts? There's more than a whiff of colonialism about the rush of Westerners and Western money into Africa. As Attah, the young Ghanaian who helped found Farmerline, put it, "There are a lot of Ivy Leaguers coming to Africa to say, 'I can solve this problem, snap, snap, snap.' They're doing good work, but little investment goes to community leaders who are doing the same work on the ground."

The Westerners I spoke to, though they pledged to hire more local executives, didn't think that the drive to help was incompatible with the desire to make money. As Poindexter put it, "There is a level of responsibility that I feel, and that I think any appropriate investor needs to have, about extraction versus contribution. I am not willing to be an extractive capitalist here, but I think that capitalism has an extremely important role to play in these communities." Helgesen—who, despite his occasional oblivious tech-dudishness, spends most of his time in very remote places trying to provide power—is unapologetic about his company's funding sources. Billionaires, he says, have the capital to make companies grow fast enough to matter. "Paul Allen didn't invest because he thought it was the easiest way to make more money," Helgesen said. "I got an awful lot of 'no's along the way from people who wanted easier money." In any event, it's not clear that other sources of funding are available, at least from the U.S.: Trump, pulling out of the Paris climate accord earlier this month, said that the country would not meet its pledge to help poor nations develop renewable energy, dismissing the plan as "yet another scheme to redistribute wealth out of the United States through the so-called Green Climate Fund—nice name."

Even when aid agencies are well funded, they haven't always delivered. Over the last decade, a strong critique of aid, ranging from William Easterly's "The White Man's Burden" to Dambisa Moyo's "Dead Aid," has laid much of the blame for Africa's continued underdevelopment on the weaknesses of sweeping programs planned from afar. Still, aid agencies and global-development banks have a useful role to play in the energy transition. It will be years before it makes financial sense for solar companies to expand to the most remote and challenging regions of the continent. As new companies launch, they will need an infusion of what Helgesen calls "ultra-high-risk capital." Private investors will supply it, he says, "but they want forty per cent of your company in return, which makes it hard to raise capital later on, because you've already sold off such a big chunk." Some aid agencies have funded private ventures in the early stages, to help them get off the ground or reach new geographic areas. U.S.A.I.D. gave Off-Grid five million dollars toward its early costs, and, over the past few years, a Dutch development agency has given the company several hundred thousand euros as it has extended into the impoverished lakes region of Tanzania, where it otherwise wouldn't have been profitable to go. Currency risks pose another problem: Poindexter told me that when she builds a Ghanaian microgrid she has invested in an asset with a twenty-year life span in a country where inflation is highly unpredictable. "We just had an election in the U.S. with huge consequences for policy," she said. "But over here every election is potentially like that." And, like anywhere in the world, national governments can make things easier by establishing clear policies. Rwanda's leaders, for instance, specified the regions in which the rapidly developing country planned to extend its grid, thereby delineating where solar would be needed most.

"African leaders used to think solar was being pushed on them," Clare Sierawski, who works on renewable energy with the U.S. Trade and Development Agency in Accra, said. "But now they all want solar. It's a confluence of things. Mostly, it's getting cheaper. And governments were tuned in to it by the Paris accord." Ananth Chikkatur, who runs a U.S.A.I.D. project in the city, had just returned from taking thirteen high-ranking Ghanaians on a trip to study solar power in California. "Renewable energy should not be considered an alternative technology," he said. "It's becoming a conventional technology now." Rwanda is not the only nation expanding its grid, and many countries are turning to large solar farms to generate power. Burkina Faso, for instance, has plans for solar arrays across its desert regions.

Distributed generation, however, is especially essential in rural areas, and it is growing fast—maybe, according to some observers, too fast. The investor Peter Bladin told me that the push for quick returns on investment could lead some companies to try to "squeeze more out of poor households" and warned about "mission drift, trying to make money off the backs of the poor in a dubious way." Earlier this year, three principals from the impact-investment firm

Ceniarth, which had put money into Off-Grid and similar companies, said that it was backing out of the industry for the time being. In an open letter, they wrote that the hype of venture capitalists and the lack of government regulation "puts consumers at risk and places a great deal of responsibility on vendors to self-police." The gush of money, they cautioned, "may be too much, too fast for a sector that still has not fully solved core business model issues and may struggle under the high growth expectations and misaligned incentives of many venture capitalists." Helgesen, unsurprisingly, disagreed with their analysis of investor over-exuberance. "It's like looking at a Palm Pilot and saying, 'This is not so great,'" he said. "Or even an iPhone 1. The iPhone 1 was a necessary step to the iPhone 7. People who have raised real money have not raised it on the premise that we'll be selling the same stuff in ten years." But he wasn't waiting for the technology to mature. "We have to think about the future, and we have to sell something people want today," he said.

Most customers I met had little interest in the fact that their power came from the sun, or that it was environmentally friendly. Since these communities weren't using power previously, their solar panels fight climate change only in the sense that they decrease pressure to build power plants that consume fossil fuel. But some observers hope that the experience in Africa—which today has more off-the-grid solar homes than the U.S.—could help drive transformation elsewhere. Already, a few dozen American cities have pledged to become one-hundred-per-cent renewable. (Pittsburgh did so the day after Trump held up its theoretically beleaguered citizens as a reason for leaving the climate accord.) The U.S. has already sunk a fortune into building its electric grid, and it may seem far-fetched to think that users will disconnect from it entirely. But, as Helgesen told me, "As batteries get better, it's going to be a lot more realistic for people to stop depending on their utility." He thinks that, in an ideal world, technological change could lead to cultural change. "The average American has no concept of electrical constraint," he said. "If we accept some modest restrictions on our power availability, we can go off-grid very quickly."

For many people in the countries I visited, solar power is creating a new hope: for electric fans. When I was there, Off-Grid Electric was expanding from the relatively cool highlands around Mt. Kilimanjaro to the scorching, humid lowlands of West Africa, and in every village we visited the message was the same: The TV is great, the light bulb is great, but can I please have a fan? Many homes are poorly ventilated; windows are expensive, and can attract burglars. Fans, however, draw a comparatively large amount of current, threatening to quickly drain the battery that a solar panel has spent the day filling. And, unlike light bulbs or televisions, fans have moving parts that easily break. "Our customers tend to make heavy use of their equipment," Off-Grid's Schreiber said. Still, she promised one village after another that fans were coming soon.

Shea Hughes, Off-Grid's product manager, is one of the employees charged with delivering on that promise. Hughes told me that he hopes to someday make Off-Grid's product powerful enough to perform industrial tasks: pumping water for irrigation, milling cacao, and so on. "I'm confident solar is capable of doing that," he said. "You just add more panels and you get to the power requirements you need. And as the price drops, well . . ." He had recently been to a consumer-electronics fair in China. "I was amazed to see the prices," he said.

For the moment, though, a workable fan would be nice. "We'd always thought a fan would take too much power for the current systems we're selling," Hughes said. "But the people in Ivory Coast were so insistent that we went back and looked at it." Because of the emerging market for super-efficient appliances, in the U.S. and elsewhere, some manufacturers had a product that, as long as you kept it set to medium, drew only eight and a half watts. (The standard incandescent light bulb that hung in American hallways for generations drew sixty.) "We've told the manufacturer to eliminate the high-speed option," Hughes said. "Now medium is high. And in our tests people are satisfied with the air speed. But they say the battery tends to run out at 3 or 4 a.m., and they typically sleep till 6 a.m. So it's not perfect, but it's getting there."

Value Meal

Tad Friend

September 30, 2019

Cows are easy to love. Their eyes are a liquid brown, their noses inquisitive, their udders homely; small children thrill to their moo.

Most people like them even better dead. Americans eat three hamburgers a week, so serving beef at your cookout is as patriotic as buying a gun. When progressive Democrats proposed a Green New Deal, earlier this year, leading Republicans labelled it a plot to "take away your hamburgers." The former Trump adviser Sebastian Gorka characterized this plunder as "what Stalin dreamt about," and Trump himself accused the Green New Deal of proposing to "permanently eliminate" cows. In fact, of course, its authors were merely advocating a sensible reduction in meat eating. Who would want to take away your hamburgers and eliminate cows?

Well, Pat Brown does, and pronto. A sixty-five-year-old emeritus professor of biochemistry at Stanford University, Brown is the founder and C.E.O. of Impossible Foods. By developing plant-based beef, chicken, pork, lamb, dairy, and fish, he intends to wipe out all animal agriculture and deep-sea fishing by 2035. His first product, the Impossible Burger, made chiefly of soy and potato proteins and coconut and sunflower oils, is now in seventeen thousand restaurants. When we met, he arrived not in Silicon Valley's obligatory silver Tesla but in an orange Chevy Bolt that resembled a crouching troll. He emerged wearing a T-shirt depicting a cow with a red slash through it, and immediately declared, "The use of animals in food production is by far the most destructive technology on earth. We see our mission as the last chance to save the planet from environmental catastrophe."

Meat is essentially a huge check written against the depleted funds of our environment. Agriculture consumes more freshwater than any other human activity, and nearly a third of that water is devoted to raising livestock. One-third of the world's arable land is used to grow feed for livestock, which are responsible for 14.5 per cent of global greenhouse-gas emissions. Razing forests to graze cattle—an area larger than South America has been cleared in the past quarter century—turns a carbon sink into a carbon spigot.

Brown began paying attention to this planetary overdraft during the late two-thousands, even as his lab was publishing on topics ranging from ovarian-cancer detection to how babies acquire their gut microbiome. In 2008, he had lunch with Michael Eisen, a geneticist and a computational scientist. Over rice bowls, Brown asked, "What's the biggest problem we could work on?"

"Climate change," Eisen said. Duh.

"And what's the biggest thing we could do to affect it?" Brown said, a glint in his eye. Eisen threw out a few trendy notions: biofuels, a carbon tax. "Unh-unh," Brown said. "It's cows!"

When the world's one and a half billion beef and dairy cows ruminate, the microbes in their bathtub-size stomachs generate methane as a by-product. Because methane is a powerful greenhouse gas, some twenty-five times more heat-trapping than carbon dioxide, cattle are responsible for two-thirds of the livestock sector's G.H.G. emissions. (In the popular imagination, the culprit is cow farts, but it's mostly cow burps.) Steven Chu, a former Secretary of Energy who often gives talks on climate change, tells audiences that if cows were a country their emissions "would be greater than all of the E.U., and behind only China and America." Every four pounds of beef you eat contributes to as much global warming as flying from New York to London—and the average American eats that much each month.

"So how do we do it?" Eisen asked.

"Legal economic sabotage!" Brown said. He understood that the facts didn't compel people as strongly as their craving for meat, and that shame was counterproductive. So he'd use the power of the free market to disseminate a better, cheaper replacement. And, because sixty per cent of America's beef gets ground up, he'd start with burgers.

A lean marathon runner with the air of a wading stork, Brown was an unlikely food entrepreneur. His older brother, Jim, said, "The idea of Pat running a company was a real surprise. The mission had always been gene mapping and finding cures for AIDS and cancer." Brown, a vegan who ate his last burger in 1976, had never spared a thought to food, considering it "just stuff to shove in your mouth." Free-rangingly curious, he lacked a C.E.O.'s veal-penned focus. "Pat gave some of the best science talks I've ever seen," Eisen told me, "and also some of the worst, because the slides wouldn't match after he started talking about something different from what he had planned."

The existing plant-based armory was unpromising; veggie burgers went down like a dull sermon. But, Brown reasoned, this was because they were designed for the wrong audience—vegetarians, the five per cent of the population who had accustomed themselves to the pallid satisfactions of bean sprouts and quinoa. "The other veggie-burger companies were just trying to be as good as the next plant-based replacement for meat, which meant they were making something no meat lover would ever put in his mouth," Brown said. To get meat-eaters to love meat made from plants, he had to resolve a scientific question, one that he decided was the most important in the world: What makes meat so delicious?

Brown assembled a team of scientists, who approached simulating a hamburger as if it were the Apollo program. They made their burger sustainable: the Impossible Burger requires eighty-seven per cent less water and ninety-six per cent less land than a cowburger, and its production generates eighty-nine per cent less G.H.G. emissions. They made it nutritionally equal to or superior to beef. And they made it look, smell, and taste very different from the customary veggie replacement. Impossible's breakthrough involves a molecule called heme, which the company produces in tanks of genetically modified yeast. Heme helps an Impossible Burger remain pink in the middle as it cooks, and it replicates how heme in cow muscle catalyzes the conversion of simple nutrients into the molecules that give beef its yeasty, bloody, savory flavor. To my palate, at least, the Impossible Burger still lacks a beef burger's amplitude, that crisp initial crunch followed by shreds of beef falling apart on your tongue. But, in taste tests, half the respondents can't distinguish Impossible's patty from a Safeway burger.

Eighteen months ago, White Castle, the nation's oldest burger chain, started selling the Impossible Slider, and sales exceeded expectations by more than thirty per cent. Lisa Ingram, White Castle's C.E.O., said, "We've often had customers return to the counter to say, 'You gave us the wrong order, the real burger.'" In August, Burger King rolled out the Impossible Whopper in all of its seventy-two hundred locations. Fernando Machado, the company's chief marketing officer, said, "Burger King skews male and older, but Impossible brings in young people and women, and puts us in a different spectrum of quality, freshness, and health."

Ninety-five per cent of those who buy the Impossible Burger are meat-eaters. The radio host Glenn Beck, who breeds cattle when he's not leading the "They're taking away your hamburgers!" caucus, recently tried the Impossible Burger on his show, in a blind taste test against a beef burger—and guessed wrong. "That is insane!" he marvelled. "I could go vegan!"

Pat Brown had built a better mouthtrap. But would that be enough?

The working title of Impossible Foods' 2019 impact report was "Fuck the Meat Industry." "I never seriously considered using it," Brown told me, "but it helps frame the mojo." Brown has a light voice, a tolerant smile, and an engaging habit of absorption; he often remarks that some scientific conundrum is "too arcane to get into," then plunges into it regardless, surfacing minutes later with a sheepish "Anyway, anyway!" as he tries to recall the topic at hand. But the mojo is conquest. "We plan to take a double-digit portion of the beef market within five years, and then we can push that industry, which is fragile and has low margins, into a death spiral," he said. "Then we can just point to the pork industry and the chicken industry and say 'You're next!' and they'll go bankrupt even faster."

Meat producers don't seem too worried that Brown will rid the earth of livestock by 2035. The three largest meatpacking companies in America have combined annual revenues of more than two hundred billion dollars. Mark Dopp, a senior executive at the North

American Meat Institute, a lobbying group, told me, "I just don't think it's *possible* to wipe out animal agriculture in sixteen years. The tentacles that flow from the meat industry—the leather and the pharmaceuticals made from its by-products, the millions of jobs in America, the infrastructure—I don't see that being displaced over even *fifty* years."

A number of alternative-protein entrepreneurs share Brown's mission but believe he's going about it the wrong way. The plant-based producer Beyond Meat is in fifty-three thousand outlets, including Carl's Jr., A&W, and Dunkin', and has a foothold in some fifty countries. Its I.P.O., in May, was the most successful offering of the year, with the stock up more than five hundred per cent; though the company is losing money, investors have noticed that sales of plant-based meat in restaurants nearly quadrupled last year. While Impossible depends on the patented ingredient heme, Beyond builds its burgers and sausages without genetically modified components, touting that approach as healthier. Ethan Brown, Beyond's founder and C.E.O. (and no relation to Pat Brown), told me, jocularly, "I have an agreement with my staff that if I have a heart attack they have to make it look like an accident."

Several dozen other startups have taken an entirely different approach: growing meat from animal cells.

Yet even Pat Brown's competitors often end up following his lead. Mike Selden, the co-founder and C.E.O. of Finless Foods, a startup working on cell-based bluefin tuna, said, "Pat and Impossible made it seem like there's a real industry here. He stopped using the words 'vegan' and 'vegetarian' and set the rules for the industry: 'If our product can't compete on regular metrics like taste, price, convenience, and nutrition, then all we're doing is virtue signalling for rich people.' And he incorporated biotechnology in a way that's interesting to meat-eaters—Pat made alternative meat *sexy*."

What's striking about Brown is his aggression. He is a David eager to head-butt Goliath. "If you could do two things of equal value for the world, and in one of them someone is trying to stop you, I would do that one," he told me. Brown doesn't care that plant-based meat amounts to less than 0.1 per cent of the $1.7-trillion global market for meat, fish, and dairy, or that meat contributes to the livelihoods of some 1.3 billion people. His motto, enshrined on the wall of Impossible's office, is "Blast ahead!" During the six months that I was reporting this story, the company's head count grew sixty per cent, to five hundred and fifty-two, and its total funding nearly doubled, to more than seven hundred and fifty million dollars. Brown laid out the math: to meet his 2035 goal, Impossible just has to double its production every year, on average, for the next 14.87 years. This means that it has to scale up more than thirty thousandfold. When I observed that no company has ever grown anywhere near that fast for that long, he shrugged and said, "We will be the most impactful company in the history of the world."

America's first commercial mock meat came out of the Battle Creek Sanitarium, in Michigan, at the turn of the twentieth century. The sanitarium was run by Dr. John Harvey Kellogg, a member of the vegetarian Seventh-day Adventist Church, who proselytized for sexual abstinence and made his eponymous cornflakes superbly bland, hoping that their ingestion would dampen lust. When Kellogg began to sell cans of Protose, an insipid mixture of nuts and gluten, he claimed that it "resembles potted veal or chicken"—meat in general, rather than any specific one.

In the seventies and eighties, soy burgers developed by MorningStar Farms and Gardenburger epitomized a peaceful life style, indicating that "no animals were harmed in the making of this patty." In 2001, Bruce Friedrich, who ran vegan campaigns at People for the Ethical Treatment of Animals, led a "Murder King" protest, trying to get Burger King to change its ways. The chain tweaked its animal-welfare policies, but kept on selling beef. Friedrich, who is now the executive director of the Good Food Institute, which advocates for meat replacements, told me, "If you're asking fast-food restaurants to pay more to compete, and to use a veggie burger that isn't very good, that's a colossal fail."

In the past decade, venture capitalists have begun funding companies that view animal meat not as inflammatory, or as emblematic of the Man, but as a problematic technology. For one thing, it's dangerous. Eating meat increases your risk of cardiovascular disease and colorectal cancer; a recent Finnish study found that, across a twenty-two-year span, devoted meat-eaters were twenty-three per cent more likely to die. Because antibiotics are routinely mixed into pig and cattle and poultry feed to protect and fatten the animals, animal ag promotes antibiotic resistance, which is projected to cause ten million deaths a year by 2050. And avian and swine flus, the most likely vectors of the next pandemic, pass easily to humans, including via the aerosolized feces widely present in slaughterhouses. Researchers at the University of Minnesota found fecal matter in sixty-nine per cent of pork and ninety-two per cent of poultry; *Consumer Reports* found it in a hundred per cent of ground beef.

For another thing, meat is wildly inefficient. Because cattle use their feed not only to grow muscle but also to grow bones and a tail and to trot around and to think their mysterious thoughts, their energy-conversion efficiency—the number of calories their meat contains compared with the number they take in to make it—is a woeful one per cent.

It's easy enough to replicate some animal products (egg whites are basically just nine proteins and water), but mimicking cooked ground beef is a real undertaking. Broadly speaking, a burger is sixty per cent water, twenty-five per cent protein, and fifteen per cent fat, but, broadly speaking, if you assembled forty-two litres of water you'd be sixty per cent of the way to a human being. Cooked beef contains at least four thousand different molecules, of which about a hundred contribute to its aroma and flavor and two dozen contribute to its appearance and texture. When you heat plant parts, they get softer, or they wilt. When you heat a burger, its amino acids react with simple sugars and unsaturated fats to form flavor compounds. The proteins also change shape to form protein gels and insoluble protein aggregates—chewy bits—as the patty browns and its juices caramelize. This transformation gives cooked meat its nuanced complexity: its yummy umami.

Mimicking these qualities was the task Pat Brown undertook in 2011, when he decided, after organizing a workshop on animal agriculture that accomplished nothing, that he'd have to solve the problem himself. He worked up a pitch, then bicycled down the road from Stanford to three venture-capital firms. His pitch had everything V.C.s like to fund: a huge market, a novel way to attack it, and a passionate founder who already talked the talk. Brown's habit of referring to "the technology that provides us with meat" made plant burgers sound like an iterative efficiency rather than like a threat to a beloved way of life. All he was doing was disintermediating the cow.

Impossible ended up taking three million dollars in seed funding from Khosla Ventures. Then Brown started hiring scientists, most of whom had no food expertise. His wife, Sue Klapholz, who trained as a psychiatrist and worked as a geneticist, became the company's nutritionist. "I had been making jewelry and doing nature photography, having this great retirement," she told me, still surprised by this turn in their lives. No one quite knew what they were doing, including Brown, who'd announce projects such as "We need every single plant-based ingredient in the world. Go!"

For alternative-protein companies, the first challenge is often producing a protein that's utterly tasteless. A flavor packet can then make it delicious. A startup called Spira, for instance, is attempting to develop algae called spirulina as a food source. "The problem is that it's a slimy goop," Surjan Singh, the company's C.T.O., told me. "And when you dry it and powderize it, it tends to biodegrade, so it tastes terrible. We're hoping to break even, eventually, where we can extract a protein isolate that's really good for you, but that tastes like as close to nothing as possible."

Impossible's first prototype burgers contained the "off-flavors" characteristic of their foundational protein, soy or wheat or pea. (Pea protein is sometimes said to evoke cat urine.) So the company's scientists had to learn how to erase those flavors, even as they were learning the subtleties of the aroma and taste they were trying to emulate.

One morning in Impossible's lab, Brown showed me a gas chromatograph–mass spectrometer, which is used to identify the molecules that appear in meat as it's cooked and to link those molecules to odors. "Some poor schmuck has their nose stuck in here for forty-five minutes," Brown said, indicating a plastic nose mold that protruded from the machine. "You have to bunny-sniff at a very high rate, often trying to characterize molecules you've never smelled before." He looked at a handwritten list from the last assay: "You might say, 'We've got to get rid of "Band-Aid," or "skunk," or "diaper pail"'—but don't judge, because all of those together make up 'burger taste.'"

Most veggie burgers are formed by an extruder, a machine that operates like a big pressure cooker, using heat and compression to replicate meat's fibrous morphology. Brown suspected that the key to a truly meaty plant burger was an ingredient. He had a hunch about heme, an iron-carrying molecule in hemoglobin (which makes your blood red), whose structure is similar to that of chlorophyll (which enables plants to photosynthesize). David Botstein, a geneticist who sat on Impossible's board, told me, "If you understand biochemistry, you understand that heme, more than anything else, is a central molecule of animal and plant life." As Brown was beginning to experiment, he pulled up clover from behind his house and dissected its root nodules, to see if there was enough heme inside to make them pink. (There was.)

In Impossible's microbiology lab, Brown told me, "An interesting, extremely speculative idea is that there's an evolutionary advantage to human beings in seeking out heme. It's a cue that means 'There's a dense source of protein and iron nearby.'" The first time that Impossible made a burger with heme, he said, "it tasted like meat, and within six months we had compelling evidence that it was the magic ingredient that gives meat its flavor."

In 2012, the company tested heme from thirty-one sources, ranging from tobacco plants to geothermal-spring water. Myoglobin from cows, the obvious candidate, oxidized too quickly (which is why ground beef goes brown in your fridge). Soy leghemoglobin performed best, so Impossible built a dozen machines to try to harvest it from the root nodules of soy. "We even rented a street sweeper and fed the soy plants in there," Brown told me. Nothing worked. "We flushed a year or more and half of our seed funding on this project I'm to blame for—the total low point," he said. They ended up manufacturing heme by genetically modifying yeast with a snippet of soy DNA. Yeast is usually white; Impossible's yeast, made in fifty-thousand-gallon tanks, is the foamy red of cocktail sauce.

Impossible's first burger, built around wheat protein, launched in 2016, at four high-end restaurants: Cockscomb and Jardinière, in San Francisco; Crossroads Kitchen, in Los Angeles; and Momofuku Nishi, in New York. An improved formulation, introduced last January, swapped out wheat for soy and was not only gluten-free but also lower in fat and cheaper to manufacture. Traci Des Jardins, the chef behind Jardinière, said, "The 1.0 version had a mushy mouthfeel, and it would adhere to surfaces and sear in a way that meat doesn't. This version has a more toothsome bounce, and it doesn't fall apart in a Bolognese sauce. The 2.0 really does behave just like beef."

Even those sympathetic to Brown's mission fret that taste and mouthfeel won't matter if the desire for meat is hardwired by evolution. Maple Leaf Foods, a Canadian company, is building a three-hundred-million-dollar facility in Indiana to make alternative proteins. But its C.E.O., Michael McCain, told me, "The human body has been consuming animal protein for a hundred and fifty thousand years, and I honestly think that's going to continue for a really long time."

Climate change, which now drives our hunt for meat substitutes, originally drove hominids to turn to meat, about two and a half million years ago, by making our usual herbivorean foodstuffs scarce. Eating animals added so much nutrition to our diets that we no longer had to spend all our time foraging, and we developed smaller stomachs and larger brains. Some scientists believe that this transformation created a powerful instinctive craving. Hanna Tuomisto, a Finnish professor of agricultural science, recently wrote, "This evolutionary predilection explains why eating meat provides more satisfaction compared to plant-based food and why so many people find it difficult to adopt a vegetarian diet."

An inborn meat hunger remains a hypothesis; meat is the object of many human urges, including the urge to construct all-encompassing theories. In the book "Meathooked," Marta Zaraska writes, "We crave meat because it stands for wealth and for power over other humans and nature. We relish meat because history has taught us to think of vegetarians as weaklings, weirdos, and prudes." The anthropologist Nick Fiddes goes further, declaring, in "Meat: A Natural Symbol," that we value meat not in spite of the fact that it requires killing animals but because it does. It's the killing that establishes us as kings of the jungle.

Ethan Brown, of Beyond Meat, suspects that nibbling plant patties doesn't exude the same macho vibe. A bearded, gregarious, six-foot-five man who played basketball at Connecticut College, he has retained a squad of athlete "ambassadors" to help dispel that perception. When I visited Ethan at the company's offices, in El Segundo, California, he pointed me to a 2009 study of Ivory Coast chimpanzees which suggested that males who shared meat with females doubled their mating success. "Men usually give women the meat first, at dinner, before the sex—you want to be a protein provider," he said. "Do you think if you take a woman out and buy her a salad you get the same reaction?"

It's worth noting that the Neanderthals, who subsisted almost entirely on meat, were outcompeted by our omnivorous ancestors. In any case, Ethan told me, meat no longer serves its original purpose, and "we can use the expanded brain that meat gave us to get us off of it." Like many alternative-protein entrepreneurs, he is a vegan; when he taste-tests Beyond's burgers, he occasionally chews a beef burger to orient his palate, then spits it out and wipes his tongue with a napkin. He has a potbellied pig named Wilbur at home that knows how to open the refrigerator: "Wilbur lives in our house to teach my kids that, from the perspective of science, the moral circle is poorly defined."

Ethan said that he launched Beyond Meat to mitigate meat's effects on "human health, climate change, natural resources, and animal welfare—we call them 'the four horsemen.'" One consequence of this compendious mission, with its attention to people's health—and to their concerns about health, warranted or not—is that Beyond, unlike Impossible, uses only ingredients taken more or less directly from nature.

For lunch, Ethan and I ate the latest Beyond Burger. Built around proteins derived from peas, mung beans, and brown rice, it was enriched with coconut oil and cocoa butter. Ethan, a self-described tough grader, rated it a 7.5 out of 10. "We've had great progress in texture and juiciness," he said, but added that the company's scientists were still working on "color transition." My burger was brown on the outside and purple in the middle, with a bloody affect encouraged by beet juice—but the fading between the two tones seemed faintly amiss. While savory, and possessed of a plausible mouthfeel, the patty was also curiously dense.

Pea protein's off-flavor was another problem to solve. Ethan said that he planned to expand his supply chain to include proteins from such plants as flax and lupine. He added, reflexively, "The best thing about pea is that it's not soy"—Impossible's chief ingredient. "I learned early on that consumers don't want a lot of soy, because they're worried about phytoestrogen, the concern being that it disrupts hormones and gives you 'man boobs.'" I observed that there was no evidence that this ever happens unless you consume soy in gigantic amounts. "I don't believe in the man-boobs theory," he said, "but who am I to question our customers?"

Ethan's scientists are skeptical of heme's efficacy. Dariush Ajami, who runs Beyond's lab, told me that he viewed it as a mere colorant, because, in collaborating with companies specializing in food chemistry, "we've never seen any flavor houses using heme as a flavor catalyzer." Ethan told me that even if heme proved to be a catalytic dynamo he wouldn't use it, or any genetically modified ingredient: "There's an evolutionary instinct, deep within us, to avoid things we don't understand." When I noted that consumers already accept many G.M.O. products—more than half the rennet used to make cheese is genetically modified, and ninety-two per cent of America's corn is G.M.O.—he conceded, "People will get used to it in the Impossible Burger." He grinned. "But will they get used to it before the burn rate gobbles the company?"

Meat producers like to point out that meat has a "clean deck": its components are few. One ag-business executive told me that consumers would, or anyway should, be alarmed by the long list of ingredients in Impossible's and Beyond's burgers: "A lot of customers think of an animal that has been around for more than a thousand years"—cows were domesticated from aurochs about ten thousand years ago—"and is just one ingredient as a natural product, versus a chemistry project of twenty-five or thirty ingredients you can't even pronounce." (Pat Brown noted, tartly, "If I gave you a poisonous mushroom, well, that's one ingredient.")

Thirty-three companies are working on a single-ingredient approach: using animal cells to grow meat in vats. The management consultants at A.T. Kearney predict that by 2040 the technique will produce thirty-five per cent of all meat. Josh Tetrick, the C.E.O. of Just, Inc., which is developing cell-based chicken nuggets and ground wagyu beef, told me that the problem with plant-based meat is that it feels ersatz: "The Silicon Valley approach of Impossible Foods and Beyond Meat is 'If we can nail taste and cost, we'll win.' But meat is about identity and authenticity. Like, I hope Tesla comes out with a pickup truck, but if they have to call it Tesla's Electric Mobility Transport Unit my friends in Alabama would never buy it."

This spring, Tetrick watched closely as I ate his chicken nugget. It tasted weirdly healthy—I missed the creamy crappiness you expect from a fast-food nugget. That's because it was mostly composed of chicken muscle cells grown in Just's lab, one floor down at the company's San Francisco headquarters. Tetrick, a charismatic vegan who started Just to save chickens' lives, knew that he had work to do: "We need to cultivate a second strain of cells, ramp up the fat program downstairs."

The cell-based approach may eventually provide meat using a tiny fraction of the land and water that livestock use. And, if companies can figure out how to grow cells on a scaffolding of mushroom or celery, or arrange them using a 3-D printer (and also surmount issues with vascularization and oxygen diffu-

and Vitamin A. In areas such as sub-Saharan Africa, where one person in five is malnourished, meat is the quickest fix. Its consumption also demonstrates to the neighbors that you can afford something other than rice, yams, or cassava. The barrier to that emblem of arrival keeps getting lower: in most places, meat is cheaper than it's ever been.

By 2050, as the world's population grows to nearly ten billion, demand for meat is expected to nearly double again. In the global-management world, this predicates what is known as "the 2050 Challenge": how do we feed all those people without hastening climate change? A five-hundred-page report, "Creating a Sustainable Food Future," released in July by the World Resources Institute, the World Bank, and the United Nations, declared that, if we stay on our present course through 2050, feeding the planet will "entail clearing most of the world's remaining forests, wiping out thousands more species, and releasing enough GHG emissions to exceed the 1.5° C and 2° C warming targets enshrined in the Paris Agreement—even if emissions from all other human activities were entirely eliminated." The chance that ten billion people will suddenly stop driving, cooling their homes, and manufacturing anything at all is, of course, zero. The report's lead author, a droopy-eyed research scholar at

Princeton University named Tim Searchinger, told me, "There were times writing it when I thought, Euthanize your children—we're all doomed."

In April, Searchinger visited Impossible's Silicon Valley headquarters, in Redwood City, hoping for better news. He tossed a notepad on a conference table, across from half a dozen Impossible executives, and looked probingly at Pat Brown. Searchinger was the fox who knows many things; Brown the hedgehog convinced of one. I'd mentioned to him that Searchinger's report detailed a raft of initiatives that humanity needed to implement to solve the 2050 Challenge, from wiser manure stewardship to increasing the global fish supply and drastically lowering the birth rate: twenty-two changes in all. "One change!" Brown had cried. "If we can just get everyone to eat plants, you don't have to disrupt everything else."

"What's the increase in your production going to be the day Burger King goes national?" Searchinger asked.

"Humongous," Brown said. He fiddled with a piece of paper, folding it into a rectangle. Impossible's rapid growth had led to a supply crunch. The company was holding meetings to determine which distributors would get less product, and had postponed launching in supermarkets. (The Impossible Burger débuts in a hundred and twenty-nine stores this fall, beginning with Gelson's locations in Los Angeles.) He went on, "That's why half the population of this building has volunteered to work in our Oakland plant." In a call-for-volunteers e-mail, Brown wrote that while the supply problem was "the biggest risk not only to our vital relationship with Burger King, [but] to our business as a whole," it was also "an epic opportunity for heroism." I'd just visited the plant—a former industrial bakery—and seen dozens of office workers in hairnets and steel-toed galoshes shadowing line workers, eager to step in.

Searchinger had brought a list of detailed questions about the company's costs and its supply chain, which the execs met with assured generalities. Brown said, "Another advantage we have over the incumbent technology is that we keep improving our product every week. The cow can't."

"How close are you on the texture issues to being able to make steak and cubed beef?"

"The level of confidence in the R. & D. team is very high," Brown replied evenly. At the moment, Impossible's steak prototypes are squishy and homogeneous, far too easy to eat. Brown announced a steak project earlier this year, then put it on hold to address the supply crunch.

Searchinger studied his list and said, "One thing that will be critical is acceptance in the developing world—finding local agricultural associations that make precursor products for you, before the local beef guys put you out of business."

"I completely agree," Brown said. North America makes up only twelve per cent of the global market for meat; he needed to wipe out livestock everywhere.

Searchinger said, "Our baseline estimate is that by 2050, to produce the beef to meet demand, we'll see a hundred and fifty-eight million hectares more pastureland in Africa alone. And the even bigger threat is from China."

Brown made a face. "To head that off, we have to be seen as successful in the U.S. and developed countries first," he said. "If we're seen as a cheap substitute, we won't get any traction in Africa."

Searchinger looked wistful. "If you could just reforest all the grazing land, 1.2 billion hectares!" he said. "Giving up all beef would be the most effective thing we could do for the planet." He has calculated that if you reduced beef consumption by three-fourths (allowing for some pastoral nomadism and dairy cows later used for beef) and reforested accordingly it would reduce global G.H.G. emissions by about twenty per cent.

"We'll take care of getting rid of all beef for you," Brown said. They smiled and shook hands.

Searchinger later told me, "Innovation from places like Impossible is the one thing that allows me to have a tiny bit of optimism." But he still believed too many complicating, countervailing things. A week after his visit, he co-wrote an op-ed for CNN that called Impossible's deals with fast-food restaurants "historic," but said that "eliminating beef is neither the goal nor realistically at stake. The point is to hold down its growth."

In June, more than a thousand people descended on the Quality Hotel Globe, in Stockholm, to discuss how to feed the world without destroying it. The annual conference of EAT, a Scandinavian nonprofit dedicated to making our food system sustainable, showcased backpacks and business beards, talk of the Global South and the Global North, and the AirDropping of dire bar graphs. There was an atmosphere of acerbic self-satisfaction, a sense that only those present understood both what it would take to save humanity and that it was probably too late. At dinner, after the chef Claus Meyer, who co-founded Noma, extolled the rhubarb on his menu for "plunging from the earth like a cold frozen fist," Pat Brown surveyed the throng and said, "If I were cynical, which of course I'm not, I'd say conferences like this are an excuse for these guys to bop around the world meeting each other."

Yet when Brown was interviewed on the main stage, wearing the outfit his comms team had specified—"NO COW T-shirt, blazer and jeans"—he was upbeat. He's become a more confident, less academic public speaker of late, having mostly learned not to point with his middle finger or end refutations with "Q.E.D." He now distilled his message to a congenial set of propositions: *Lecturing people doesn't work. This is a technology problem. And we've solved it.* He left his provocative "I ♥ GMO" water bottle in his backpack.

Offstage, however, he couldn't resist disputation. Watching a panel discussion in which a British cattle rancher lauded "regenerative grazing," Brown stuck out his tongue and murmured, "I am so tempted to shout out, 'This is bullshit!'" The rancher's ideas were premised on the increasingly popular practice of "grass-feeding" cattle, and further shaped by the theories of the Zimbabwean rancher Allan Savory, who believes that herds of livestock that are ushered to a new pasture as soon as they've cropped the grass can reverse desertification and make grasslands a carbon sink. To Brown's chagrin, the EAT crowd seemed more receptive to this dream of Eden than to his unrepentant bovicide.

While all cattle graze on grass for much of their lives, at least ninety-five per cent of American beef cattle spend their last four to six months being fattened on grain at feedlots. Because cattle "finished" on grass gain weight half as fast as they do on grain, they are kept alive longer; for that reason, and because the microbes in their bellies process grass more thoroughly, the cows belch out forty-three per cent more methane. Grass = gas. When a Costa Rican at Brown's table at dinner proudly announced, "One of the priorities of our government is decarbonizing cattle ranching," Brown said, "You can't decarbonize cattle ranching. It's impossible. You just need to get rid of those cows!"

At a meeting in the hotel's lobby, Lindiwe Majele Sibanda, a Zimbabwean scholar and policy advocate who co-chairs the Global Alliance for Climate-Smart Agriculture, politely told Brown that his plan didn't apply to her continent. "Ninety per cent of Africans are not eating meat in quantity," she said. "For most smallholders, it's a goat or a chicken. We use livestock for dowries, for diversity of diet, and as a store of wealth. They are literally cash cows."

Brown had told me repeatedly that he wasn't trying to displace poor farmers' goats, but he replied, "Even those goats and those chickens are taking a big toll on biodiversity. They're eating the grasses and shrubs and bugs that wild animals would otherwise be eating."

"I have yet to see scientific evidence that goats and chickens have pushed out other species," Sibanda said. "Remember, you're looking at arid and semiarid areas, so when you say, 'Meat is bad for the environment,' I say, 'Which environment? The thing that grows best here is goats!'"

"The global biomass of goats and sheep is more than two-thirds that of all wild animals," Brown said.

"Disadvantaged people have their own systems of livelihood—"

"We're not attacking farmers who are raising goats! We're just trying to remove the economic incentive for covering the earth with livestock."

They shook hands and rose without regret. Afterward, Sibanda told me, "You're selling the environmental argument to us, but it's the northern countries—right?—that are responsible for the majority of the damage. In the south, the feeling is 'How can my fifty grams of meat cause a problem?'"

Brown said, "She cares about many of the same things we do, obviously, but we were almost from different universes." He added that he wished he had a short film to show "what the world would be like in 2035 on its present course, and what it would be like if we eliminate animal ag." In the second scenario, he said, "the canonical poor farmer with his goat, or whatever, would get to keep it. But he would also get the benefits of averting catastrophic climate change and of our eliminating the biggest drain on his freshwater sources and his land—which is his neighbors raising cows. People need to see 'How does it improve *my* life?'" He sighed. "It's all so complicated and indirect."

When Pat Brown was twelve, and he and his six siblings were living with their parents in Taiwan, he figured out that his father, Jim, was in the C.I.A. He didn't tell anyone, because he didn't want to blow his father's cover or impede his mission of keeping an eye on China. "There's this real misconception about the C.I.A., that it's the dirty-deeds arm of the U.S. government," he told me. "When my dad joined, he'd been a P.O.W. in World War Two, in Belgium, where he ended up weighing ninety-something pounds, and he came out of it with a well-developed sense that there are bad people in the world who need to be watched."

The family was uprooted with Jim Brown's postings: to Paris, Taipei, Washington, D.C. This itineracy, Brown came to feel, made him a resourceful citizen of the world. Brown's younger brother, Richard, a neurobiologist who works at Impossible studying how we perceive taste and odor, said that the family was Catholic, but guided less by doctrine than by curiosity and fairness: "We were driven by 'What is intellectually the most interesting thing to work on, and what is of the most public service?'" Brown was a fractious student; a generation later, he might have been given a diagnosis of A.D.H.D. "In Taiwan, I would get F's, F's, F's for conduct," he said. "I was intrinsically not into anyone having authority over me—I was kind of an asshole. Most of the things of value that I learned I learned on my own."

In college, at the University of Chicago, Brown loved pure mathematics, but felt that it was too removed from public service. So he majored in chemistry. He became a vegetarian the summer after he graduated, spurred by his younger sister Jeanne, whose animal-welfare arguments convinced nearly everyone in the family to stop eating meat. That same year, Brown met Sue Klapholz, and began an M.D.-Ph.D. program at Chicago; afterward, he did a residency in pediatrics. The couple married in 1982 and six years later relocated to Stanford, where Brown became an associate professor and an investigator at the Howard Hughes Medical Institute. They had three children and brought them up as vegetarians.

They still live in the cedar-shingled faculty-housing condominium they moved into more than thirty years ago, now accompanied by a deaf, senile rescue mutt named Sebastian. The rooms, a riot of wooden and ceramic animals, call to mind Kafka's observation as he admired fish at an aquarium: "Now I can look at you in peace; I don't eat you anymore." Brown seems almost angry that when Impossible Foods goes public he'll likely become a billionaire. "We've got it so good here," he told me one morning, as he sat with Klapholz in their back garden over bagels and blackberries, watching juncos flit overhead. "Why would we want to change the way we live?"

Every other arrangement, though, has always been up for grabs. "I don't know anyone more passionate than Pat—and it's hurt him," Suzanne Pfeffer, his former department chair at Stanford, said. "We'd tease him about not hitting the Send button on e-mails to the dean or the N.I.H." Joe DeRisi, a leading malaria researcher who once worked in Brown's lab, showed me a photo he keeps on his phone from those days: the first slide in a presentation Brown gave at Howard Hughes, which said "Eating meat, publishing in *Nature*, and other asinine things you dumb f***s keep doing." "I thought, Man, do I admire that," DeRisi said. "What I learned from Pat was 'You have a certain amount of time on the planet—you should work on important stuff.'"

In 1995, Brown's lab published pioneering work on the microarray, a method of determining which genes are being expressed in a given cell. The technique proved hugely useful in distinguishing normal tissue from cancerous tissue and identifying a given cancer; it established, for instance, that there isn't one kind of breast cancer but six. In 2001, he co-founded the Public Library of Science, a nonprofit publisher of open-access science journals that competed, with some success, with the commercial journals that offended his principles by limiting access to their trove of knowledge.

At Impossible, Brown second-guesses himself in ways he never had to as a scientist. He loves the office—"It gives me a burst of happiness when I come in"—but hates having to compartmentalize information and to suppress his instinct for combat. "My favorite thing to do is to get into an argument, but my superego can't snooze through the day the way it used to," he told me. Still, he can't resist interrogating norms that strike him as defective. At a recent meeting to consider promotions for ten staffers, Brown derailed the agenda by questioning the whole idea of tiered titles. After half an hour, Impossible's new president, Dennis Woodside, the former C.O.O. of Dropbox, said, in gentle disbelief,

"Last week, we were very close to promoting eight or nine people, and now we're going to take everyone's titles away?" Unruffled, Brown said, "Is there a way to have a more sensible system that wasn't invented for I.B.M.?"

Brown's brother Richard said, "Pat optimistically holds to the belief that people are rational and can be convinced by evidence. Some of the frustration he feels is that food is different—there's so much subjectivity to it." Brown remains mystified, for instance, by Americans' eagerness to add protein to their diets when they already consume far more than is necessary. Nonetheless, he beefed up the protein in his burgers. "There are things we do that are effectively just acknowledging widespread erroneous beliefs about nutrition," he said. "For the same reasons, we initially used only non-G.M.O. crops, which was essentially pandering. We're not trying to win arguments but to achieve the mission."

He is equally baffled by challenges from people who agree with his goals but question his methods. In 2017, the environmentalist organizations ETC Group and Friends of the Earth attacked Impossible, claiming that heme was potentially unsafe and that its patty "implicates the extreme genetic engineering field of synthetic biology, particularly the new high-tech investor trend of 'vat-itarian' foods." Brown published a comprehensive response, in which he pointed out that "your own bloodstream right now contains about as much heme as 300 pounds of Impossible Burgers."

When Impossible undertook the required animal testing to get F.D.A. approval of heme as a color additive, People for the Ethical Treatment of Animals promptly strafed the company for feeding soy leghemoglobin "to a total of 188 rats in three separate tests, killing them, and cutting them up." PETA spitefully added that "the Impossible Burger is probably the unhealthiest veggie burger on the market." Brown told me he was wounded by the attacks: "With a lot of fundamentalist religious groups, it's bad if you're a nonbeliever. But if you're a *heretic*—that's a capital crime."

The spread of livestock is largely responsible for the ultimate in the unethical treatment of animals; since 1970, the world's wild animal populations have diminished by an average of sixty per cent. But PETA, in its zeal, often fails to grapple with the nuances of means and ends. For instance, it opposes eating chicken, pointing to the abuses of factory farming. American broilers, chickens raised for meat, are bred and confined in ways that make them more than four times larger than broilers were in the nineteen-sixties; as a result, they often collapse from their own weight. Jacy Reese, in

"The End of Animal Farming," noted that "consuming smaller animals leads to far more suffering per calorie because it takes far more animals." By comparing the number of days that various kinds of livestock spend in factory-farm conditions, Reese determined that eating chicken is nineteen times worse than eating beef. But it's vastly better for the environment—poultry production has about one-eighth the climate impact of beef production.

Believing you're right doesn't salve the bruises from these ethical struggles. Sue Klapholz told me, "Our mission was too important not to do the animal testing, but Pat and I would never want to do it again. Our youngest son had a pet rat, and they're very smart animals that like to have toys. I wouldn't even swat a mosquito—I'm that kind of vegan. The protest was personally shattering to me, as a longtime PETA supporter." She looked out the window. "I feel like I lost a friend."

In Stockholm, Pat Brown had breakfast with Solina Chau, an energetic Hong Konger who is the co-founder of Horizons Ventures. The firm, underwritten by one of Asia's richest men, Li Ka-shing, has led two rounds of investment in Impossible. Over coffee and avocado toast at the Grand Hotel, Chau was trying to revise Brown's plan for introducing his burgers into China. Brown said that he envisioned telling the central government, "'I want to help you solve your biggest national-security problem.' Because China is the biggest meat consumer in the world"—between 1961 and 2013, the average Chinese person's meat intake went up more than fifteenfold—"but it's completely dependent on imports," chiefly from Brazil and Germany.

Chau had told me she didn't think Impossible should attempt to eradicate meat in China, or anywhere else: "There's not enough supply to feed future demand, so it's a coexistence scenario." She suggested to Brown that Impossible partner with the tech-friendly city of Shenzhen: "You must align your interest with the local government, and they will do your work for you and protect the investment. And they'd help you with the regulatory issue!" Because heme is a novel ingredient, Impossible's burgers require regulatory approval in both Europe and China, which Brown told me will take "probably two years in Europe and eighteen months to infinity in China." Chau's way would be slower, but safer.

Brown waggled his head: he'd think about it. He was well aware that a Chinese company could entice him into a joint venture and then hijack Impossible's intellectual property. However, he told Chau, "it's just a risk you take. Either you go there and reach some accommodation that's not complete exploitation, or you go there and maybe they exploit you and you end up with nothing, or you don't go there and you *definitely* end up with nothing." Impossible has explored a way to keep its heme-production process from being bootlegged. Nick Halla, the executive in charge of new markets, told me, "We'd send the buckets into China rather than the recipe, just the way Coca-Cola sends in the syrup."

Brown assured Chau, "We're not going to *give* it away." Yet his instinct is to do exactly that, with companies around the world. "In five or ten years," he told me, "I'd love to give small entrepreneurs free access to our technology, with the idea that they'd pay us royalties once they got to a million dollars in revenue. The way I'd pitch it as a business is 'Now you have a million new employees who are basically working for free.'" Such a plan would cut into Impossible's profits, but, he said, "the animal industry will be worth three trillion dollars in ten years, and if we have a small fraction of that we'll be one of the most successful companies on earth. And if we tried to have all of it, and we controlled the world's food supply, we would guarantee being the most hated company in history."

Brown sees himself as a guide rather than as a micro-manager—"I have no idea if the company paid taxes last year. The C.E.O. is supposed to know that, I guess"—but he is determined to retain control. When Google made an early offer to buy the company, he said, he turned it down "in less than five seconds, because we would have just been one of their suite of nifty projects." And he made it a condition of his deal with Khosla Ventures that Impossible couldn't be sold without his approval to any of about forty "disallowed companies"—meat producers and agricultural conglomerates.

Those companies, which like to say that they're in the business of providing whatever protein consumers want to eat, have finally begun to respond to the plant-based boom. Nestlé offers an Incredible plant-based burger overseas and is about to release an Awesome one in the U.S., and Kellogg just announced a plant-based line called Incogmeato. Many of these new products seem aimed less at meat-eaters than at flexitarians, a dignifying name for the wishy-washy: Perdue's "Chicken Plus" nugget mixes chicken with cauliflower and chickpeas, and Tyson Foods is releasing a burger that blends beef with pea protein.

The agribusiness giant Cargill recently invested in Puris, which supplies Beyond Meat with pea protein, and in two cell-based startups. Brian Sikes, who runs

Cargill's protein-and-salt group, told me that "plant-based is part of the solution" to the 2050 Challenge, "and potentially cell-based is, too."

Though Sikes repeatedly assured me that Cargill's purpose is "to be leaders in nourishing the world," the company recently said that—like many agricultural conglomerates—it would miss its target of removing deforestation from its supply chain by 2020. And the environmental group Mighty Earth just excoriated Cargill as "The Worst Company in the World." When I asked Sikes if he'd learned anything from Impossible and Beyond, he said, "They're master marketers. They've made us realize that we need to tell the story of traditional animal protein better."

Samir Kaul, Brown's original investor at Khosla Ventures, told me, "There have to be ways to partner with the large food companies," but Brown remains skeptical. "If Tyson called us, we wouldn't go into it with the naïve idea that they want to help us," he said. "The best outcome for them, given their sunk costs, would be to slow us down." He allowed, cautiously, that "if Tyson shut down their meat-production operations and broke all their artificial-insemination rods and melted them down and turned them into hoes—well, that would get my attention."

A few months ago, in Washington, D.C., I visited the National Cattlemen's Beef Association, which lobbies on behalf of American cattle producers and feeders. Five of the N.C.B.A.'s employees sat across from me in leather chairs at a long conference table, surrounded by paintings of cowboys performing their manly duties, and explained why Pat Brown was misguided. Danielle Beck, a senior lobbyist, said, "Consumers like locally grown, supporting the small rancher—we have a good story to share, and our product is superior. So I don't think we need a Plan B."

"It comes down to taste," Ed Frank, who runs policy communications, said.

"Ed and I tried the Impossible Burger for our podcast," Beck said, referring to a 2018 episode called "We Tried Fake Meat So You Don't Have To!" She made a face: "Salty. Odd aftertaste."

"We faced a moral and ethical dilemma. What if it was as good as ground beef? What would we say then?" Frank said. "Fortunately, it wasn't, so I was able to sleep at night." I noted that Impossible has since put out a much improved burger—had they tried it? Frank and Beck shook their heads and looked away.

Meanwhile, local ranchers' groups have convinced twelve state legislatures to pass laws that prohibit words such as "meat" and "burger" from being used on labels for anything that's not "harvested" from carcasses. In July, a law went into effect in Arkansas that forbids the makers of plant-based meat even to use the term "veggie burger." The laws' alleged intent is to avoid "customer confusion," but most people have no trouble grasping that almond milk doesn't gush from an almond's udders. The laws' actual intent, of course, is competitive hindrance. Mark Dopp, of the North American Meat Institute, told me that when Impossible Foods has to put "bioengineered" on its labels, in 2022, once a federal labelling law takes effect, "that will be a challenge for them. I'm sure they'll try to escape it."

In fact, Impossible will label itself as bioengineered this fall, when it goes on sale in supermarkets. "We're totally transparent," Pat Brown said, adding, "I'd love to have them have to put labels on their meat that say 'Processed in a slaughterhouse,' with a symbol of a friendly bacterial cell smiling and saying, 'Contains aerosolized fecal bacteria!'"

While the lobbyists at the N.C.B.A. acknowledged that beef has some environmental liabilities, they said that those concerns would soon be mitigated by the same American ingenuity that has "productized" every inch of the cow. After sixty-four per cent of the animal is turned into meat, including beef hearts sold to the

Middle East, tongues to Asia, and tripe to Mexico, eighteen volleyballs can be made from the hide, and other remnants are used to produce bone china, gelatine, dog food, ink, nail-polish remover, laundry pretreatments, and antifreeze.

I observed that, despite all these efficiencies, the magazine *Science* had recently identified giving up meat and dairy as the most powerful environmental act any individual could make. "There are more reports like that than we care to see," Colin Woodall, the N.C.B.A.'s senior vice-president of government affairs, said ruefully. "We just go back to the two-per-cent number from the E.P.A." By the association's reading of a 2019 E.P.A. report, only 2.1 per cent of America's greenhouse gases come directly from beef production. "Is two per cent really going to change climate change?" Woodall said. "No. A lot of people like to throw rocks at us, but they do so while driving down the road at seventy miles per hour in an air-conditioned car."

The N.C.B.A.'s math doesn't account for nitrous-oxide emissions from manure-covered pastures or emissions from producing crops for feed and from manufacturing the beef itself, all of which raise the figure to 3.8 per cent. More significantly, the E.P.A.'s accounting, like many such assessments, fails to factor in the G.H.G. impact of animal agriculture's land use. According to the

World Resources Institute, if Americans replaced a third of the beef in their diets with legumes, it would free up a land area larger than California, much of which could be reforested (at great expense, and if the owners of the land were so inclined).

In most of the world, beef production is vastly less efficient than it is in America. Frank Mitloehner, a professor in the department of animal science at the University of California, Davis, who is often cited by pro-meat forces, acknowledged, "We have way too much livestock in the world—it poses a serious risk to our ecosystems." By incorporating American know-how abroad, he added, "we could feed everyone in the developing world with one-quarter of the current global herds and flocks."

Sciencing the cow to make this possible, the N.C.B.A. suggested, was where everyone should be focussing their efforts. Colin Woodall proudly reported, "Since 1977, we can produce the same amount of beef with one-third fewer cattle." In the past two decades, the dressed weight of a cow—the amount of beef that ends up for sale—has increased ten per cent. Woodall noted that agronomists are working on new corn varieties and seed additives to reduce methane, as well as nitrification inhibitors to diminish the nitrous oxide given off by manure. However, he said, "we're never taking cattle completely off of grass, so it really comes down to: what are the new tools to put more meat on that animal?"

While the Impossible Burger is still trying to match the flavor of beef, in certain respects it's begun to improve upon the original. Celeste Holz-Schietinger, one of the company's top scientists, told me, "Our burger is already more savory and umami than beef, and in our next version"—a 3.0 burger will be released in a few months—"we want to increase the buttery flavor and caramelization over real beef."

Richard Brown said, "Early on, we had two goals that were fully aligned: to be identical to a burger from a cow, and to be much better than a burger from a cow. Now they're somewhat at odds, and we talk about the chocolate-doughnut problem. What if what people really like in a burger is what makes it taste like a chocolate doughnut, so you keep increasing those qualities—and suddenly you're not making a burger at all?"

Rob Rhinehart thinks that Brown should double down on doughnut. Five years ago, Rhinehart created Soylent, a wan, nutritive sludge that allows you to keep playing Mortal Kombat as you replenish; he now runs MarsBio, an accelerator for companies working on bioreactors and engineered microalgae. "There's all these comical efforts to make new food look like the old food," he said. "I want Impossible Foods to do something totally new. Alien meat! Or a burger that tastes like a human—a brain burger!"

Brown is drawn to such flights of fancy. He told me, "There's reason to doubt that the handful of animals we domesticated thousands of years ago provide the most delicious meats possible. We could choose a meat flavor better than beef or chicken or pork, and call it a brontosaurus burger—or anything you like. It would be super fun to make übermeat!" He added, regretfully, "But it has to be a side project, for now, because the more sure way to crush the chicken producers is to make the best version of chicken."

One morning in June, Impossible's chief science officer, David Lipman, took me through the test kitchen. As nine scientists in lab coats and hairnets looked on, I drank a glass of Impossible Milk, which had the consistency, color, fat, and calcium content of dairy milk. The only issue was that it tasted like water. "We have to do more work to give it dairy flavor," Lipman said, optimistically.

The flavor scientist Laura Kliman made me a tasty fish paella. The recipe for Impossible's anchovy-flavored broth is about eighty per cent similar to its recipe for the Impossible Burger. "Once we cracked the code on meat flavor," Kliman said, "if you change a few of the ratios and ingredients, it's not that hard to get fish or pork or chicken."

Next up was Impossible Steak Flavor—a beaker full of red juice. A scientist named Ian Ronningen poured it into a saucepan, turned on the gas, and began swirling the juice with a metal spatula. As it reduced and turned brown, he said, "Now you're getting a change of flavor."

His colleague Allen Henderson softly confided, "We feel that we have sufficiently recapitulated the multiple chemistries of cooked beef."

Ronningen bent over the bubbling goo, wafted the steam toward his nose, and said, "I'm starting to get that really wonderful fat note."

"Ah, yes," Lipman said, doing some wafting. "There's an animalic quality. It's more musky than a burger."

"And we get these grizzled pieces, just like a steak," Ronningen said. "If we have a deflavored protein, which we're good at, we can take this flavor and put it on a textured protein base." He took the pan off the heat and we dipped pieces of bread into the gritty juice. It was literally the sizzle, not the steak—but it was delicious.

Brown told me it was "time to double down on steak, for mission reasons." He planned to use another chunk of the three hundred million dollars he'd just raised to accelerate his R. & D., hiring ninety more scientists. Small teams would immediately begin work on chicken nuggets and melty cheese for pizza. He also planned projects to spin proteins into structural fibres, and to pursue a general methodology for stripping plant proteins of their off-colors and off-flavors.

After years of focus, Brown was beginning to return to his preferred mode of swashbuckling inquiry. He yearns to pursue a project that gripped him early in Impossible's development: using RuBisCo, the most abundant protein in the world, as his staple ingredient. RuBisCo is an enzyme used for photosynthesis that's found in the leaves of plants like soy and alfalfa; by Brown's calculations, it would enable him to meet the world's protein requirements using just three per cent of the earth's land. But no one produces RuBisCo at scale: to do so requires processing huge quantities of leaves, which tend to rot in storage, and then isolating the enzyme from indigestible cellulose. However, Brown said, "for a year, our prototype burgers used RuBisCo, and it worked functionally better than any other protein, making a juicy burger." He folded a napkin smaller and smaller. "We *will* build a system for producing protein from leaves."

Though Brown longs to transmute leaves into loaves and fishes, the more immediate concern is the drive-through at fast-food restaurants. Chipotle and Arby's have declared that they have no plans to serve plant-based meats, and Arby's went so far as to develop a mocking rejoinder: the "marrot," a carrot made out of turkey. Other chains have lingering concerns. One is price: Impossible's burgers, like Beyond Meat's products, cost about a dollar more than the meats they're intended to replace. At White Castle, the Impossible Slider sells for a dollar ninety-nine, one of the highest prices on the menu. "Honestly, that's the biggest barrier to the new product for college kids, and for our customers who can only afford to pay three dollars for a meal," Kim Bartley, White Castle's chief marketing officer, said.

Early on, Brown believed that his burger would be cheaper than ground beef by 2017. His original pitch claimed, in a hand-waving sort of way, that because wheat and soy cost about seven cents a pound, while ground beef cost a dollar-fifty, "plant based alternatives can provide the nutritional equivalent of ground beef at *less than 5%* of the cost." But establishing a novel supply chain, particularly for heme, proved expensive. The company has increased its yield of the molecule more than sevenfold in four years, and, Brown said, "we're no longer agonizing over the impact of heme on our cost." He now hopes to equal the price of ground beef by 2022.

Plant-based meat won't become a shopping-cart staple unless it achieves price parity, and some observers worry about how long that's taking. Dave Friedberg, the founder of the Production Board, an incubator for alternative-protein companies, noted the expense of heme and texturized soy protein. "I'm concerned that we're *never* going to get to the price of ground beef," Friedberg told me. "And to sell people a product that's *not* meat, and charge more for it, won't shift the world to a new agricultural system."

Shifting the world to a new agricultural system is not part of a fast-food chain's business model. So the chains question whether plant-based will prove to be a trend, like spicy food, or merely a fad, like rice bowls. Lisa Ingram, the White Castle C.E.O., told me she was agnostic on animal ag. Eradicating it by 2035 "is Pat's view of the world," she said, "and every customer gets to decide if they agree. If they do, then in 2035 we'll sell the Impossible Slider and the Impossible Chicken Slider and the Impossible Fish Slider. If they don't, then we're going to sell the Impossible Slider as part of our menu just as long as people want to buy it."

Right now, they do. In July, Impossible announced that, after tripling production at its Oakland factory and signing a deal to make its burgers at plants belonging to a meat-processing behemoth called the OSI Group, it was no longer restricting deliveries to any of its distributors. The company planned to increase production fourfold by the end of the year. It was once more blasting ahead.

Yet, the greater its progress, the wider the gap between what Brown hopes to do and what his investors expect to gain—between idealism and market value. Vinod Khosla, at Khosla Ventures, has assured Brown, "If we never make a penny from our operations in Africa, I'm fine with that." But you won't find this promise in any of Khosla's contracts. Bart Swanson, who sits on Impossible's board, suggested that any potential conflict is not imminent, adding, "By the time we go into Africa, I hope I'm alive." Swanson is fifty-six. Brian Loeb, an investor at Continental Grain, a large agricultural-products holding company that invested in Impossible in 2016, said, "The industry-wide conversation now is around 'Can plant-based meats get to five per cent of the market?' "

During my last visit to Impossible Foods, Brown admitted that he was somewhat at the mercy of his investors. "I was more naïve than I wish I'd been early on in terms of how my control gets affected by repeated rounds of funding," he said, as we sat in Yam, a small conference room near his desk. "I don't have the hard power to say no, if someone wants to buy us. I have a reasonable amount of soft power, to the extent that I can convince our investors that they'd be missing out on continued growth if they sold. The best defense we have is doing well—and if we're not doing well then who cares if we get sold to Tyson Foods?"

"Do you, deep down, believe that nobody else is approaching the problem correctly?" I asked.

"I'm worried about how it sounds, but yes," he said. "Nobody else has caught on to the fact that this is the most important scientific problem in the world, so their results are just a reheated version of veggie burgers from ten years ago, maybe with a little lipstick on them. And cell-based companies are just taking the same technology cows have used to grow meat for a thousand years and making it *less* efficient." His impatience was plain. It struck me that while, as a scientist, Brown welcomes searching questions and alternative ideas, as a missionary he believes that searching questions and alternative ideas waste time—time we simply don't have.

I wondered whether it had occurred to him that he had essentially devised a tortuous work-around for human selfishness. "Yes," he said slowly. "I do find that interesting. Strategically, a hamburger is hugely symbolic. But it's also completely trite and ridiculous. If you'd told me ten years ago that I'd be totally focussed on burgers, I'd have thought, *Well, that's not a life I want.*"

After a moment, he returned to the question of whether he had any true partners. "I'm aware of our investors' feelings, and to some extent disappointed by them," he said. "People aren't used to doubling in size and impact every year—that's a very steep and unrelenting curve, and even venture investors are incredibly conservative. They realize that something far short of our goal is a massive investment success for them. If they were *completely* confident, they would be backing trucks up to Impossible Foods loaded with billion-dollar bills." He grinned, and went on, "But they're wrong! Kodak and the horse-and-buggy industry thought they'd just coexist with the new technology, too. I only picked 2035 because it seemed like something you could plausibly achieve, something that other people could at least see a path to. *I* would have picked sooner."

Trailblazers

Nicola Twilley

August 26, 2019

Before Terry Lim handed me an aluminum flask filled with a blend of gasoline and diesel and asked me to set fire to the Tahoe National Forest, he gave me a hard hat, a pair of flame-resistant gloves, and a few words of instruction. "You want to dab the ground," he said. "Just try to even out the line."

The line was a low ridge of flame, no more than a foot high, creeping toward us through the forest. In front of it, the ground was springy, carpeted with a dense layer of pine needles and studded with tufts of grass. Specks of sunlight shimmered in the deep, almost kaleidoscopic green, bouncing off lime-colored ferns and conifer boughs. A foot-long alligator lizard skittered in front of me, pausing to pump out a couple of quick pushups before vanishing into the brush. Beyond the line, the ground was black and silent. Silhouettes of large trees loomed out of a sallow gray haze.

The lit cannister of fuel I was holding, known as a drip torch, had a long, looped neck that emitted a jaunty quiff of flame. I took a deep breath, and ducked my way through the scrub to the far end of the line. Then I walked back, dotting the tip of the torch's neck to the forest floor a few feet in front of the flames, as if I were tapping out a message in Morse code. The dots and dashes ignited small fires, which joined up so rapidly that at one point I set fire to my boots. A swift, panicky battering with my gloved hands smothered the flames before any damage was done.

The main fire was advancing into the wind, so it moved slowly and stayed close to the ground. But my new flames had the wind at their back and quickly jumped across the gap separating them from the original front, transforming the line's ragged edge into a wall of flame. It was mesmerizing and thrilling, and I couldn't wait to do it again. As the afternoon wore on, I began setting my ignitions farther away from the line, in order to consume the forest faster. I started to anticipate how terrain would affect the pace of fire: open stretches of pine needles caught instantly, but I learned to place my dabs in tight clusters near saplings and denser shrubbery.

I wasn't really supposed to be setting the forest on fire. That was the job of the United States Forest Service crew whose work I was there to observe. Their task was to carry out a prescribed burn—a carefully controlled, low-intensity fire that clears duff and deadwood, reducing the risk of a catastrophic wildfire. But the crew were temporarily occupied by what they called "a slop-over event": a rogue ember had leaped across a trail that acted as a firebreak at one edge of the burn, sparking a half-acre blaze so hot that standing within a few feet of it made my chest hurt. While the crew used chainsaws and hoes to create a new firebreak, it fell to me to insure that no part of the line got ahead of the rest. If flames are allowed to break ranks and surge forward, they can whirl around and start running with the wind, burning more intensely and smokily than the prescription allows.

It took the team more than an hour to fully contain the slop-over. Then they returned to the line with their drip torches. By the end of the day, they had set fire to a hundred and twenty acres of forest. As Lim walked me out of the woods, through the gray-gold twilight of the burn zone, he gave a satisfied sigh. "See, now that's nice," he said. "The trees have breathing room."

The contrast between that day's prescribed burn and the uncontrolled blaze that the crew had rushed to extinguish epitomizes California's spiralling problem with fire. Throughout the twentieth century, federal policy focussed on putting out fires as quickly as possible. An unintended consequence of this strategy has been a disastrous buildup in forest density, which has provided the fuel for so-called "megafires." The term was coined by the Forest Service in 2011, following a series of conflagrations that each consumed more than a hundred thousand acres of woodland.

Megafires are huge, hot, and fast—they can engulf an entire town within minutes. These fires are almost unstoppable and behave in ways that shock fire scientists—hurling firebrands up to fifteen miles away, forming vortices of superheated air that melt cars into puddles within seconds, and generating smoke plumes that shroud distant cities in apocalyptic haze. Centuries-old trees, whose thick bark can withstand lesser blazes, are incinerated and seed banks beneath the forest floor are destroyed. Without intervention, the cinder-strewn moonscape that megafires leave behind is unlikely to grow back as forest.

Six of the ten worst fires in California's history have occurred in the past eighteen months, and last year's fire season was the deadliest and most destructive on record. More than a hundred people were killed, and more than seventeen thousand homes destroyed. Experts have warned that this year's fire season could be even worse, in part because record-breaking rains early this year spurred the growth of brush and grasses, which have since dried out, creating more fuel. Governor Gavin Newsom proclaimed a wildfire state of emergency in March, months before fire season would normally begin.

The tools and techniques capable of stopping megafires remain elusive, but in the past few decades a scientific consensus has emerged on how to prevent them: prescribed burns. When flames are kept small and close to the ground, they clear the leaf litter, pine needles, and scrub that fuel wildfire, and consume saplings and low-level branches that would otherwise act as a ladder conveying fire to the canopy. With the competing vegetation cleared out, the remaining trees grow larger, developing a layer of bark thick enough to shield them from all but the hottest blazes. California's state legislature recently passed a bill earmarking thirty-five million dollars a year for fuel-reduction projects.

"And yet no one is actually burning," Jeff Brown, the manager of a field station in the Tahoe National Forest, told me when I visited him there recently. Although prescribed burns have been part of federal fire policy since 1995, last year the Forest Service performed them on just one per cent—some sixty thousand acres—of its land in the Sierra Nevada. "We need to be burning close to a million acres each year, just in the Sierras, or it's over," Brown said. The shortfall has several causes, but, some fifteen years ago, Brown set himself the almost impossible task of devising a plan for the forest he helps maintain that would be sophisticated enough to overcome all obstacles. Now he is coördinating an urgent effort to replicate his template across the Sierra Nevada.

The Sagehen Creek Field Station, where Brown is the manager, lies twenty miles north of Lake Tahoe, in the eastern Sierra Nevada. It was established in 1951 to conduct fishery and wildlife research, and is part of the University of California, Berkeley. Its amenities include a dozen radio-linked meteorological towers, snowpack sensors, tree-sap monitors, and a stream-depth gauge. It is not open to the public, but some twenty small red cabins are occupied by an ever-changing assortment of visiting researchers, student field-trippers, and even artists-in-residence.

When I drove there, in May, there were still patches of snow in the shade, but the banks of Sagehen Creek were dotted with the first buttercups of spring. I followed a rutted dirt road for a couple of miles through the forest, arriving at a simple shingled cottage, where Brown lives with Faerthen Felix, the station's assistant manager. From here, they help oversee the Sagehen Experimental Forest, nine thousand acres of mountain meadows, alkaline fens, and pristine streams surrounded by dense stands of Jeffrey and lodgepole pine.

Brown, who is in his mid-sixties, is a former competitive triathlete, ski patrolman, and river-rafting guide, and he has the rugged look and expansive manner of a lifelong outdoorsman. When I visited, he was taking two filmmakers on a tour of the station. He led us out into a clearing and unrolled a map on the forest floor. In the distance, three young does picked their way through the undergrowth. Behind us was a shed with an underground window onto the next-door stream, for the observation of spawning trout. Over the decades, dozens of insect, bird, and other forest-dwelling species have been studied and monitored at Sagehen, and the station's records constitute one of the longest-running and most detailed data sets on the Sierra. "We're the best-inventoried forest in the western United States," Brown told me.

As he led us through the trees, Brown pointed out that we were following an old railroad bed. Sagehen was clear-cut in the mid-nineteenth century to help build the railways and mines of the gold-rush era. (Sutter's Mill, where the first gold was discovered, in 1848, is less than a hundred miles away.) After loggers felled the large trees, smaller ones became fuel for locomotives, and the eastern slopes of the Sierra are so dry that there are still stacks of cordwood left over from the eighteen-eighties. Nearby, Brown bopped up and down on pine needles that coated the ground. "See this?" he said. "These go down ten inches deep in places."

When Brown and Felix arrived at Sagehen, in 2001, they saw their responsibility as straightforward: to keep this assiduously catalogued patch of wild Sierra forest unchanged, for future generations of researchers. Only gradually did they grasp that the forest they had inherited was in terrible shape. During their first summer at the station, there were three big wildfires nearby, and Brown realized that all that dry wood and all those pine needles could easily go up in flames. Then, in 2004, scientists who had conducted research at Sagehen gathered for a belated celebration of its fiftieth anniversary. Several had not returned in decades, and expressed shock at how dense the forest had become.

The local district ranger at the time was worried, too, and asked Brown whether she and her team could help reduce the forest's fuel load by doing some thinning—something the Forest Service does either by sending in loggers with chainsaws or by using a backhoe-like machine called a masticator, which shreds anything in its path. Brown was horrified at the suggestion. Like many staunch environmentalists, he was suspicious of the agency, because part of its remit is to generate revenue by logging timber like a crop. "To my mind, the Forest Service was the enemy, because if you cut down one tree you were doing something wrong," he told me.

Elsewhere in the Sierra Nevada, conditions were much the same—overstuffed forests, stripped of big old trees and filled with smaller ones crammed together—and global warming amplified the risk of disaster with each passing year. The average temperature on a summer day in California is 2.5 degrees Fahrenheit hotter than it was in the nineteen-seventies, and in the same period there has been a fivefold increase in the acreage consumed by wildfire. Fire seasons have been getting longer and more severe since the nineteen-eighties. Brown realized that doing nothing was no longer an option.

When the conquistador Juan Rodríguez Cabrillo sailed three ships along the coast of California, in September, 1542, and became the first European to set foot in the state, he reported seeing a great pall of smoke drifting over the landscape. As the ethnobotanist M. Kat Anderson has documented, indigenous tribes traditionally set fire to the forest at a variety of intervals, for a variety of reasons: to create better habitat for elk; to encourage the growth of edible or useful plants, such as mushrooms or chia; and to minimize the risk of fire. Precontact California burned constantly but rarely disastrously. In her book "Tending the Wild," Anderson writes, "Legends about destructive fires reflect the almost universal belief among California Indian tribes that catastrophic fires were not a regular, natural occurrence but rather a rare punishment."

In 2004, one of Brown's colleagues at Berkeley, a fire scientist named Scott Stephens, came to Sagehen and took samples from the stumps of huge trees cut down during the gold-rush era. Examining tree rings and scorch marks, Stephens was able to construct a record of fires dating back to the sixteen-hundreds. His findings confirmed that, in pre-Colonial times, Sagehen burned regularly. Those fires sometimes occurred naturally, from lightning strikes, but they were also deliberately set by Native Americans. The consensus now is that the entire Sierra Nevada burned every five to thirty years.

"The Washoe tribe used to hang out here in the summer, and then light it on fire in the fall, on their way out for the winter," Brown told me. "Especially near the creek—they wanted fresh willow shoots in the spring for basket-making." At Sagehen, some of the drier, south-facing slopes seem to have burned as often as every two years. Not only did the forest's native species evolve to survive fire; several of them actually require it in order to thrive. Lodgepole pinecones do not open until heated by fire. Black-backed woodpeckers dine almost exclusively on seared beetle larvae.

Brown began to see the outlines of an opportunity to reduce Sagehen's risk of a catastrophic wildfire, by working with the Forest Service and scientists at Berkeley to figure out how to implement prescribed burns. At the local Forest Service office, an eager young silviculturist, Scott Conway, was assigned to the project. When I talked to Conway, he recalled, "Somebody told me, kind of under their breath, 'Sagehen is never going to happen, don't get involved.' And, of course, I immediately took that as a challenge."

There were plenty of reasons to suppose that Brown's attempt would fail. One was the mutual mistrust between the Forest Service and environmentalists who object to public land being used as a lumberyard. After the passage of the National Environmental Policy Act, in 1969, conservationist groups became adept at using its protections of threatened species and habitats as a basis for lawsuits to bring logging to a halt.

In the early nineties, "The Sierra in Peril," a Pulitzer Prize-winning series of reports that appeared in the Sacramento *Bee*, spurred Congress to commission studies on California's forest ecosystems. As a result, the Forest Service revised its policies to allow prescribed fire as well as thinning. However, the agency had very little experience in designing and conducting prescribed burns in the American West. The Sierra Nevada's mountainous terrain and dry, Mediterranean climate make controlling even a planned fire challenging, and a century's worth of fire suppression had left forests so flammable that the smallest spark might trigger an inferno.

Brown and the rest of the Sagehen planning team decided to pursue a strategy that had recently been developed by a Forest Service scientist at its Rocky Mountain Research Station. Affectionately known as SPLAT, for Strategically Placed Landscape Area Treatment, the technique involves clearing rectangular chunks of forest in a herringbone pattern. This compels any wildfire to follow a zigzag path in search of fuel, travelling against the wind at least half the time. The SPLATs function as speed bumps, slowing the fire enough that it can be contained, while allowing the Forest Service to get away with treating only twenty to thirty per cent of any given landscape.

The SPLAT technique had been tested only in flat grasslands in Utah, and adapting it to the mountainous topography of Sagehen proved tricky. When fire travels uphill, it preheats the ground in front of it, often doubling its velocity; fire usually moves downhill more slowly, but a lit pinecone rolling down a slope can easily ignite new areas. Topography also affects other factors that determine the pace of a fire, such as wind speed, rainfall, and soil-moisture levels. Scott Stephens and one of his doctoral students embarked on a multiyear study to gather all the landscape data needed to model fire behavior at Sagehen.

Adapting the SPLATs to Sagehen's terrain took four years. Then, just as the plan was being finalized, a paper was published documenting the unexpected decline of the American pine marten at Sagehen. The marten, a member of the weasel family, is not endangered, but its population levels are seen as a useful proxy for forest health. Soon, the Sagehen planning team heard from Craig Thomas, the director of the environmental group Sierra Forest Legacy, which has a long history of litigation against the Forest Service. Thomas asked them to redesign the project, with an eye to protecting marten habitat.

Thomas, a small-scale organic farmer in his seventies, told me that he was astonished when the Sagehen group, especially the Forest Service, seemed open to the idea. "Instead of getting their backs up, they jumped in with both feet," he said. Conway recalled his own response a little differently. "I was, like, really?" he said. "It meant a bunch of complexity, and making this project, which was already really too long, much, much longer." Still, as Thomas recalls, Conway "went away and read every marten ecology paper in existence by the time the next phone call happened. And I went, Ah, this is somebody I think I want to work with."

So in 2010 the team, which had now been working together for six years, began planning all over again, this time with an even larger group of collaborators and a more expansive goal. "It started as science, but it became diplomacy," Brown told me. "How could we get all these people—groups that didn't trust each other, were actively suing each other—to a consensus on what was best for the forest?"

Brown secured grants, hired a professional facilitator, and brought together loggers, environmental nonprofits, watershed activists, outdoor-recreation outfits, lumber-mill owners. Sometimes there were upward of sixty people at meetings. Scientists from all over the region presented the latest findings on beaver ecology or the nesting behaviors of various bird species. To categorize Sagehen's diverse terrains—drainage bottoms with meadows and those without, north- and south-facing slopes, aspen stands with conifer encroachment—working groups hiked almost every yard of the forest.

Arriving at a consensus took years of discussion, but, in the end, the strategy the team decided on turned out to mimic the way fire naturally spreads. For instance, fire burns intensely along ridges and more slowly on north-facing slopes. Martens, having adapted to these conditions, rely on the open crests to travel in search of food and mates, while building their dens in shadier, cooler thickets. Following the logic of fire would create the kind of landscape preferred by native species such as the California spotted owl or the Pacific fisher—a mosaic of dark, dense snags and sunlit clearings, of big stand-alone trees and open ridgelines connecting drainages. Conway then led an effort to formulate a detailed implementation plan whose treatments varied, acre by acre, according to the group's predictions. Some areas were to be left as they were, some were to be hand-thinned with a focus on retaining rotting tree trunks, and some were to be aggressively masticated and then burned.

Typically, a Forest Service project takes two months to plan. Sagehen had been in the works for nearly a decade, but Brown eventually achieved the impossible: a plan that everyone—environmentalists, scientists, loggers, and the Forest Service—agreed on. Then, three days before the group was due to sign off on the plan, there was yet another hitch: in one of the units of Sagehen that were scheduled to be burned, a Forest Service employee discovered a nesting pair of goshawks—raptors that are federally protected as a sensitive, at-risk species.

This time, it was the conservationists who compromised. "I could have said, 'O.K., this area is now off limits, and if you don't believe me I'll sue your ass,'" Craig Thomas recalled. But, after some discussion, he agreed to stick with the plan. He knew that burning might make the birds leave or fail to fledge young, but, he told me, "the collaboration effort and what we had accomplished together mattered more."

When the Sagehen Forest Project tested its fire regimen on two five-acre plots, the results were striking: a bespoke application of thinning followed by a prescribed burn reduced fire risk just as efficiently as the Forest Service's standardized SPLATs, while also preserving more wildlife habitat and producing a higher yield of usable timber. The remaining trees seemed to respond well to fire, too; sensors that monitor levels of ethylene gas, which plants exhale when they're under stress, showed that the forest relaxed almost immediately post-burn.

But, despite the success of the project, enormous challenges remain. The Forest Service struggles to muster the resources and the staff necessary to burn safely. The California Air Resources Board restricts prescribed burns to days when pollution is at acceptable levels and the weather likely to disperse emissions from fire. In practice, this means that burning can occur only during a few weeks in the spring. In summer and autumn—the seasons when forests would burn naturally—the state's air usually falls foul of the Clean Air Act. These are also the months that are most prone to uncontrollable wildfires, whose smoke is far more damaging to human health than that from prescribed fire. But, perversely, because wildfires are classified as natural catastrophes, their emissions are not counted against legal quotas.

The window of time available for prescribed burns is further reduced by the stringent requirements of staffing, weather, and conditions on the ground, so that, in effect, there are just a few days each year when the Forest Service can set fires—nowhere near enough time to burn at the required scale. Even at Sagehen, large tracts of forest that should have been treated with fire remain untouched. When I made a second visit there and hiked through the forest with Brown and Faerthen Felix, he gestured ruefully as we passed through an area that seemed reasonably uncluttered. "We thinned this section years ago," he said. "We just haven't been able to burn, so it's a mess."

He pointed a few hundred feet ahead, to a couple of piles of spindly logs, two stories high. They represented another challenge. "These aren't big enough to go to a mill to be processed into boards," Brown said. "Ideally, we'd chip them and drag them down the road to burn for fuel and power, but the math doesn't add up." Traditional logging fells the biggest, most salable trees, but those are the ones that Sagehen's strategy is designed to spare. Thinning produces timber that has no value as lumber. Brown was resigned to simply burning these woodpiles, but air-quality restrictions had prevented him from doing even that. So the logs just sat there, increasing the risk of wildfire.

Brown has begun working with a group of researchers at U.C. Santa Cruz to imagine the outlines of a timber industry built around small trees, rather than the big trees that lumber companies love but the forest can't spare. In Europe, small-diameter wood is commonly compressed into an engineered product called cross-laminated timber, which is strong enough to be used in multistory structures. Another option may be to burn the wood in a co-generation plant, which produces both electricity and biochar, a charcoal-like substance used to replenish soil. Brown has also been talking to a businessman who hopes to burn waste wood to heat an indoor greenhouse-aquaculture operation. His vision is to provide organic vegetables and shrimp to buffets in Las Vegas, and then to interest California's cannabis farmers in using shellfish-dung-enriched biochar as fertilizer.

Throughout California, creative efforts are being made to tackle the obstacles that have slowed implementation of the Sagehen plan and now hamper its replication elsewhere. Regional air-quality officials have been brought into collaborative projects, in the hope that they will permit more flexibility. New state legislation has allocated millions of dollars to hire full-time burn crews, and will also require California's air board to quantify emissions from wildfires, in order to reverse the incentive against prescribed fire. To help entrepreneurs build business plans for monetizing small-diameter timber, Forest Service scientists are trying to quantify how much of it will be removed from forests.

Across the region, the Forest Service is devising projects to thin and burn on the Sagehen model. Meanwhile, Brown has helped launch the largest forest-restoration venture yet undertaken in California: the Tahoe-Central Sierra Initiative. It encompasses an enormous swath of forest that extends as far north as Poker Flat, level with Chico, and as far south as the American River, level with Sacramento. Brown's goal is to return fire to three-quarters of a million acres in the next fifteen years.

Achieving this will require a radical acceleration of the process that took place at Sagehen. Scott Conway has been exploring ways of using artificial intelligence to synthesize satellite data and aerial laser imaging into precise, three-dimensional maps of the more than a million acres that make up the Tahoe National Forest. With a grant of $1.3 million dollars from the Moore Foundation and the support of Silicon Valley startups, he has begun work on creating an open-access platform currently called the California Forest Observatory. Information that required years of on-the-ground counting and analysis at Sagehen—tree diameter, forest structure, fuel load—should soon be almost instantly accessible. Currently, the fire-risk map used by the California Department of Forestry and Fire Protection doesn't include weather data and hasn't been updated to show burned areas since 2005. The prototype Forest

Observatory will incorporate fresh satellite imagery on a daily basis.

Perhaps Sagehen's most important legacy is cultural: persuading the Sierra's warring stakeholders to conceive of forest management in ways they had previously rejected. Three of California's national forests have recently mandated allowing wildfire to spread in areas where it will be beneficial. Forest Service employees will have to file paperwork to justify putting out a fire that has started, where previously any decision not to extinguish a fire was ground for disciplinary investigation.

Attitudes among conservationists have evolved, too. In July, I joined Craig Thomas, the former director of Sierra Forest Legacy, for a hike along Caples Creek, in the Eldorado National Forest, just south of Lake Tahoe. "I would take those out," he said, pointing at two lovely little cedars nestled in the shade of an enormous sugar pine, their crowns just grazing its lower branches. They posed an existential threat to the larger tree, offering fire a fast track up to the canopy, and a lack of sunshine and nutrients had left them stunted. Thomas, a man who once spent much of his time suing the Forest Service, told me that he recently became certified to operate a chainsaw.

The Illilouette Creek wilderness area, in Yosemite National Park, is encircled by granite peaks that create a natural firebreak. Because it is so unlikely that any fire could spread beyond them, the National Park Service, in 1972, made the decision not to suppress wildfire within the basin's fifteen thousand acres. Since then, thanks to more than a hundred and fifty lightning ignitions, almost every acre, excepting bare rock and the creek itself, has burned at least once—some in small, pocket blazes, some in larger, more intense conflagrations. The resulting landscape provides a glimpse of what California's forests ought to look like—how they will look if Brown's Sagehen strategy succeeds.

In June, I visited Illilouette with Katya Rakhmatulina, a doctoral student who works with Scott Stephens studying the hydrological effects of wildfire. On a two-mile hike to one of three monitoring stations she maintains there, we passed perhaps only a hundred and fifty feet of what most people would consider picture-postcard Sierra Nevada forest—dark-green, conifer-packed woods with a rust-colored carpet of fallen pine needles. The rest was a surprising patchwork of landscapes: rush-filled meadows, crisscrossed with fallen logs; large, sunny grasslands punctuated by a few big trees; copses of young pines and willows; and recently burned expanses, where the ground was brownish black, spattered with delicate pink flowers and adorned with carbonized trunks, gleaming and sculptural.

Rakhmatulina was going to the station to rewire some cables that had been detached by bears. While she attempted to reboot the station's instrumentation, she told me about her research and the ways that fire affects groundwater supply. Having more trees in the landscape depletes water resources—like having more straws in a drink. Furthermore, pine needles and bark on the forest floor can form a resinous layer that prevents snowmelt and rainwater from sinking in and building up groundwater reserves.

More than sixty per cent of California's water supply originates in the Sierra Nevada, so anything that can preserve and increase that resource ought to be of immense value to the state's residents. Brown says that he sees California's water utilities and agribusiness as future converts to his cause and imagines a day when forest restoration could be paid for by a couple of extra cents on everyone's water bill.

I left Rakhmatulina to her tangle of wires and wandered back through the basin. Long vistas extended in all directions, allowing views of snow-covered mountains. The "forest" felt more like a lightly wooded park—it has an average of fifty trees per acre, compared with the four to five hundred that are typical elsewhere in the Sierra Nevada—and I began to realize that saving these forests will require a profound adjustment in our sense of what nature looks like here. The dark, dense, wild forests of European fantasy translate, in the drier conditions of California, to a landscape that is both dying and deadly—but how many of us are ready to make that perceptual shift? The picnickers, hikers, and mountain bikers who fill the parking lots of the Sierra Nevada each weekend, and the wealthy summer-home owners who prize the privacy of Lake Tahoe's emerald shores, will have to learn to appreciate more open, meadowlike environments. Logging jobs that have been lost could be replaced by new careers in fire management. Californians will have to forge a new relationship with their forest, and see the Sierra more as its native inhabitants once did—as a landscape that should be tended like a garden rather than harvested as a crop or protected as a wilderness.

Afterword

Elizabeth Kolbert

Like millions of other Americans, I first learned about climate change in the summer of 1988. For its day, it was a scorcher: Yellowstone National Park burst into flames; the Mississippi River ran so low that almost four thousand barges got backed up at Memphis, and, for the first time in its history, Harvard University shut down owing to heat. It was on an afternoon when the mercury in Washington, D.C., hit ninety-eight degrees that James Hansen, then the head of NASA's Goddard Institute for Space Studies, told a Senate committee that "the greenhouse effect has been detected and is changing our climate now." Speaking to reporters after the hearing, Hansen went a step further: "It is time to stop waffling so much and say that the evidence is pretty strong that the greenhouse effect is here."

Hansen's warning, which you will have encountered earlier in this volume, was certainly not the first. A report to President Lyndon Johnson in 1965 noted that the effect of burning fossil fuels was likely to be "deleterious from the point of view of human beings." Another report, prepared for the Department of Energy in 1979, predicted that even a relatively small increase in temperature could lead to the ultimate "disintegration" of the West Antarctic ice sheet, a process that would raise global sea levels by sixteen feet. A third report, also from 1979, found that, as carbon accumulated in the atmosphere, there was no doubt that the climate would change and "no reason to believe" that the change "will be negligible." But, for some reason, when Hansen spoke up, on that sweltering afternoon in June, the story of climate change shifted. The *Times* ran its article at the top of page one, under a three-column headline: "GLOBAL WARMING HAS BEGUN, EXPERT TELLS SENATE." The following year, Bill McKibben published "The End of Nature," first as a *New Yorker* piece under the rubric "Reflections," and then, in longer form, as a book.

Had the words of either man been heeded in the intervening three decades, the world today would be a very different place—incalculably better off in innumerable ways. Instead, during that interval some two hundred billion metric tons of carbon have been spewed into the atmosphere. (This is roughly as much CO_2 as had been emitted from the start of the Industrial Revolution to that point.) Meanwhile, trillions of dollars have been sunk into coal-burning power plants, oil pipelines, gas pipelines, liquid-natural-gas export terminals, and a host of other fossil-fuel projects that, in a saner world, would never have been constructed. And global temperatures, as everyone can by now attest—though some still refuse to acknowledge—have continued to rise, to the point where the sweltering summer of 1988 no longer stands out as particularly hot. The nineteen-nineties were, on average, warmer than the eighties, the aughts hotter than the nineteens, and the past decade hotter still. Each of the past five years has ranked among the warmest on record.

The New Yorker has run dozens of pieces on climate change. All might be described as "reflections" on this fundamental disconnect. Even as the consequences—rising seas, fiercer droughts, longer wildfire seasons, more devastating storms—have become daily news, global carbon emissions have continued to increase. In 2019, they reached a new record of ten billion metric tons. Emissions in India rose by almost two percent, and in China by more than two percent. In the United States, they actually dropped, by about 1.5 percent. On November 4, 2019, the Trump Administration formally notified the United Nations that it planned to withdraw from the Paris climate accord, negotiated by the Obama Administration back in 2015. The very next day, a group called the Alliance of World Scientists released a statement, signed by eleven thousand researchers, warning that "the climate crisis has arrived and is accelerating faster than most scientists expected."

"Especially worrisome," the statement continued, were "irreversible climate tipping points," the crossing of which "could lead to a catastrophic 'hothouse Earth,' well beyond the control of humans."

What will the Earth look like thirty years from now? To a discomfiting extent, the future has already been written. There's a great deal of inertia in the climate system; as a result, we've yet to experience the full effects of the CO_2 that's been emitted to date. No matter what happens during the next few decades, it's pretty much guaranteed that glaciers and ice sheets will continue to melt, as temperatures and sea levels continue to rise.

But to an extent that, depending on your perspective, is either heartening or horrifying the future—and not just of the next several decades but of the next several millennia—hinges on actions that will be taken by the time today's toddlers reach adulthood. What's technically referred to as "dangerous anthropogenic interference with the climate system" and colloquially known as "catastrophe" is warming so dramatic that it's apt to obliterate whole nations (such as the Marshall Islands and the Maldives) and destroy entire ecosystems (such as coral reefs). A host of scientific studies suggest that a temperature increase of two degrees Celsius (3.6 degrees Fahrenheit) or more would qualify. A great many studies suggest that warming of 1.5 Celsius (2.7 degrees Fahrenheit) would be enough to do the trick. At current emissions rates, the 1.5-degree threshold will be crossed in about a decade. As Drew Shindell, an atmospheric scientist at Duke University, told *Science:* "No longer can we say the window for action will close soon—we're here now."

So how hot—which is to say, how bad—will things get? One of the difficulties of making such predictions is that there are so many forms of uncertainty, from the geopolitical to the geophysical. (No one, for example, knows exactly where various "climate tipping points" lie.) That being said, I'll offer three scenarios.

In one scenario—let's call this "blue skies"—the world will finally decide to "stop waffling" and start to bring emissions down more or less immediately. In the U.S., proponents of the Green New Deal have proposed a "ten-year national mobilization" in order to meet a hundred percent of the country's power demand "through clean, renewable, and zero-emission energy sources." Such a timetable is obviously fantastically ambitious, but not for this reason infeasible. According to a report by the International Energy Agency, using technologies now available, offshore wind turbines could provide the country with twice as much electricity as it currently uses, and according to some (admittedly partisan) estimates, weaning the U.S. off fossil fuels would create tens of millions of jobs.

Bending the emissions curve globally is an even more formidable challenge. Leaders of many developing nations point out the injustice in asking their countries to forgo carbon-based fuels just because richer nations have already blown through the world's carbon budget. India, which will soon overtake China as the world's most populous country, gets three-quarters of its electricity from coal, and that proportion has in recent years been growing. Still, it's possible to imagine (at least as an intellectual exercise) that global emissions could peak in the next decade or so. Were this to happen, the increase in global temperatures could be held to less than two degrees Celsius. The world in 2050 would still be hotter than it is now, but it would also be less polluted, less given over to vast concentrations of oil wealth, and, in all likelihood, more just. As Narasimha Rao, a professor at Yale's School of Forestry & Environmental Studies, put it in the *Times*, it's hard to see how serious global-emissions cuts could take place without "increased attention to equity."

Alternatively, global emissions could continue to grow through the middle of the century and, along with them, global inequity. In this scenario, by 2050 a temperature increase of 2 degrees Celsius would, for all intents and purposes, be locked in. Developed nations would have constructed storm-surge barriers to keep out the sea and erected border walls to keep out refugees. They would also have started to air-condition the outdoors. Developing nations, meanwhile, would have been left to fend for themselves. To a certain extent, all of this is already happening. A study published in 2019 by Noah Diffenbaugh and Marshall Burke, both of Stanford, found that in the past fifty years warming had slowed economic growth in those parts of the world which have emitted the least carbon, perhaps by as much as twenty-five percent. "Not only have poor countries not shared in the full benefits of energy consumption, but many have already been made poorer (in relative terms) by the energy consumption of wealthy countries," the two wrote. Qatar, one of the world's hottest countries and also one of the richest, already cools its soccer stadiums and its outdoor malls.

In a third scenario, global warming could by 2050 produce global conflict that draws in poor nations and rich ones alike. This, too, already seems to be happening to a certain extent. A significant body of research suggests that the Syrian civil war was caused, at least in part, by a drought that pushed more than a million people out of their villages. The war, which, as of this writing, rages on, has, over the years of bloodshed, involved the U.S., Russia, Saudi Arabia, Iran, and Turkey. Future droughts in the Middle East are apt to be even more severe and prolonged; meanwhile, other volatile regions—like the Horn of Africa—are also suffering from water shortages probably exacerbated by climate change. It doesn't seem that it would take too many more Syrian-scale conflicts to destabilize large swathes of the globe. At the very least, climate change "will endanger the stability of the international political order and the global trading networks upon which American prosperity rests," Michael Klare, an expert on resource competition and a professor at Hampshire College has written. "As conditions deteriorate, the United States could face an even more perilous outcome: conflict among the great powers themselves."

If all these scenarios appear to be either too unrealistic or too unpleasant, I invite readers to write their own. Here's the one stipulation: it must involve drastic change. At this point, there's simply no possible future that averts dislocation. Billions of people will have to dramatically change the way they live or the world will change dramatically or we will see some combination of the two. My experience reporting on climate change, which now spans almost twenty years, has convinced me that the most extreme outcomes are, unfortunately, among the most likely. As the warnings have grown more dire and the consequences of warming more obvious, emissions have only increased that much faster. They are now tracking the highest of the so-called pathways studied by the Intergovernmental Panel on Climate Change. If this continues, the I.P.C.C. projects that, by the end of this century, global temperatures will have risen by almost eight degrees Fahrenheit. Let's just say that at that point no amount of outdoor air-conditioning will be sufficient.

A few years ago, I interviewed James Hansen for a video project that I was working on. Hansen retired from NASA in 2013, but he has continued to speak out about climate change—and to get arrested protesting projects like the Keystone XL pipeline. He was blunt about the world's failure. When I asked him if he had a message for young people, he said, "The simple thing is, I'm sorry we're leaving such a fucking mess."

Acknowledgments

We're grateful for the dedication, judgment, and skill of the team at Ecco, not least Dan Halpern, Denise Oswald, Martin Wilson, Sonya Cheuse, Caitlin Mulrooney-Lyski, Miriam Parker, Meghan Deans, Norma Barksdale, Allison Saltzman, and Dominique Lear; to Eric Simonoff, our wise matchmaker, and his agile aide, Jessica Spitz; to some terrifically helpful colleagues at *The New Yorker*, including Carol Anderson, Elizabeth Barber, John Bennet, Fabio Bertoni, Nicholas Blechman, Alan Burdick, Antonia Burdick, Leo Carey, Mengfei Chen, Willing Davidson, Deirdre Foley-Mendelssohn, Ann Goldstein, Jessica Henderson, Jessie Hunnicutt, Leily Kleinbard, Anya Kordunsky, Cressida Leyshon, Pam McCarthy, Erin Overbey, Natalie

Raabe, Katherine Stirling, Nick Trautwein, Dorothy Wickenden, and Daniel Zalewski, not to mention all the contributors in this volume. Big thanks, finally, to Clare Sestanovich, who did so much of the hard stuff and made it seem easy.

Contributors

Burkhard Bilger has been a staff writer at *The New Yorker* since 2001. He was previously a writer and deputy editor for *The Sciences*, where his work helped earn two National Magazine Awards, and a senior editor at *Discover.* He is the author of "Noodling for Flatheads: Moonshine, Monster Catfish, and Other Southern Comforts."

Dexter Filkins was awarded a 2009 Pulitzer Prize as part of a team at the New York *Times.* He has received two George Polk Awards, three Overseas Press Club Awards, and is the author of "The Forever War," which received a National Book Critics Circle Award. He joined *The New Yorker* in 2011.

Jonathan Franzen has contributed to *The New Yorker* since 1994. He is the author of five novels, including "Purity," "Freedom," and "The Corrections." His collections of non-fiction include "Farther Away" and "The End of the End of the World."

Ian Frazier is the author of twelve books, including "Great Plains," and, most recently, "Hogs Wild: Selected Reporting Pieces." He has written for *The New Yorker* since 1974.

Tad Friend, a staff writer at *The New Yorker* since 1998, is the author of the memoir "Cheerful Money" and the essay collection "Lost in Mongolia."

Christine Kenneally, formerly a senior contributor at BuzzFeed News, is the author of "The Invisible History of the Human Race" and "The First Word: The Search for the Origins of Language."

Tom Kizzia is the author of "Pilgrim's Wilderness: A True Story of Faith and Madness on the Alaska Frontier" and "The Wake of the Unseen Object: Travels through Alaska's Native Landscapes." He wrote extensively about rural Alaska for the *Anchorage Daily*

News; his work has also appeared in the Los Angeles *Times* and the Washington *Post.*

Eric Klinenberg is the Helen Gould Shepard Professor of Social Science and the Director of the Institute for Public Knowledge at New York University. He is the author or co-author of five books, including "Heat Wave: A Social Autopsy of Disaster in Chicago" and "Palaces for the People: How Social Infrastructure Can Help Fight Inequality, Polarization, and the Decline of Civic Life."

Elizabeth Kolbert, a staff writer at *The New Yorker* since 1999, is the author of "The Prophet of Love: And Other Tales of Power and Deceit," "Field Notes from a Catastrophe," and "The Sixth Extinction: An Unnatural History," which won the 2015 Pulitzer Prize for nonfiction.

Bill McKibben is an author, environmentalist, and scholar-in-residence at Middlebury College. A contributing writer at *The New Yorker*, he is author of seventeen books, including "The End of Nature" and "Falter: Has the Human Game Begun to Play Itself Out?"

Fen Montaigne is the author or co-author of five books and helped launch *Yale Environment 360*, where he is a senior editor. He has reported for *National Geographic* from every continent except Australia.

David Owen, a staff writer at *The New Yorker* since 1991, is the author of more than a dozen books, including "Volume Control: Hearing in a Deafening World."

Kathryn Schulz, a staff writer at *The New Yorker* since 2015, won the 2016 Pulitzer Prize for feature writing.

Michael Specter has been a staff writer at *The New Yorker* since 1998. Previously, he worked at the New York *Times* as a senior correspondent in Rome and the bureau chief in Moscow. He is an adjunct professor of bioengineering at Stanford University, and the author of "Denialism: How Irrational Thinking Hinders Scientific Progress, Harms the Planet, and Threatens Our Lives."

Ben Taub has been a staff writer at *The New Yorker* since 2017. He was awarded the Pulitzer Prize in 2020, and has received a National Magazine Award, an Overseas Press Club Award, and two George Polk Awards.

Nicola Twilley, a regular contributor to *The New Yorker*, is the co-host of the Gastropod podcast.